Beyond Bibliometrics

Beyond Bibliometrics:

Harnessing Multidimensional Indicators of Scholarly Impact

edited by Blaise Cronin and Cassidy R. Sugimoto

The MIT Press
Cambridge, Massachusetts
London, England

MIT Press books may be purchased at special quantity discounts for business or sales promotional use. For information, please email special_sales@mitpress.mit.edu.

This book was set in Stone Sans Std and Stone Serif Std by Toppan Best-set Premedia Limited, Hong Kong. Printed and bound in the United States of America.

Library of Congress Cataloging-in-Publication Data
Beyond bibliometrics : harnessing multidimensional indicators of scholarly impact / edited by Blaise Cronin and Cassidy R. Sugimoto.
pages cm
Includes bibliographical references and index.
ISBN 978-0-262-02679-6 (hardcover : alk. paper) — ISBN 978-0-262-52551-0 (pbk. : alk. paper) 1. Bibliometrics. 2. Bibliographical citations—Evaluation. 3. Scholarly publishing—Evaluation. 4. Scholarly electronic publishing—Evaluation. 5. Scientific literature—Evaluation. 6. Research—Evaluation—Statistical methods. 7. Communication in learning and scholarship—Technological innovations. I. Cronin, Blaise, editor of compilation. II. Sugimoto, Cassidy R., editor of compilation.
Z669.8B49 2014
010. 72'7—dc23

2013027449

10 9 8 7 6 5 4 3 2 1

Contents

Preface

The etymology of *bibliometrics* consists of the Greek words for "book" (*biblos*) and "measure" (*metron*). In recent decades the subject matter of concern to bibliometricians has expanded well beyond books to include scholarly journals and journal articles, authors and institutions, bibliographic references, citations, acknowledgments, patents, and much more. With the advent of the Web, electronic journals, digital libraries, citation databases, and social media, we are now much better positioned to see what scholars produce (from monographs to videos, from datasets to tweets), where they publish, how they communicate with their various audiences, how their work is received, critiqued, and used, whether and in what ways their diverse contributions are acknowledged, how influential they are within and across different intellectual communities, and whether or not their *oeuvre* has an impact over time, and, if so, precisely what kind of impact. The *we* above refers, of course, to any interested party, be it an individual scholar, university administrator, budget director, federal funding agency, or national research council, curious to know more about the immediate or long-term outcomes and effects of research, whether personal or institutional in nature, funded or unfunded.

Such is the variety of digital trace elements that can be tracked and measured using currently available tools and platforms that the term *bibliometrics* can seem a mite anachronistic when placed alongside younger lexical cousins, such as *webometrics* and *scientometrics*. But in defense of the term *bibliometrics* we would merely note that both of these neologisms themselves have quite specific and limiting referents. For that reason, we have decided to stick with *Bibliometrics* (albeit prefaced with *Beyond*) in the title of this book, it being a term that has both pedigree and currency. We see it as the logical starting point for any serious discussion of the newer metrics—often grouped under the rubric *alternative metrics*—currently being proposed or already being developed and tested in the context of research assessment, faculty evaluation, or resource allocation exercises.

Bibliometrics, once the preserve of a small population of information scientists and mathematicians, is now a sprawling, still fast-growing specialty, a multidisciplinary mix of methodologists, conceptualists, policy analysts, software developers, evaluators, and application specialists, along with a sprinkling of critical theorists. The heady enthusiasm that sometimes characterizes the field should not, however, cause us to lose sight of the fundamentals. Algorithmic and statistical refinement matters not a jot if the metrics in question do not measure what they claim to measure or are applied inappropriately: fitness for purpose needs to be demonstrated, not presumed. Despite the many technical advances recorded in the literature—think, for instance, of all the papers published in the last five years alone on Jorge Hirsch's *h*-index and its numerous derivatives—issues relating to validity (and, no less, reliability) continue to plague blibliometrics.

Do the tools we use and the indicators we favor measure what we claim or believe them to measure, and, if so, are those measures reliable—that is, capable of producing consistent, and, ideally, transparent results? Are we not sometimes so seduced by the incrementing sophistication of our procedures (data capture and cleaning, weighting, normalization, multivariate analysis, modeling, visualization) that the technical tail could almost be said to be wagging the disciplinary dog? Predictably, opinions are cleft and debate on the pros and cons of evaluative bibliometrics is as eristic as ever. But as the stakes rise, by which we mean that the progressive institutionalization of metrics-based assessment in both academe and science policy circles has become an ineluctable trend, so, too, does the need to explore the ramified ethical and cultural consequences of relying on metrics of one kind or another to capture putative evidence of research quality, scholarly impact, and academic influence.

Beyond Bibliometrics showcases a diversity of recent advances in metrics-based research, both theoretical and applied, and also presents a number of penetrating critiques of the subject. The chapters in this collection reflect the Janus-faced nature of (biblio)metrics research: the proliferation of techniques and applications demonstrates that the field is flourishing, yet critics highlight the considerable potential for misuse and abuse. It is our hope that *Beyond Bibliometrics* will help promote critical reflexivity within the field and foster a more enlightened appreciation of the pros and cons of metrics-based assessment among relevant communities of practice.

Acknowledgments

We are grateful to Jylisa Doney and Andrew Tsou for their assistance with bibliographic checking.

I History

1 Scholars and Scripts, Spoors and Scores

Blaise Cronin

Est modus in rebus.
Horace, *Satires* 1.1.106

Communicating Science

In medieval Europe, scientific news traveled at a measured pace, as slow or as fast as the peripatetic scholars who journeyed from one center of learning to another. The medium was the messenger. With the introduction of postal services in the sixteenth and seventeen centuries the speed of dissemination picked up. Letter writing gradually became the dominant form of exchange between gentlemen scientists such as the chemist Robert Boyle, the statistician William Petty, and the other grandees of the Scientific Revolution's early days. Their correspondence was often channeled through what might today be termed clearinghouses, there being copied for wider distribution and subsequent reading at local scientific gatherings (Manten, 1980). In Paris, Friar Marin Mersenne, variously referred to as the "mailbox of Europe" (Hatch, 2000, p. 265) or "chief philosophical intelligencer of his time" (Dear, 2000, p. 668), was the energetic embodiment of a communications hub; he corresponded with many of the leading minds on the continent, including Descartes, Galileo, Huygens, and Pascal. In London, Henry Oldenburg performed a similar role with industry and distinction.

Members of these "invisible colleges"—informal networks of natural philosophers, unaffiliated with formal institutions of learning (e.g., Crane, 1972; Lomas, 2002; Wagner, 2008)—would assemble at private dwellings, coffee shops, or taverns to conduct their business and exchange news of the latest scientific findings. From such loose arrangements evolved the national academies of the seventeenth century, the Royal Society in London and the Académie Royale des Sciences in Paris. Shortly after their establishment, both bodies created in-house journals, the *Philosophical Transactions* and *Journal des Sçavans*, respectively. These prototypical scientific journals—learned might be a more accurate qualifier given the initial breadth of subject coverage (Waterman, n.d.)—superseded the communication of experimental reports via correspondence, formalizing a process that until then had depended largely on individual agency and personal motivation. The aforementioned Henry Oldenburg was appointed Secretary to the Royal Society and became the first editor of the *Transactions*, collating and editing the many reports that flowed to him from near and far. In Paris, Denis de Sallo, the founding editor of *Journal des Sçavans*, which began as a weekly publication, fared less well; he was ousted from his position after only a few months because of his controversial book-reviewing practices (Waterman, n.d.).

Not only did the two academies spur the development of the scholarly journal as we have come to know it today, but they also introduced and systematized the process of third-party peer review. Over time, enlightened amateurism ceded to a system in which learned societies, society fellows, journal editors, and outside subject experts, through distributed collective action, conferred legitimacy on experimental reports and vouchsafed the integrity of the truth claims put forward by individual authors. That system, albeit in modified form, is still with us today.

Much, it need hardly be said, has changed in the realm of scholarly communication since the mid-17th century. But the key components of the system post "etherization" (Koku, Nazer, & Wellman, 2001, p. 1754)—that is, the scientific article (today very formal and structured in its presentation of experimental results), the journal of record (now as likely to be commercial as societal in nature), and the process of peer review (massively scaled up and more complex than in Oldenburg's time)—remain central to the effective conduct of science. Across the centuries, the discursive character and overall architecture of the scholarly journal article have changed greatly, as Gross, Harmon, and Reidy (2002) demonstrate in their meticulously comprehensive biography of this doughty publication genre.

The gradual incorporation of bibliographic references into scholarly articles is a good illustration of the kind of paratextual changes that have taken place. Citing the works of scholars who have materially influenced one's thinking is a normatively mandated aspect of academic writing, but one that was slow to establish itself in consistent fashion across disciplines. Another relatively recent and now near-ubiquitous addition to the feature set of the scholarly journal article is the acknowledgments section (Costas & Van Leeuwen, 2012; Cronin & Franks, 2007). This easily overlooked species of microattribution, typically located at the end of a paper, captures a miscellany of second-order contributions made by trusted colleagues, technicians, and others. Together, references and acknowledgments—which Heffner (1981) referred to aptly as indicators of "subauthorship collaboration"—provide a cumulating ledger of contributions, major and minor, made by scholars and researchers to the commonwealth of knowledge. As such, they—references in particular—have become essential raw material for bibliometricians, allowing us to move beyond productivity measures (publication counts) to indicators of impact (citation counts) and, more recently, indicators of influence (acknowledgment counts).

Changes in the shape, content, and interactivity of the traditional journal article, and even the scholarly monograph, are imminent as commercial interests (see, for example, Elsevier's Article of the Future project[1]) and also the scientific community itself (see the Force 11 [Future of Research Communications and e-Scholarship] manifesto[2]) seek to take full advantage of the capabilities of digital technologies and networked infrastructures in an effort to improve the transparency, efficiency, and overall effectiveness of the multi-billion-dollar primary communication system.

Crediting Influence

The progressive institutionalization of bibliographic referencing behavior, coupled with the development in the mid-1950s of the first experimental citation indexes to the literature of science, created the preconditions necessary to support large-scale multidimensional citation and cocitation analyses (Cronin & Atkins, 2000). As the size, depth, and reliability of commercially available citation databases (e.g., Web of Science, Scopus) grew, so, too, did the sophistication of the techniques being developed to monitor, measure, and visualize various aspects of the scholarly communication process. Bibliometrics—defined parsimoniously as the quantitative analysis of publications, authors, and bibliographic references—gradually became a significant site of experimentation and research activity in its own right, something much more than a set of methods to be used by a handful of quantitatively inclined information scientists.

It is no exaggeration to say that the advent of the Internet and World Wide Web turbocharged research in the field. For example, the affordances of massive datasets and the easy manipulability of usage statistics have greatly expanded the appeal of bibliometrics and extended the ways quantitative techniques can be applied to the inputs, outputs, and processes of science and scholarship. Today, the field is a kaleidoscope of overlapping domains—informetics, scientometrics, cybermetrics, webometrics, influmetrics, digimetrics are just some of the neologisms—and is populated by researchers from almost every disciplinary background. It also has all the paraphernalia (journals, professional societies, conferences, prizes, curricula, research centers, etc.) that one associates with a mature academic specialty. Interest in bibliometrics, broadly construed, has never been greater, as corporations, universities, and funding bodies seek to identify robust indicators of research performance, whether at the macro (country), meso (discipline), or micro (program) level.

In the United Kingdom, by way of illustration, the government's Research Excellence Framework (REF), successor to the Research Assessment Exercise (RAE), which began in the mid-1980s, will focus on impact more broadly than heretofore: unlike the RAE, REF will seek evidence of not only the scientific/scholarly impact of research but also evidence of downstream social, economic, and cultural benefits.[3] Scholars, as Thelwall (2012, p. 430) notes, may now be able to make the case that they have "an impact on the world that is insufficiently represented by the citations that their publications receive." For example, academics who spend a considerable proportion of their time advising and mentoring doctoral students should soon be able to demonstrate the long-term fruits of their often-hidden labors by using tools that reveal and quantify academic genealogy (Russell & Sugimoto, 2009; see also Sugimoto, chapter 19, this volume). Researchers whose ideas have commercial significance can draw on diverse indicators of impact from patent citation data to trade and industry press coverage. Lewison (2005), for example, describes five alternatives/complements to conventional citation indexes that can be used to track the overall diffusion and impact of biomedical research, namely, references to research that appear in international standards, national policy documents, clinical guidelines, textbooks, and newspapers. The message seems to be that Web of Science and Scopus do not tell the whole story.

The Web has engendered a variety of corpora, data types, and, somewhat less concretely, "genres of invocation" (Cronin, Snyder, Rosenbaum, Martinson, & Callahan, 1998, p. 1326) that can be mined to reveal heretofore largely invisible traces of interaction and influence. Blog posts and tweets about a scholar's ideas, two instances of what has been termed "polymorphous mentioning" (Cronin et al., 1998, p. 1320), can now be incorporated into the impact portfolio of individuals or groups alongside established indicators: the Total Impact web application is an early prototype of what such a system might look like.[4] We are no longer limited to capturing data about formal publications and citations (the scripts and spoors, respectively, in the title of this chapter). Rather, the evaluator's net can be cast more widely to trawl for novel or overlooked indicators—alt(ernative) metrics to use the term of art (Priem, 2010; see also chapters 14, 17, and 16, by Priem, Haustein, and Bar-Ilan et al., respectively, in this volume)—of scholarly engagement and impact that are generated automatically in the digital communication environment.

We are moving from a reliance on exclusively *citation*-based metrics to the use of multidimensional, *usage*-based metrics (Kurtz & Bollen, 2010). The easier data capture is, the richer the picture of use and performance that can be produced. This applies not only to scholarship. To take but one illustration, in the highly competitive world of professional soccer, information management and data-mining tools are used routinely to assess the contributions made and value added by individual team members during every game over the course of a season: "The network of cameras around the ground misses not a trick; every move, every step is tracked and transformed into frequencies, averages, ratios and correlations. Metrics are beginning to replace mystique" (Cronin, 2009, p. 3). Signs of a similar trend can be observed in the groves of academe.

Developments in online and open-access publishing are inducing transparency across the various stages of the scholarly communication process. The black box of traditional scientific publishing is being opened up to scrutiny, and orthodoxies, such as anonymous peer review (single- or double-blind) and citation-based evaluation, are being questioned by a growing band of scientists and scholars who reflect the values of a different Zeitgeist (e.g., Kravitz & Baker, 2011) and are increasingly disillusioned with the status quo. This extract from the Force 11 Manifesto (Bourne et al., 2011, para. 3) captures the mood of the moment, in some quarters at least: "We see a future in which scientific information and scholarly communication more generally become part of a global, universal and explicit network of knowledge; where every claim, hypothesis, argument—every significant element of the discourse—can be explicitly represented, along with supporting data, software, workflows, multimedia, external commentary, and information about provenance." Force 11 is a self-described community of scholars, librarians, archivists, publishers, and research funders that has arisen organically to bring about change in the scholarly communication process. One of its aims is to make the full range of a scholar's contributions more visible than is the case at present, to counteract what might be called the "neglect of silent evidence" (Taleb, 2010, p. 103).

Evaluating performance, allocating credit, and distributing pecuniary rewards will become more complicated because citations are now only one among many indicators (and potential indicators) of scholarly influence, visibility, and impact, broadly construed. How does one factor into the academic reward system data on mentoring successes, the inclusion of a scholar's work on syllabi, the frequency with which a researcher is acknowledged by his or her peers? And how, to take an admittedly contrived example, does one establish the degree of equivalence between a positive review on Faculty of 1000 (F1000), six complimentary tweets, four document downloads, and two citations in *Nature*? This is not a novel issue: in reality it is a variant of the old question, should a citation from a Nobel laureate be weighted the same as one from a doctoral student, one from *Scientific American* the same as one coming from a minor journal? It also throws into relief the problem of relying on indicators that are incommensurable—yet one more illustration of the all-too-prevalent phenomenon that Hadley Arkes has termed "the ritual of empty exactitude" (Arkes, 2010, p. 131).

The idea of currency exchange rates being applied to symbolic capital markets may not be quite as preposterous as it sounds. It is conceivable that next-generation academics, comfortable with the notion of networked participatory scholarship (Veletsianos & Kimmons, 2012), could and would be happy to be judged on more than formal published output. Faculty evaluation and research assessment exercises might incorporate data on the overall use of scholarly publications (e.g., broadcast media mentions; number of times a work is included on reading lists) or solicit evidence of social media–based community engagement and peer interactive communication (e.g., discussing ideas, debating issues, sharing resources). Pierre Bourdieu (1988) seemed to anticipate current thinking when he spoke of "journalist-academics who, despite the contempt in which the more academically successful affect to hold them, are invested with the power of celebration and criticism afforded by privileged access to the daily and weekly press and are thereby able to exercise quite real effects within the field itself" (p. 77). Were he writing today, Bourdieu would likely not limit himself to his original example of *Le Nouvel Observateur*. The contemporary public intellectual, no less than the assiduous self-promoter, has a battery of communicative options, both institutional and personal, at his or her disposal, ranging from conventional broadcast channels and newspapers to a wide range of social media. One can but hope that this does not result in what Dalrymple (2005), writing about contemporary art in the United Kingdom, cautions against, namely, a culture "that confers the reward of eminence on those who use self-advertisement and vulgarity . . . as their means to obtain it" (p. 152).

Genres Galore

In years past, a relatively small number of scholarly publication genres existed: the monograph, journal article, and conference paper being exemplars. Each of these canonical forms was associated with an identifiable (corresponding) author (occasionally aided by one or more coauthors); each was fixed in character, the text immutable on the printed page; each existed in isolation, unlinked, other than via inert bibliographic references, to prior or contemporaneous work. The situation today is different. Papers published in online or hybrid journals are located in cyberspace alongside myriad other papers and a miscellany of digital objects (from datasets to peer commentary) of potential relevance, connected in real time through hyperlinks in a way that was inconceivable only a few decades earlier. Alongside the historically dominant genres, we now have a congeries of emergent publication forms: micro, mega, molecular, mutable, and mobile scripts. What was fixed is now mutable; what was static is now mobile; what was monolithic is now modular. Today, the scholarly communication system is less linear, less rigid, and less opaque than before; both the process and the end products are being transformed, slowly if inexorably.

A powerful illustration of the general trend is the so-called data explosion or data revolution: Big Data in short (Blatecky, 2012). It is not just high-energy physicists, astronomers, volcanologists, geneticists, and climatologists who are struggling to capture, clean, curate, analyze, share, and repurpose large-scale datasets; humanists and social scientists, too, are trying to figure out ways to manage and exploit the proliferation of digital datasets that have become a striking epiphenomenon of 21st-century scholarship. In some fields, data matter as much if not more than text, a point made succinctly by Bourne (2005) in the form of a rhetorical question: "Will a biological database be different from a biological journal?" If the lines are indeed blurring, and some domains are moving toward a post-narrative future, then we need to think systematically about the ways we cite data so that (a) readers are able to access the primary data in order to validate, replicate, or extend the original work as required, and (b) data creators receive full acknowledgment and due attribution for their endeavors. In order for data to be citable and for credit to be allocated equitably, be it to an individual, group, or institution, data elements must, of course, be uniquely identifiable. In Borgman's words (2012, p. 4): "For people to invest in making data discoverable, they should receive credit for creating, cleaning, analyzing, sharing, and otherwise making data available and useful." Therein lies one part of the data challenge. Other nontrivial issues include archiving and preservation, privacy protection, and data licensing.

Multiauthorship is now commonplace, and in some areas (e.g., biomedical research) the single-author paper is a virtual anachronism. Hyperauthorship raises a host of questions about (a) what it means to be an author, (b) how credit should be distributed across the various contributors to a paper, and (c) how responsibility or ownership of a collective work should be determined (Cronin, 2002, 2012). It was hard enough with only three or four coauthors to settle on an equitable weighting/credit allocation formula, but when the population of coauthors is literally in the hundreds, it becomes intractably difficult, if not meaningless, to grant slivers of individualized credit to each named individual. The flipside of the credit allocation coin is citation counting. How do we distribute symbolic capital equitably when there are multiple authors? This issue has been a source of considerable debate for decades—various approaches such as full counting and fractional counting have been proposed (e.g., Long & McGinnis, 1982)—and the problem is exacerbated by the rise in coauthorship across almost all fields of scholarly and scientific endeavor (Gazni, Sugimoto, & Didegah, 2011). Defining a contributor is no easier than defining a contribution; in fact, it is almost as difficult as counting angels on a pinhead.

Texts typically go through multiple versions, becoming in the process quasi-organic objects that link to related materials (e.g., supplementary online data), connect to other resources (e.g., videos of an experiment), and undergo continual refinement as a result of pre- and postpartum feedback from colleagues and/or open peer commentary at various points along the now much more transparent creation-dissemination value chain (dynamic versioning). They are, in other words, mutable and mobile in nature, capable of morphing into megascripts that "trap all the phases of script writing which are currently captured but held in different systems" (Davenport, 1993, p. 39). Texts can be short and sweet: a tweet is a microscript.

Of late, the term *nanopublication* is being used to denote any core, highly granular scientific claim (e.g., "malaria is transmitted by mosquitoes") made in a traditional, full-length scientific paper (Groth, Gibson, & Velterop, n.d.). Nanopublications relating to the same claim can be harvested, mined, and evaluated in aggregate to identify the degree of community-wide consensus on any given topic. Thus, the scientific/scholarly paper in its entirety need not be the primary or exclusive focus of attention; instead, specific claims embedded therein become the unit of analysis. In sum, the journal article is capable of being deconstructed, reconstituted, and networked with other digital objects with which it has kinship, thematic, or functional ties in a way that Oldenburg could not have imagined.

It was this kind of thinking that motivated the LiquidPub Project.[5] The title derived from the concept of liquid modernity, introduced by the sociologist Zygmunt Bauman in books such as *Liquid Love* (Bauman, 2003) and *Liquid Times* (Bauman, 2007). We live, he argues, in a moment when social, structural, and affective bonds have loosened such that we need to be flexible and capable of operating under conditions of uncertainty and impermanence. Liquid publications, to extend his metaphor, are considered to be "evolutionary, collaborative, and composable scientific contributions" (LiquidPub, 2012, para. 2). They go beyond the conventional journal article to include "blogs with interesting ideas, scientific experiments, comments on somebody else's paper, reviews, slides, videos, demos, even data" (Baez & Casati, n.d., p. 1). The concept has also been applied to monographs: liquid books are "collaborative, evolutionary, possibly open-source and multi-faceted" in nature (Casati & Ragone, 2009, p. 1). The advocates of change are growing in number and new modes of scholarly communication (e.g., blogs, social bookmarking sites, wikis) are gaining traction within the academic community.

Symbolic Capitalism

The myriad trace elements generated by the ever-growing number of digital systems and services on offer are increasingly amenable to automated real-time capture and systematic analysis. Policymakers, scientists, scholars, publishers, research administrators, and funding bodies now have at their fingertips an unprecedented trove of data of varying granularity on (a) human information seeking and communication behaviors, (b) the use of digital assets and services, and (c) the ways professional contributions are valued and reputations constructed. Digital analytics allows not just companies but also institutions of higher education to track movements in their brand or reputation over time, based on, inter alia, the frequency of media mentions. If we view individual academics (or research teams) as microbrands, there is no in-principle reason why a scholar's presence—"buzz" in the vernacular (Hoover, 2012)—could not be tracked via media mentions, in addition to conventional indicators of impact such as citations (see Cronin & Shaw, 2002). In the commercial world this is already reality. An individual's Klout score[6] is based on (i) true reach (the number of people you influence), (ii) amplification (how much you influence people), and (iii) network (the influence of those in your true reach). Individuals are rated on a scale of 1–100, which, as the company notes, is basically your social credit score. It is not hard to see the potential benefits to marketers of this metric; equally, it is not hard to see how such a system could be abused or simply become a time-consuming distraction.

We are, it is argued, moving away from a "citation-fetishizing article monoculture" (Priem, as quoted in Howard, 2012, para. 16) to a culture of multiples; think in terms of a matrix of metrics (frequencies, ratings, etc.) derived from social media such as Twitter, ResearchGate, CiteULike, F1000, Academia.edu, and Mendeley. The prospector's tools and the evaluator's armamentarium will indubitably become more powerful in the years ahead, but questions relating to the validity and reliability of different kinds of bibliometric and post-bibliometric indicators will continue to be raised and debated with vigor. The old aphorism attributed to Albert Einstein, that not everything that can be counted counts and not everything that counts can be counted, certainly holds true in the context of evaluative bibliometrics. Exactly what kinds of impact or influence do "altmetrics *events*" (e.g., downloading a file; social bookmarking) attest to, singly and in combination (Haustein & Siebenlist, 2011; Priem, Piwowar, & Hemminger, 2012)? Do citations and newer forms of mentioning correlate with one another and with established bibliometric indicators, and if so to what extent and under what conditions (Bollen, Van de Sompel, Smith, & Luce, 2005; Shuai, Pepe, & Bollen, 2012)? Are altmetrics indicators "mostly orthogonal to citation," as some early empirical evidence suggests (Priem, Piwowar, & Hemminger, 2012, para. 1)? Do downloads predict citations (Watson, 2009)? Are tweets harbingers of citations (Eysenbach, 2011)? A flurry of correlational analyses can confidently be expected.

The New Metrics

Changes in the tools and platforms that support scholarly exchange and publication are giving rise to a new wave of metrics that can, with greater or lesser confidence, be used in research evaluation exercises, alongside more established (if still contested) indicators. As I predicted, "There will soon be a critical mass of web-based digital objects and usage statistics on which to model scholars' communication behaviors—publishing, posting, blogging, scanning, reading, downloading, glossing, linking, citing, recommending, acknowledging—and with which to track their scholarly influence and impact, broadly conceived and broadly felt" (Cronin, 2005, p. 196). By way of analogy, think of the *Annales* school in France in the 1950s, with its commitment to the notion of *l'histoire totale*. Today we are witnessing the emergence of something similar in the world of performance evaluation: total bibliometrics. However, much work, in terms of prototyping and validity testing, remains to be done in order to even get close to creating anything like a standard set of performance indicators with across-the-board applicability such as are to be found, for example, in the world of business.

The proliferation of alternative indicators should serve agencies of control and narcissists equally well (Wouters & Costas, 2012). Imagine a business intelligence dashboard that visualizes an array of key performance indicators for research groups and academic departments, pulling data from multiple sources, and in real time. Instead of data on, say, market share, production volumes, regional sales, and revenue targets, the metrics in this case might include categories such as the following: scholarly outputs (papers, conference presentations, number of doctoral students graduated, published datasets, patents, etc.); measures of esteem (prizes, fellowships, teaching awards and honors); funded research activity (number, value, and source of grants); citation indicators (the impact factor of the journals in which scholars publish; how often authors' papers are cited, where they are cited and by whom; data citation rates); acknowledgment frequency; number of views/downloads; frequency of bookmarking; news coverage and social media mentions. The ability of such a system to generate all-source comparative performance data with which to compute an individual's or group's composite impact factor would surely appeal to a certain class of administrative elite. Welcome to the world of "cognitive capitalism" (De Angelis & Harvie, 2009); welcome to the mandarinate of measurement.

By the same token, highly competitive or narcissistic scholars, keen to enumerate their accomplishments, establish indicators of their relative market worth, enhance their professional visibility, or just make themselves feel good, might well appreciate having multicomponential indicators of their productivity, impact, and professional salience—akin to a value-added "live CV" (Priem, Piwowar, & Hemminger, 2012) or, to take things a step further, a Q score as used in the media and marketing industries[7]—available at their fingertips. Reputation management, or career grooming, may thus become an inescapable and not especially edifying corollary of scholarly life as researchers actively attempt to manipulate the system to their own advantage. This concern has been aired repeatedly with regard to citation behavior (e.g., Corbyn, 2008; Davis, 2011), most presciently perhaps by Franck (1999), who raised the specter of "citation cartels" (p. 54), and it will be heard loudly again as the mapping, measuring, and monitoring of academic performance become ever more sophisticated in contemporary science and scholarship. Indeed, in a perverse reversal of values, visibility may become an end in itself. As Jensen (2007, para. 38) has observed, "scholarly invisibility is rarely the path to scholarly authority," and that is truer than ever for the increasingly open and interactive world of Web 2.0 and beyond. But, one might ask, why shouldn't researchers be able to more efficiently track the beneficial effects and influence their professional endeavors have on both their peer community and society at large? And wouldn't it be natural in an era of public accountability and open government for federal funding agencies to want to track the immediate and longer-term effects of the programs and projects they fund (see Lane and colleagues, chapter 21, this volume, for more on this subject)? Let a hundred flowers bloom, as the saying goes.

Well before the emergence of alternative metrics, fears were being expressed about the ways citation analysis could be used as an instrument of control and as a means of disciplining scholarly discourse. Sosteric (1999), in a spirited polemic that predated the advent of social media, foresaw "an Orwellian surveillance net" and the "cyberbnating of the academy" (p. 4)—language triggered in part by Cronin and Overfelt's (1994) assertion that citations and other performance indicators could be used diagnostically "to help shape individual faculty productivity profiles" (p. 71). Sosteric was not the only one to foresee some of the unintended social consequences of using digital analytics to evaluate research performance and scholarly impact. Hicks and Potter (1991), for instance, noted "new possibilities for producing hierarchies of difference and categories of normal/abnormal scientific behavior" as a byproduct of advances in automated citation analysis (p. 475). Sociotechnically speaking, the new media shape metrics as much as the new metrics shape media use. In short, media and metrics are co-constitutive.

Publication counts, citation scores, and impact factors do not necessarily tell the whole story, nor should they be expected to. A scholar's work may well have a range of impacts over time—in different contexts, with different audiences, and for different reasons—and traditional bibliometric indicators may not fully reflect these multivalent contributions. But with new-age metrics come age-old concerns relating to the features and functionality of the platforms, from Mendeley to Google Scholar, that generate the raw data being used to determine scholars' "true contributions" (Bourne et al., 2011, para. 6). Among the issues that will need to be addressed are the following: (a) the durability of the different platforms and toolsets; (b) the transparency of the underlying algorithms and assumptions; (c) the reliability of the data, in terms of consistency and completeness; (d) the validity of the indicators being developed (at the most basic level, determining whether they measure what they purport to measure; see the chapters by Day, Furner, and Gingras); and (e) the susceptibility of these tools to gaming—"coercive citation" being a case in point (Wilhite & Fong, 2012)—whether by individual scholars or cliques intent on maximizing their professional visibility across a multiplicity of media and contexts. In addition, there is the rather more mundane matter of time; researchers and scholars have only so much time to manage their online presence and only so much time to attend to the work of others. As has been noted, "Only a certain amount of research can be said to have impact without making the very notion of impact meaningless" (Schroeder, Power, & Meyer, 2011, para. 5). Academe is not Lake Wobegon, though we may sometimes wish it were.

Since the early days of commercial citation indexes there has been continual debate within the wider scientific community on issues relating to the utility, usability, ethicality, reliability, and validity of citation indexes for research evaluation purposes, the last of these most particularly and most persistently (e.g., MacRoberts & MacRoberts, 1989; Seglen, 1992). This extract from the Force 11 Manifesto makes clear the distrust felt by many toward the use of, and growing reliance on, journal impact factors by authors, editors, and publishers: "We need to acknowledge the fact that notions such as journal impact factors are poor surrogates for measuring the true impact of scholarship . . . we need to derive new mechanisms that allow us to more accurately measure *true contributions*" (Bourne et al., 2011 para. 6; italics added). We can expect much more discussion as to what exactly is meant by "true contributions" and continuing pushback on the use of the journal impact factor, which Vanclay (2012) compares with phrenology, in research evaluation exercises, as the limitations of the indicator become more widely acknowledged (see Lozano, Larivière, & Gingras, 2012). In the Web 3.0 world many more elements will be available for building impact, reputation, and authority metrics, as this detailed and prescient list (see below) drawn up by Jensen (2007) indicates, in the process making impact factors seem like very blunt instruments.

- Prestige of the publisher (if any).
- Prestige of peer prereviewers (if any).
- Prestige of commenters and other participants.
- Percentage of a document quoted in other documents.
- Raw links to the document.
- Valued links, in which the values of the linker and all his or her other links are also considered.
- Obvious attention: discussions in blogspace, comments in posts, reclarification, and continued discussion.
- Nature of the language in comments: positive, negative, interconnective, expanded, clarified, reinterpreted.
- Quality of the context: What else is on the site that holds the document, and what's its authority status?
- Percentage of phrases that are valued by a disciplinary community.
- Quality of author's institutional affiliation(s).
- Significance of author's other work.
- Amount of author's participation in other valued projects, as commenter, editor, etc.
- Reference network: the significance rating of all the texts the author has touched, viewed, read.
- Length of time a document has existed.
- Inclusion of a document in lists of "best of," in syllabi, indexes, and other human-selected distillations.
- Types of tags assigned to it, the terms used, the authority of the taggers, the authority of the tagging system. (Jensen, 2007)

Indeed, it may not be farfetched to imagine an array of radar charts depicting the relative strengths of a population of scholars or research groups on a battery of traditional and newer performance indicators (e.g., peer-reviewed publication count, citation rank, mainstream media visibility, social media salience, mentoring prowess): performance tracking at a glance.

Conclusion

How, in the light of the foregoing, does one go about creating "a nuanced, multidimensional view of multiple research impacts at multiple time scales" (Priem, Piwowar, & Hemminger, 2012, para. 3)? Is consensus even possible, given the antipathy in many quarters toward the culture of accountability, metrification, and monetization being imposed on the academy (Burrows, 2012; Cronin, 2000)? To what extent do "true contributions" differ within and across disciplines and epistemic cultures? Might such contributions also differ across time, as policy mandates and administrative expectations change, or as the conduct and processes of science itself change? Do scientists define "true contributions" in the same way as funding agencies and policymakers? To what extent do scholars themselves agree on the validity, utility, and appropriateness of alternative metrics in peer and program evaluations? How many different kinds of alternative metrics, with what weights and normalizations, can and should be used in evaluating individuals or research groups? Is there an upper bound, a point at which the evaluative process collapses under its own weight, where triviality trumps transparency? What, more concretely, are the trade-offs between ease of use of "metric assemblages" (Burrows, 2012, p. 357) and both the overhead and opportunity costs of developing, fine-tuning, and managing a metrics-rich assessment effort (Thelwall, 2012)?

Pandora's box has been opened and the challenge will be to harness the proliferation of alternative metrics intelligently to the assessment of scholarly impact while simultaneously guarding against abuses—be they inadvertent, opportunistic, or engineered—of the system, whether by scholars themselves or those who mandate and superintend the ever more complex processes of performance evaluation that have become an inescapable feature of the higher education landscape.

Notes

1. http://www.articleofthefuture.com
2. http://www.force11.org
3. http://www.ref.ac.uk
4. http://impactstory.it
5. https://dev.liquidpub.org/svn/liquidpub/papers/deliverables/LiquidPub%20paper-latest.pdf
6. http://klout.com/home
7. http://www.qscores.com/Web/Index.aspx

References

Arkes, H. (2010). Pornography: Settling the question in principle. In J. R. Stoner & D. M. Hughes (Eds.), *The social costs of pornography* (pp. 127–142). Princeton, NJ: Witherspoon Institute.

Baez, M., & Casati, F. (n.d.). Liquid journals: Knowledge dissemination in the Web era. Retrieved from http://wiki.liquidpub.org/mediawiki/upload/9/9b/Liquid-journal-proposal_v0.13.pdf.

Bauman, Z. (2003). *Liquid love: On the frailty of human bonds*. Cambridge: Polity Press.

Bauman, Z. (2007). *Liquid times: Living in an age of uncertainty*. Cambridge: Polity Press.

Blatecky, A. (2012). Opening remarks by project sponsors. In National Research Council of the National Academies, *The future of scientific knowledge discovery in open networked environments* (pp. 3–5). Washington, DC: National Academies Press.

Bollen, J., Van de Sompel, H., Smith, J. A., & Luce, R. (2005). Toward alternative metrics of journal impact: A comparison of download and citation data. *Information Processing & Management, 41*(6), 1419–1440.

Borgman, C. L. (2012). Why are the attribution and citation of scientific data important? In National Research Council of the National Academies, For attribution: Developing data attribution and citation practices and standards (pp. 1–8). Washington, DC: National Academies Press.

Bourdieu, P. (1988). *Homo academicus*. Cambridge: Polity Press.

Bourne, P. (2005). Will a biological database be different from a biological journal? *PLoS Computational Biology, 1*(3), e34. doi:10.1371/journal.pcbi.0010034

Bourne, P. E., Clark, T., Dale, R., De Waard, A., Herman, I., Hovy, E., et al. (Eds.). (2011). *Improving future research communication and e-scholarship*. Retrieved from http://www.force11.org/white_paper.

Burrows, R. (2012). Living with the h-index? Metric assemblages in the contemporary academy. *Sociological Review, 60*(2), 355–372.

Casati, F., & Ragone, A. (2009). Liquid book: Reuse and sharing of multifacet content for evolving books. Retrieved from http://wiki.liquidpub.org/mediawiki/upload/f/f2/LiquidBook.pdf.

Corbyn, Z. (2008). Researchers play dirty to beat REF. *Times Higher Education*. Retrieved from http://www.timeshighereducation.co.uk/story.asp?storycode=400516.

Costas, R., & Van Leeuwen, T. (2012). Approaching the "reward triangle": General analysis of the presence of funding acknowledgments and "peer interactive communication" in scientific publications. *Journal of the American Society for Information Science and Technology, 63* (8), 1647–1661.

Crane, D. (1972). *Invisible colleges: Diffusion of knowledge in scientific communities*. Chicago: University of Chicago Press.

Cronin, B. (2000). Knowledge management, organizational culture and Anglo-American higher education. *Journal of Information Science, 27*(3), 129–137.

Cronin, B. (2002). Hyperauthorship: A postmodern perversion or evidence of a structural shift in scholarly communication practices? *Journal of the American Society for Information Science and Technology, 52*(7), 558–569.

Cronin, B. (2005). *The hand of science: Academic writing and its rewards*. Lanham, MD: Scarecrow Press.

Cronin, B. (2009). *Stickmen: Reflections on the goalie's eccentric art*. Bloomington, IN: AuthorHouse.

Cronin, B. (2012). Collaboration in art and science: Approaches to attribution, authorship, and acknowledgment. *Information & Culture, 47*(1), 18–37.

Cronin, B., & Atkins, H. B. (Eds.). (2000). *The web of knowledge: A Festschrift in honor of Eugene Garfield*. Medford, NJ: Information Today & American Society for Information Science.

Cronin, B., & Franks, S. (2007). Trading cultures: Resource mobilization and service rendering in the life sciences as revealed in the journal article's paratext. *Journal of the American Society for Information Science and Technology, 57*(14), 1909–1918.

Cronin, B., & Overfelt, K. (1994). Citation-based auditing of academic performance. *Journal of the American Society for Information Science, 45*(2), 61–72.

Cronin, B., & Shaw, D. (2002). Banking (on) different forms of symbolic capital. *Journal of the American Society for Information Science and Technology, 53*(13), 1267–1270.

Cronin, B., Snyder, H. W., Rosenbaum, H., Martinson, A., & Callahan, E. (1998). Invoked on the Web. *Journal of the American Society for Information Science, 49*(14), 1319–1328.

Dalrymple, T. (2005). *Our culture, what's left of it: The mandarins and the masses*. Chicago: Ivan Dee.

Davenport, E. (1993). *Risks and rewards and electronic publishing: A case study of information science in the United Kingdom using a qualitative methodology*. Unpublished doctoral dissertation, University of Strathclyde, UK.

Davis, P. (2011). Gaming the impact factor puts journal in time-out. Retrieved from http://scholarlykitchen.sspnet.org/2011/10/17/gaming-the-impact-factor-puts-journal-in-time-out.

De Angelis, M., & Harvie, D. (2009). "Cognitive capitalism" and the rat-race: How capital measures immaterial labour in British universities. *Historical Materialism, 17*(3), 3–30.

Dear, P. R. (2000). Marin Mersenne. In W. Applebaum (Ed.), *Encyclopedia of the scientific revolution: From Copernicus to Newton* (pp. 668–670). New York: Garland Publishing. http://mey.homelinux.org/companions/Wilbur%20Applebaum%20(edt)/ENCYCLOPEDIA%20OF%20THE%20SCIENTIFIC%20REVOLUTIO%20(714)/ENCYCLOPEDIA%20OF%20THE%20SCIENTIFIC%20REVOLUTIO%20-%20Wilbur%20Applebaum%20(edt).pdf.

Eysenbach, G. (2011). Can tweets predict citations? Metrics of social impact based on Twitter and correlation with traditional metrics of scientific impact. *Journal of Medical Internet Research, 13*(4). http://www.jmir.org/2011/4/e123.

Franck, G. (1999). Scientific communication—a vanity fair? *Science, 286*(5437), 53–55.

Gazni, A., Sugimoto, C. R., & Didegah, F. (2011). Mapping world scientific collaboration: Authors, institutions, and countries. *Journal of the American Society for Information Science and Technology, 63*(2), 323–335.

Gross, A. G., Harmon, J. E., & Reidy, M. (2002). *Communicating science: The scientific article from the 17th century to the present.* Oxford: Oxford University Press.

Groth, P., Gibson, A., & Velterop, J. (n.d.). The anatomy of a nano publication. Retrieved from http://www.w3.org/wiki/images/c/c0/HCLSIG$$SWANSIOC$$Actions$$RhetoricalStructure$$meetings$$20100215$cwa-anatomy-nanopub-v3.pdf.

Hatch, R. A. (2000). Correspondence networks. In W. Applebaum (Ed.), *Encyclopedia of the scientific revolution: From Copernicus to Newton* (pp. 263–267). New York: Garland Publishing. Retrieved from http://mey.homelinux.org/companions/Wilbur%20Applebaum%20(edt)/ENCYCLOPEDIA%20OF%20THE%20SCIENTIFIC%20REVOLUTIO%20(714)/ENCYCLOPEDIA%20OF%20THE%20SCIENTIFIC%20REVOLUTIO%20-%20Wilbur%20Applebaum%20(edt).pdf.

Haustein, S., & Siebenlist, T. (2011). Applying social bookmarking data to evaluate journal usage. *Journal of Informetrics, 5*(3), 446–457.

Heffner, A. G. (1981). Funded research, multiple authorship, and subauthorship collaboration in four disciplines. *Scientometrics, 3*(1), 5–12.

Hicks, D., & Potter, J. (1991). Sociology of scientific knowledge: A reflexive citation analysis *or* science disciplines and disciplining science. *Social Studies of Science, 21*(3), 459–501.

Hoover, E. (2012). Colleges, ranked by "media buzz." *Chronicle of Higher Education.* Retrieved from http://chronicle.com/blogs/headcount/colleges-ranked-by-media-buzz/29880?sid=pm&utm_source=pm&utm_medium=en.

Howard, J. (2012). Scholars seek better ways to track impact online. *Chronicle of Higher Education.* Retrieved from http://chronicle.com/article/As-Scholarship-Goes-Digital/130482/#disqus_thread.

Jensen, M. (2007). The new metrics of scholarly authority. *Chronicle of Higher Education.* Retrieved from http://chronicle.com/article/The-New-Metrics-of-Scholarly/5449.

Koku, E., Nazer, N., & Wellman, B. (2001). Netting scholars: Online and offline. *American Behavioral Scientist, 44*(10), 1752–1774.

Kravitz, D. J., & Baker, C. I. (2011). Toward a new model of scientific publishing: Discussion and a proposal. *Frontiers in Computational Neuroscience, 5*(55). Retrieved from http://www.frontiersin.org/Computational_Neuroscience/10.3389/fncom.2011.00055/abstract.

Kurtz, M. J., & Bollen, J. (2010). Usage bibliometrics. *Annual Review of Information Science & Technology, 44*, 3–64.

Lewison, G. (2005). Beyond SCI citations—new ways to evaluate research. *Current Science, 89*(9), 1524–1530.

LiquidPub. (2012). Liquid publications: Scientific publications meet the web. Retrieved from http://project.liquidpub.org/liquid-publications-scientific-publications-meet-the-web-1.

Lomas, R. (2002). *The invisible college: The Royal Society, freemasonry and the birth of modern science.* London: Headline.

Long, J. S., & McGinnis, R. (1982). On adjusting productivity measures for multiple authorship. *Scientometrics*, *4*(5), 379–387.

Lozano, G. A., Larivière, V., & Gingras, Y. (in 2012s). The weakening relationship between the impact factor and papers' citations in the digital age. *Journal of the American Society for Information Science and Technology*, *63*(11), 2140–2145.

MacRoberts, M. H., & MacRoberts, B. R. (1989). Problems of citation analysis: A critical review. *Journal of the American Society for Information Science*, *40*(5), 342–349.

Manten, A. A. (1980). The growth of European scientific journal publishing before 1850. In A. J. Meadows (Ed.), *Development of scientific publishing in Europe* (pp. 1–22). New York: Elsevier.

Priem, J. (2010). Alt-metrics: A manifesto. Retrieved from http://altmetrics.org/manifesto.

Priem, J., Piwowar, H. A., & Hemminger, B. M. (2012). Altmetrics in the wild: Using social media to explore scholarly impact. *arXiv:1203.4745v1*.

Russell, T., & Sugimoto, C. R. (2009). MPACT family trees: Quantifying genealogy in library and information science. *Journal of Education for Library and Information Science*, *50*(4), 248–262.

Schroeder, R., Power, L., & Meyer, E. T. (2011). Putting scientometrics in its place [v0]. Retrieved from http://altmetrics.org/workshop2011/schroeder-v0.

Seglen, P. O. (1992). The skewness of science. *Journal of the American Society for Information Science*, *43*(9), 628–638.

Shuai, X., Pepe, A., & Bollen, J. (2012). How the scientific community reacts to newly submitted preprints: Article downloads, Twitter mentions, and citations. *PLoS ONE*, *7*(11), e47523. Retrieved from http://www.plosone.org/article/info%3Adoi%2F10.1371%2Fjournal.pone.0047523.

Sosteric, M. (1999). Endowing mediocrity: Neoliberalism, information technology, and the decline of radical pedagogy. *Radical Pedagogy*, *1*(1). Retrieved from http://www.radicalpedagogy.org/Radical_Pedagogy/Endowing_Mediocrity__Neoliberalism,_Information_Technology,_and_the_Decline_of_Radical_Pedagogy.html.

Taleb, N. N. (2010). *The black swan: The impact of the highly improbable*. New York: Random House.

Thelwall, M. (2012). Journal impact evaluation: A webometric perspective. *Scientometrics*, *92*(2), 429–441.

Vanclay, J. K. (2012). Impact factor: Outdated artefact or stepping-stone to journal certification? *Scientometrics*, *92*(2), 211–238.

Veletsianos, G., & Kimmons, R. (2012). Networked participatory scholarship: Emergent techno-cultural pressures toward open and digital scholarship in online networks. *Computers & Education*, *58*, 766–774.

Wagner, C. S. (2008). *The new invisible college: Science for development*. Washington, DC: Brookings Institution.

Waterman, S. (n.d.). Literary journals. In *Encyclopedia of Life Support Systems*. Retrieved from http://www.eolss.net/Sample-Chapters/C04/E6-87-04-03.pdf.

Watson, A. B. (2009). Comparing citations and downloads for individual articles at the Journal of Vision. *Journal of Vision (Charlottesville, Va.), 9*(4). Retrieved from http://www.journalofvision.org/content/9/4/i.

Wilhite, A. W., & Fong, E. A. (2012). Coercive citation in academic publishing. *Science, 335*(6068), 542–543.

Wouters, P., & Costas, R. (2012). *Users, narcissism and control—tracking the impact of scholarly publications in the 21st century*. Amsterdam: SURFfoundation.

2 History and Evolution of (Biblio)Metrics

Nicola De Bellis

Introduction

A number of standard definitions of bibliometrics and cognate fields have long been available:

> [Bibliometrics]: the application of mathematics and statistical methods to books and other media of communication. (Pritchard, 1969, p. 349)
>
> [Scientometrics]: the quantitative methods of the research on the development of science as an informational process. (Nalimov & Mulchenko, 1971, p. 2)
>
> [Informetrics]: the study of the application of mathematical methods to the objects of information science for the description and analysis of its phenomena, the discovery of its laws and the support of its decisions. (Nacke, 1979, p. 220)[1]

These definitions point to the fact that biblio/sciento/informetrics is about the application of mathematics and statistical tools to an increasingly elusive set of objects: books, science, information. We would probably agree on what counts as a book if we were asked to pin down a concrete example, but a similar consensus would be difficult, if not impossible, to achieve in the case of scientific or informational items. What counts as scientific, in fact, depends on compliance with agreed-on criteria of demarcation between science and non-science; at the same time, claiming that something qualifies as information is overtly noninformative since information is everywhere. The nice thing is that, despite the broad range of definitions, we do not have to worry about potential sources of uncertainty, for the true object of most current research in the field boils down to the quantitative analysis of published scholarly literature, notably journal articles and the network of their bibliographic connections. This makes sense. From the mid-17th century onward, at least in the natural and biomedical sciences, peer-reviewed journals have been the most efficient channels of reliable, certified knowledge. In addition, starting from the first half of the 20th century, journal articles have undergone a process of standardization relative to structure, format, and style that makes them ideal candidates for the automatic extraction of metadata preliminary to any quantitative analysis.

Contemporary bibliometrics is a deeply specialized and extensively institutionalized area of inquiry with strong practical implications in the science policy arena. Bibliometric measures, such as the number of publications and citations, are widely used as indicators of research performance (see Lane and colleagues, chapter 21, this volume). These measures can tip the balance of power in favor of one particular scientist or institution when it comes to promotion, funding, tenure, and the like. Additionally, sophisticated bibliometric maps can be drawn, portraying the structure and dynamics of scientific territories with unusual resolution. Bibliometrics is not rooted in a particular knowledge domain. By handling information patterns that belong to the knowledge transfer process generally, it holds the key to unlock the hidden structure of science and to "indicate" the occurrence of epistemic values in potentially any scholarly field: most productive, most cited, most collaborative, most "whatever" authors, groups, institutions, and countries.

How did we get to the point where scientific information is a measurable entity and the measure itself is a tool for managing the social forces involved in the production of new, supposedly better, scientific information? The sections that follow provide a critical overview of the intellectual background to bibliometrics across three deeply interconnected historical domains: (a) the emergence of an "objective" study of social facts in early positivist and functionalist philosophies; (b) the birth and rise of citation indexing and analysis, which supplied social scientists and technocrats with an evidence-building machine to capture unobtrusive indicators of research performance; and (c) the discovery and formalization of the mathematical structure of information processes—the so-called bibliometric laws—and their subsequent integration into the mainstream of deterministic and probabilistic modeling, which enabled bibliometricians to play the counting game according to widely shared mathematical rules.

The basic assumption is straightforward, although its consequences are not: bibliometrics is not just applied mathematics but a social science, sharing with other social sciences both the "dealing-with-people" part of the job and the propensity to use conceptual devices that, by modifying the very objects they are supposed to investigate, lend themselves to a wide range of theoretical and practical uses, including the justification of political arguments and decisions. In what follows, the emphasis will be on the three fundamental operations that, long before the advent of bibliometrics as a research field, prepared the discursive ground for a reflective study of science in terms of countable units: defining the object, producing the evidence, and setting up the rules of the game.

Defining the Object: Science as a "Good as It Is" Social System in the Positivist and Functionalist Traditions

It would be tempting, if not convenient, to reduce the history of bibliometrics to the history of statistics and the progressive colonization of scientific territories by descriptive and inferential data-handling techniques: lotteries, dice rolling, astronomical positions, crop yields, census and mortality tables, and, eventually, books and journal articles. But the big picture entails more than that. First and foremost, it involves the emergence of a reductionist attitude toward science whereby the full range of social and mental activities bolstering the generation of new knowledge ends up distilled in formal relationships between observable, easily collectible, and countable units. Such a history cannot be traced back to an exact starting point, the mythical origin date that no professional historiography seeks anymore. Nevertheless, some converging lines of perspective can be identified in conjunction with particular episodes or figures. Let's draw a conventional boundary at the point where, in the aftermath of the 16th–17th century Scientific Revolution, scholars became increasingly familiar with Galilei's idea that the *Book of Nature* is a sort of mathematical treatise writ large. Mathematics and physics, at that time, were the role model for other sciences, but reductionism could take various forms, the most relevant being those contingent on the calculation of probabilities and the biological functions of the human body.

It did not take long for 18th-century French scientists and *philosophes* to recognize that the mathematics of uncertainty, triggered by Blaise Pascal and Pierre de Fermat's correspondence in the 1650s, could be put to more vital uses than the mastering of chance games. In medical sciences, the association between number of favorable hits and scientific value came up on probabilistic grounds as soon as physicians were confronted with the task of determining, with irrefutable arguments, the efficacy of two alternative treatments: by administering each of them to equal groups of patients under exactly the same conditions, the more effective would reveal itself in the number of favorable outcomes. Moral and political sciences, too, were expected to benefit strongly from the calculation of probabilities, even more so in times of moral collapse and political upheaval. During the French Revolution, Nicolas de Condorcet, a politically committed member of the Académie des Sciences, set out to shape social sciences on the model of physico-mathematical disciplines. His project revolved around the *mathématique sociale*, a quantitative approach to human affairs aimed at the formulation of firm rules for value judgments on the shaky grounds of such controversial tasks as establishing the advantages/disadvantages of an election method or weighing the strength of our reasons to believe/disbelieve something. He maintained that, for the determination of such rules to be possible, general laws of behavior had to be inferred from observation by replacing individual values with mathematically treatable average values (Condorcet, 1793/1994). This kind of mathematical reductionism encountered the opposition of scholars who held human passions to be immeasurable: for the *idéologues* Destutt de Tracy and Pierre Cabanis, the main drivers of individual and aggregate human behavior could not be captured by mathematical formulas and if basic explanatory principles of social phenomena were to be found, human physiology was the right place to look for them.

Both mathematical and physiological reductionism were discredited by the subsequent evolution of sociology, but they did not lose seductive power among scholars heading for new avenues of research. The former resurfaced very soon, during the first half of the 19th century, under the guise of the *physique sociale* envisioned by the Belgian astronomer Adolphe Quetelet. In the post-Napoleonic era, nation-states started amassing statistics of every kind on people and their habits: enumerating and classifying groups of individuals for statistical purposes, especially in connection with deviant conditions (crimes, suicides, diseases, etc.), amounted to exercising a more effective social control over them. To a certain extent, statistical categorizing helped invent these groups, while the insistence on averages and dispersions objectified the idea of a "normal" status of beings and things (Hacking, 1990). For Quetelet, the scientific study of society followed from the mechanical extension of the law of error in astronomy to physical, moral, and intellectual qualities. The empirical frequencies of sizes and weights as much as those of spiritual propensities in human populations, in fact, revealed the universal occurrence of the bell-shaped Gaussian or normal distribution. In a given state of society, under the influence of certain causes, regular effects are expected, which oscillate around a fixed mean point. Thus, by calculating mean values over large populations it was possible, in Quetelet's view, to outline the true profile of the "average man," on which any political intervention should rest: the average man is in a nation what the center of gravity is in a body; he is "the centre around which oscillate the social elements" (Quetelet, 1842, p. 8). Yet surprisingly, even in Quetelet's averaging eyes, moral and intellectual qualities, such as criminal propensity, courage, and intellectual prominence, were not so easy to "normalize" since they required the indirect assessment of the effects referable to those qualities, namely, the number of criminal, courageous, or intelligent deeds. But no moral calculus could possibly be expected to achieve a degree of precision comparable to that of geometry: "How can we ever maintain, without absurdity, that the courage of one man is to that of another as five is to six, for example, almost as we should speak of their stature?" Or that "the genius of Homer is to that of Virgil as three to two?" (Quetelet, 1842, p. 9).

Similarly to Quetelet's mythical notion of the "average man," the physiological model of human behavior worked much like a scientifically productive nonscientific construct. To begin with, it channeled the assumption, shared among others by August Comte and Herbert Spencer, that society resembles a living organism, with its complex network of functional relationships among individual parts and the inexorable evolution through definite stages of growth and differentiation. In Comte's philosophical system, having done away with theological and metaphysical modes of knowledge, the human mind had finally reached the true "positive spirit," which consists of "substituting the study of the invariable Laws of phenomena, for that of their so-called Causes . . . in a word, in studying the *How* instead of the *Why*" (Comte, 1880, p. 34). The human mind—he argued—cannot be a direct subject of observations. We cannot observe other people's inner thoughts any more than we can observe our own from the outside. The best we can manage, then, is to analyze the mind's products, its concrete achievements following the translation of ideas into action. Specifically, the mental operations underlying the discovery of scientific truths in different sciences are reflected by the concrete deployment of their research methods. So a general philosophy of science, the science of science par excellence, ensues from the synthesis of the research methods that have proved successful in each individual area of inquiry. It is an empirical science of science, standing aloof from both mathematical and physiological reductionist attitudes. Émile Durkheim brought to logical completion Comte's and Spencer's functionalist approach. He claimed that social phenomena are—or should be tackled as—"objective facts" that can be observed and measured just like natural events, given the appropriate level of abstraction. Their study, accordingly, should be pursued through empirical data collection and statistical reasoning, and their explanation must refer to other social facts instead of narrowing to psychological or biological determinants. Statistical classification enables the observer to isolate a social fact, for instance a current of opinion, from its individual manifestations in real-world phenomena, such as high birth or suicide rates somehow ascribable to that current of opinion. Thus statistical averages instead of descriptions of the individual occurrences provide an expression—an indicator, we would say—of a certain state of the "collective soul" (*l'âme collective*) (Durkheim, 1895/1964).

By the turn of the 20th century, the idea that one of the highest forms of mental activity (i.e., engaging in scientific research) is the natural expression of a well-formed, "good as it is" social system, conveniently addressed through the tangible traces of its existence, was taken for granted by scholars of diverse origins. As to the formal definition of what a scientist is and how scientific excellence can be detected, the solution was already available in the conceptual space of Darwin's theory of evolution. According to Francis Galton, the most reliable way to identify a "man of science" as such was "to take the verdict of the scientific world as expressed in definite language" (Galton, 1874, p. 3). Reputation among peers, therefore, was the key indicator since "high reputation is a pretty accurate test of high ability" (Galton, 1869, p. 2). For quite some time, though, the only material evidence of high reputation was limited to an individual's appointment to prestigious academic positions or the inclusion in biographic dictionaries and encyclopedias. In the wake of Galton, Alphonse de Candolle and James McKeen Cattell issued early compilations of qualitative and quantitative data on prominent scholars identified according to their reputation. Like Galton, Candolle (1873) wished to make a point in the nature-nurture debate, emphasizing the environmental conditions for the occurrence of genius against the alleged predominance of hereditary factors. Cattell (1906) pushed the line even further and used the statistical analysis of peer ratings by experts to rank about a thousand of the scholars listed in his directory. As the size and ramification of scientific domains grew beyond manageable limits, however, the inadequacy of the old reputational criterion became clear, and turning to published records of scientific discoveries was the natural next step. On this basis, pioneering scientometric analyses were carried out, with very different purposes and from different theoretical frameworks, by scholars as well as librarians from the second half of the 19th century.

In their spare time, natural scientists started looking back at the progress made by their research field in terms of publication growth patterns. They did so either by applying simple descriptive tools, as in the landmark picture of the history of comparative anatomy drawn by Cole and Eales (1917), or venturing into curve-fitting exercises, as in Wilson and Fred's (1935) investigation of the "biological properties" of the literature on nitrogen fixation by plants. Librarians, for their part, carried out early literature-centered quantitative analyses with practical goals. Detached from the academic role-play and pressured by tightening budgets and physical space constraints, they were in the right place to consolidate the bond between published literature and objectified quality criteria, which they did using citation rates as proxies of journal quality (Gross & Gross, 1927). Above the fray, the cosmic eye of Paul Otlet foresaw the librarian of the future deeply involved with the systematic collection and classification of measures applicable to any kind of document. These were not just output estimates, but also content-related measures aimed at determining "the place, time and, insofar as the readers are concerned, the probability for texts to be read, hence for exerting their action on the society" (Otlet, 1934, p. 16).

By the end of the 19th century, the basic ingredients were ready for the precooked construct of science as a social institution amenable to factual analysis of its actors and products. At around the same period, the seeds were planted also for the opposing stance, namely, the antireductionist argument in favor of the irreducibility of human sciences to the cognitive style of the natural sciences. A heated debate broke out in Germany, the homeland of historical and philological studies, on the nature and methodology of the *Geisteswissenschaften* (human sciences). John Stuart Mill's received view that human sciences could achieve the same level of certitude and generality of the natural sciences building on the firm ground of individual psychology met with strong opposition from the anti-positivist camp. Several variations took form on the theme that the human sciences' distinctive feature is the uniqueness of their object and/or the specificity of their method. To be properly pursued, then, human studies required some sort of "artistic induction" (Hermann von Helmoltz), the ability to empathetically understand the inner world of lived experience (Wilhelm Dilthey), or the deployment of an a priori conceptual framework (Wilhelm Windelband, Heinrich Rickert, Georg Simmel, Max Weber). A sharp disagreement persisted on the role psychology should play in the foundational process, but in the end neo-Kantian ideas got the better of competing positions, and Max Weber's insistence on interpretative understanding (*Verstehen*) as critical to the assessment of subjective meaning in human actions marked indelibly the epistemic borderline between the social and the natural sciences for the years to come (Anderson, 2003).

Bibliometrics chose for itself the reassuring mantel of mathematics and the hard sciences, thereby dodging any possible identity crisis. Comte's commitment to the *How* instead of the *Why* showed the way. In one direction, Durkheim's model of morphological analysis and his emphasis on the nature, number, and interrelations among parts as crucial to sociological analysis would spawn a number of different research programs. The most relevant here is the British line of structuralism that, through the work of Alfred Radcliffe-Brown and Siegfried Nadel, developed into social network analysis, a very popular approach in recent quantitative studies. On a parallel track, consistent with Durkheim's and Talcott Parsons's insistence on internalized norms as critical to the equilibrium of social systems, Robert Merton explored the normative structure of science in terms of functional requirements that scientometricians would immediately endorse: scientific progress is possible only insofar as scientists, more or less consciously, abide by a set of rules delimiting what can be accepted as true scientific behavior from what cannot. These norms imply that scientists pursue universal knowledge by checking their hypotheses against shared criteria of logical validity and empirical verifiability; they engage in scientific research for its own sake, not for personal interest; and above all they do not keep the findings to themselves: communication in peer-reviewed scientific literature is integral to the functional equilibrium of the system and so is the practice of acknowledging prior relevant art in the form of bibliographic citations. Citing, specifically, is the same as peer reviewing, just on a smaller scale. Hence bibliographic citations are atomic components of the cognitive and reward system of scientific communication (Merton, 1942/1973). Defining the scientist as a paper-delivering professional whose reputation is also dependent on the network of mutual citation interlinkages in published literature was the best (or worst) service done by social studies to the burgeoning field of scientometrics.

Producing the Evidence: Citation Indexes and the Quantity-Quality Connection

Many bibliometric exercises and rankings count citations like pebbles in a jar: in perfect positivist style, each citation forms a solid and indivisible manifestation of endorsement. The more citations, the heavier the jar; the heavier the jar, the higher its impact potential. This peculiar form of evidence is context-free and theory-independent, a privilege that contemporary epistemologies do not grant to the factual evidence gathered in the process of justification or falsification of any scientific theory. To collect such a privileged form of evidence, a citation index is necessary, which lists the documents cited in the reference section of a selected body of (mostly periodical) literature next to the sources from which the citing references originated. The *Science Citation Index* (SCI) was the first interdisciplinary citation index to the journal literature available for large-scale scientometric studies. It was devised by Eugene Garfield during the 1950s and published on a regular basis since 1964 by his firm, the Institute for Scientific Information (ISI)—now Thomson Reuters—in Philadelphia.

The original mission of the SCI was not citation counting but improving literature searches. The birth of the index, in fact, would be unimaginable outside the efforts undertaken during the 1950s by computer and information scientists to find new solutions to an old problem: how to extract automatically and as quickly and efficiently as possible, from large text corpora, the appropriate index terms to be used for retrieving specific documents meeting the user's search criteria. Landmark solutions to the problem by Peter Luhn, Gerard Salton, and Karen Spärck Jones came from the analysis of language's statistical properties, resulting in such effective schemes as the "vector-space" model and the "term frequency–inverse document frequency" measure.

Garfield's out-of-the-box solution came instead from his acquaintance with structural linguistics, specifically from the idea that the complexity of scientific language could be reduced to a manageable set of structural units. Where to look in order to find such basic units was the crucial issue. Structural linguistics emphasized the importance of metatext in written communication—that is, text whose function is not to communicate a conceptual content but to introduce and locate it. A particular kind of metatextual relationship is also the one existing between the text of a journal paper and the concepts not exposed discursively but hinted at or summarized by the bibliographic references to the documents containing them. Plus, in a particular type of journal paper—review articles in the natural and life sciences—the metatextual relationship between texts and citations is predominant because almost every sentence in a review is backed up by a bibliographic reference, and the sentence itself introduces and partly anticipates the cited document's conceptual content with "an unusually definitive indexing statement" (Garfield, 1983, p. 7). So, by virtue of their status as "concept symbols" (Small, 1978), bibliographic citations could effectively complement, and sometimes even replace and outperform, words and subject headings as indexing units in a literature retrieval system. Even more, they could disclose an entire network of bibliographic relationships among documents and authors, a network susceptible to further analysis, both quantitatively and qualitatively, by sociologists and historians interested in tracking down the origins of an idea (Garfield, 1955).

Notwithstanding its initial vocation, from the very beginning the SCI's path to glory took a sharp detour from the bibliographic highway toward science policy quarters. In post–World War II science there was little or no room for the amateur and the genial dilettante. "Big science," the science behind the Manhattan Project, the Hubble space telescope, the antibiotics industry, and similar undertakings, was largely a collaborative effort, exceeding traditional academic and institutional boundaries, massively funded by government money and private capital as well. "Medium" and "small" science, too, carried out by academic research groups mostly at the local level, were facing new managerial challenges due to the hyperspecialization of research fields and the obligation, on the part of funding bodies, to account for the investment of public money. After the Sputnik affair it became clear in the West that science and technology were a matter of national welfare whose destiny could not be left to chance. They required monitoring and strategic planning, which science studies could help to accomplish by treading the same path as in Russia, where two influential scientometric schools had been established during "Khrushchev's Thaw" by Gennady Dobrov in Kiev and Vassily Nalimov in Moscow. A key role in channeling Eastern "Red" notions of science policy into the Western provinces was played by the British scientist John Desmond Bernal, whose work (1939/1967) triggered a knock-on effect of cross-fertilization between the two scientometric traditions on the opposite sides of the Cold War.

It was not until later in the 20th century that the SCI's potential in science administration gained momentum. Being cited by other authors is not simply a matter of intellectual lineage. When the score gets high, it is likely that the cited document is exercising an impact on the citing sources. Insofar as science is perceived as a cumulative enterprise, impact amounts to positive influence, contributing to knowledge advancement. This forward-pushing potential, in turn, is the hallmark of scientific quality. Thus, backed up by Merton's assumptions regarding the reward structure of science, citations retrieved and counted via the SCI formed the source of empirical evidence for the quantification of quality in research evaluation.

Unfortunately, the status-granting privilege worked also the other way around. Citations not only allowed quick and dirty estimations of the relative standing of documents and authors, but the very ability to provide them in a mechanical fashion by a market-leading, privately owned database enhanced the database's status to that of an additional filtering device, a sort of invisible gatekeeper acting alongside peer review at the level of journal subscription recommendations and individual publishing strategies. Being listed among the SCI sources would enhance the symbolic power and financial status of a journal to the point that libraries acquired it; scientists, accordingly, started submitting their manuscripts preferably to journals processed by ISI. The Journal Impact Factor, introduced by Garfield and Irving Sher during the mid-1970s, would further consolidate this trend. Originally meant to normalize citation counts at the journal level for supporting the SCI source selection process, it turned into a widely misused shortcut for attaching a crude score to the scientific merit of journals and even individual papers and authors.

What happened next is still too "live" to grant the privileges of a true historical perspective. The SCI citation data officially entered the evaluation game by inclusion in the National Science Foundation's *Science Indicators Reports 1972*. Initially, theory lagged behind practice as scholars were trying to make sense of the theoretical background of bibliometric indicators (Elkana, Lederberg, Merton, Thackray, & Zuckerman, 1978). Within a few years, however, things changed dramatically. After the launching of the journal *Scientometrics* in 1978 and the establishment of the Information Science and Scientometric Research Unit (ISSRU) at the Hungarian Academy of Sciences, Budapest, scientometrics blossomed into a full-fledged discipline, with its own apparatus of research facilities consisting of scientific conferences and societies, specialized reviews and monographs, and dedicated research centers and programs in United States and

Europe. The marriage between citations and scientometrics has been further cemented, from the mid-1970s onward, by the evolution of citation analysis in the five main directions outlined below.

Extension and Fragmentation of the Empirical Base

ISI citation data are full of noise, especially when it comes to institutional affiliations. Hence, leading scientometric agencies have been working, since the 1970s, preferably with in-house, enriched, and cleaned-up versions of the SCI. Each version forms a slightly different empirical base for the studies carried out by its users. The situation has deteriorated over the last two decades. After the acquisition of the ISI by a subsidiary of the Thomson Corporation, in 1992, the SCI and sister indexes were merged into the *Web of Science* portal hosted on the *Web of Knowledge* platform. They ruled the market until 2004, when Elsevier's *Scopus* and *Google Scholar* were released as competing tools for multidisciplinary citation searching. Hypertext technologies and markup languages have enormously facilitated the creation and multiplication of citation counting facilities in commercial and open-access databases of scholarly literature. Even the traditional elective affinity of citation indexes with the scientific article in English language has been questioned, over the last decade, by the extension of citation indexing to sources in different languages (e.g., the *Russian Science Citation Index*, the *Indian Citation Index*) and document types different from the journal paper, such as books and scientific datasets (e.g., Thomson Reuters' *Book Citation Index* and *Data Citation Index*). But more evidence is not necessarily better evidence: because of the differences in the source selection and indexing methodologies, the use of different citation indexes can lead to sharply different bibliometric profiles for the same collection of documents depending on the discipline or research field to which they belong.

Construction and Refinement of Citation-Based Indicators of Research Impact

The logic of development, in this regard, has been driven by the assumption that, through elementary algebraic manipulations, it is possible to "normalize" raw citation counts so as to distill pure citedness from confounding factors, such as size, age, self-citedness, multiple authorship, and above all, field-related citation attitudes. Furthermore, if Markov chain models are called in, it is possible to account for the relative standing of documents, authors, and journals in terms of the prestige of the citing sources. Understanding the conditions for unbiased comparisons of "like with like" has been, over the past three decades, the internal puzzle-solving activity underpinning the paradigm of evaluative bibliometrics. Early exemplary solutions to the problem include Francis Narin and Gabriel Pinski's PageRank-like algorithm for weighting the value of citations according to that of the citing sources (Narin, 1976); Tibor Braun and colleagues' normalized output and citation impact indicators for cross-field and cross-country comparisons (Schubert & Braun, 1986); Ben Martin and John Irvine's converging partial indicators (Martin & Irvine, 1983); and Anthony van Raan and colleagues' work on the empirical and methodological refinements of bibliometric indicators at the mesolevel of the research group or institution (Moed, Burger, Frankfort, & van Raan, 1985).

Investigation of the Meaning of Citations in the Context of Citing Texts and Citing Behaviors

Science historians, philosophers, and information scientists have long been aware that there is more to citations than settling an intellectual account. Merton himself had pointed out some possible rifts in the normative structure—for instance, the rich-get-richer phenomenon allowing elite scientists to accumulate disproportionate amounts of recognition by virtue of their acquired social status (the "Matthew effect"). Constructivist sociologists stressed the fictional nature of published papers' formal accounts and the rhetorical, essentially power-driven mission of cited references (Latour, 1987). The act of citing, accordingly, has been conceptualized as a process embedded in the social dynamics of scientific research, and citations have been downgraded to complex, unpredictable results of the interplay between private dispositions and social constraints, a status not compatible with a single grand theory (Cronin, 1984). Reaction to this awareness has taken three main forms. Some have tried to identify citer motivations within selected samples of authors and texts (for a review see Bornmann & Daniel, 2008). Others have been looking for the best predictors of citation scores through some form of regression analysis, a practice initiated by Stewart (1983) and Baldi (1998). Others have denied altogether the possibility and utility of a theory of citation based on citing behavior (van Raan, 1998; Wouters, 1999). Unfortunately, the microlevel streams of research appear no longer to attract a significant number of original contributions: a citation analysis "in context" for science policy purposes does not (yet) exist.

Development of Increasingly Refined Techniques for Drawing Bibliometric Maps of Science

Garfield's original idea was to investigate whether and to what degree "historiographs" (i.e., maps of the bibliographic connections between nodal papers retrieved via the SCI) could support the science historian's job in reconstructing the intellectual background of an important discovery (Garfield, Sher, & Thorpie, 1964). Next came cocitation analysis, with Small and Griffith's pioneering work at ISI during the 1970s. The road from surface to structure was built: no longer the thin bibliographic ties between documents visible to the naked eye, but the clusters of documents emerging from the application of multivariate techniques for dimension reduction (e.g., factor analysis, multidimensional scaling) and classification (e.g., cluster analysis) to groups of highly cocited papers (Small, 1973). Stable associations of highly cocited documents, in fact, were construed as markers of the intellectual boundaries of a knowledge domain. During the next decades, progress in the field was unrelenting. Cocitation analysis was extended to other mapping units (authors, journals, subject categories); the mapping arsenal was further enriched by the application of network analytic techniques to journal intercitation networks; hybrid approaches relying on both citation and full-text data were experimented with; and dramatic advances in computer visualization along with more advanced and scalable algorithms for data analysis turned maps into fine-grained, colorful, browsable representations of dynamic scientific territories ranging from emergent research fields to the global system of science (Börner, 2010). In the 1980s, a complementary line of research on science mapping was started by French sociologists at the École Nationale Supérieure des Mines, Paris: coword analysis, namely, the statistical analysis of word pairs from the title, abstract, or full text of scientific documents. Inspired by Latour's Actor-Network theory, coword analysts sought to identify common patterns of associations between words in order to shed light on the mechanisms controlling the production of literary texts (Callon, Law, & Rip, 1986). Recurring clusters of words, in their opinion, reflected the power relationships among the actors involved in scientific communication. Curiously, then, bibliographic citations and texts of scientific papers started telling sharply different bibliometric stories: the former in a functionalist/positivist style, the latter from a constructivist standpoint. The two poles would never reconcile in practice, even though a mathematically sophisticated attempt to optimize them in the framework of entropy statistics and information theory has been performed by Loet Leydesdorff (2001).

Extension of Bibliometric Techniques to Cyberspace

With the advent of the Internet and World Wide Web, cited references became linked nodes in the network of digitally available scientific documents. The Web itself manifested a citation-like organization, with hyperlinks between web pages formally similar to bibliographic citations in journal papers, suggesting the natural extension of bibliometric techniques to its communication structure (e.g., Thelwall, Vaughan, & Björneborn, 2005). Within the new field of cybermetrics, commercial or in-house-built web engines are used as surrogates of citation indexes to retrieve relevant data, and web server logfiles are a valuable source of countable footprints left by scholars accessing online documents. As a result, impact (webo)metric indicators have been devised similar to the Impact Factor and normalized citation scores; the cognitive and social structure of the hyperlink network in specific domains has been addressed through hyperlink network analysis; and both the Internet and the Web have been treated as special cases of heterogeneous, self-organizing networks to be modeled using the advanced mathematical tools of complex network analysis. Furthermore, in its latest stage of evolution, the Web 2.0 is a place where scientists publicly disclose much more of their pre- and post-publication activities than ever before. They form visible colleges through dedicated social networks, sharing data and experimental workflows as well as bibliographic records; they issue live comments and engage in endless conversations in blogs and tweets; and they rate other scientists' works outside the citation game (see Priem, chapter 14, this volume). All these activities are traceable and supply unique open information to complement the traditionally non-open procedures of peer reviewing and retrieving citation counts from privately owned databases. In the long run, a true leap forward in bibliometrics will also depend on the ability of next-generation metrics to take advantage of this enlarged empirical base for throwing new light on the dark regions of scientific communication.

Setting Up the Rules of the Game: Mathematical Life in a Skewed World

> My approach will be to deal statistically, in a not very mathematical fashion, with general problems of the shape and size of science and the ground rules governing growth and behavior of science-in-the-large. . . . The method to be used is similar to that of thermodynamics. . . . One does not fix one's gaze on a specific molecule called George, traveling at a specific velocity and being in a specific place at some given instant; one considers only an average of the total assemblage in which some molecules are faster than others. (Price, 1963, p. viii)

When the British historian of science Derek John de Solla Price announced the goal and methods of his scientometric research program, he had already issued his famous exponential law: whatever indicator of the growth of science one considers from the mid-17th century onward, be it, for instance, the number of publications or the number of universities, its normal growth rate is exponential, that is, it multiplies in equal periods of time by a constant factor. The exponential pattern was not homogeneous across all research fields and would not go on forever, but it appeared as the first bold materialization of a physical law in the study of science, mirroring the success of the Malthusian model of population growth in demography. Meanwhile, Price had obtained the machine printouts of the SCI 1961 data, which disclosed the network of citation links in a very large portion of scientific literature. He regarded citations as genuine markers of the intellectual linkages knitting together the fabric of new knowledge. Unlike the sheer distribution of publications, the distribution of citations reflected the merit structure of science, which turned out to be no less undemocratic (Price, 1965). Working out the details of such distributions, Price set the stage for the solution of three critical problems faced by scientometricians in the years to come: mapping out the "invisible colleges" of informally connected, highly cited scientists driving innovation on the "research front" (a task subsequently carried out through the analysis of coauthorship networks and cocitation analysis); pursuing the correlation between quantity and quality in the publication game—the more productive scientists also being the most cited (an idea later incorporated in *h*-type indexes); and investigating the deep regularities of citation practices at the disciplinary level, such as the "immediacy effect" in physical and biomedical sciences (a prerequisite for the field-based normalization of citation-based indicators).

Zooming in and out of the citation network, however, revealed the paradoxical nature of any quantitative inquiry into science. You can discover meaningful statistical patterns by watching the system from above, just looking at the global behavior of randomly moving actors, but you cannot get hold of the main driving force behind scientific progress—that is, the individual creativity of the "specific molecule called George, traveling at a specific velocity and being in a specific place at some given instant." At the heart of such impossibility, beyond the psychological and sociological factors accounting for the unpredictability of individual behavior, there is also a fundamental mathematical gap, a structural unevenness in the distribution of bibliometric qualities undermining the plain application of standard statistical tools to real science situations. This story goes back to the early decades of the 20th century, when three basic empirical regularities in the field of information science were revealed by Alfred Lotka, Samuel Bradford, and George Zipf.

Lotka (1926), Bradford (1934), and Zipf (1936) employed crude mathematical formulations, often referred to as "laws," to express the empirical relation between sources and the items they deliver in three areas: authors producing papers in a given field, journals producing papers on a given subject, and texts producing words with a given frequency. Their common denominator is a striking inequality in the pattern of information processes under observation: a few authors are responsible for most of the scientific literature in a given research field; a few scientific journals publish the majority of the papers relevant to any given subject; and a relatively small number of recurrent word units govern individual linguistic behaviors in scientific communication. In all three cases, the basic graph structure for the distribution of individual productivity values yields a hyperbolic or J-reverse or power law function, with a long tail of scattered values falling down toward regions of high concentration. Similar regularities also emerged in practical contexts. Donald Urquhart's analysis of interlibrary loans of journals from the Science Museum Library in 1956 revealed that less than 10 percent of titles accounted for approximately 80 percent of the requested items; likewise, in the late 1960s, a study of journal citation patterns in the SCI led to Garfield's "Law of Concentration," a generalization of Bradford's law stating that the core literature for all scientific disciplines involves a basic set of no more than 1,000 journals. In both cases, the same selective mechanism seemed to be at work, which aligned journal and library usage phenomena with a more general cumulative advantage process reflecting the social stratification of scholarship (Bensman, 1985).

Skewness is ubiquitous in natural and social phenomena alike. Statisticians knew it long before Lotka, Bradford, and Zipf. Indeed the realization that, contrary to Quetelet's belief, skewed instead of bell-shaped patterns of events are the norm in empirical datasets was at the core of the dramatic intellectual shift leading, in the second half of the 19th century, to the development of the tools and concepts of modern inferential statistics. The revolution's first phase began in England with the work of Galton and the British mathematicians moving around the circle of the journal *Biometrika*, including Walter Weldon, Karl Pearson, William Gosset (better known by his pen name "Student"), and George Udny Yule. Their technical accomplishments marked the switch from a mechanistic to a probabilistic view of experimental science: no matter how far from "normality" most empirical datasets appear to the naked eye, the real thing in investigating their structure is not the set of numbers resulting from necessarily inaccurate measurements, but the scatter of values according to a theoretical probability distribution expressed in mathematical form. It is the equation that makes the measurements significant by linking each individual outcome with its probability of occurrence, and the equation is univocally identified by abstract, unobservable "parameters" estimated from the data themselves. Pearson had worked out an entire family of such "skew distributions" supposedly fitting any conceivable empirical dataset. His solutions turned out to be inaccurate in several respects, but they pushed forward the idea that randomness is integral to observational phenomena. The only way to cope with it is to allow uncertainty into the picture from the beginning in the shape of sound mathematical models (Salsburg, 2001).

Lotka's, Bradford's, and Zipf's empirical regularities granted the right of citizenship to skewness in the realm of library and information science and prepared the ground for many mathematical treatments of informetric phenomena. Starting in the 1950s, but more persistently from the 1970s onward, several scholars tried to work out more rigorous mathematical versions of the original statements in order to test their goodness-of-fit with data gathered in several distinct subject domains. Yet the most interesting developments occurred when the structural similarity in the patterns of events described by those regularities became apparent and their relationship with analogous patterns in adjacent fields, such as Pareto's law in economics, was further explored. As a result, it was proved that, under proper assumptions, the three bibliometric "laws" along with other hyperbolic distributions occurring in nature and society are mathematically equivalent and can be adequately explained (or reduced) in mathematical terms by means of general stochastic (e.g., Price, 1976) or deterministic (e.g., Egghe, 2005) principles.

In the meantime, Lotka/Bradford/Zipf-type datasets appeared in many fields of the social sciences, from economics and sociology to linguistics, arousing the suspicion that normal and hyperbolic distributions conceal basic and irreducible differences in the structure of the events. The former seemed more adequate to represent natural phenomena deriving from the aggregation of many random, independent patterns of behavior; the latter appeared more consistent with social processes where randomness is constrained by the appearance of outliers that cannot simply be wiped out or otherwise domesticated, but have to be accounted for in a suitable mathematical framework. Following an early suggestion by Benoît Mandelbrot, some authors even jumped to the conclusion that power law distributions cannot be managed properly within the Gaussian paradigm and that a new framework was necessary for coping with such weird objects as unstable means, infinite variances, and unstable confidence intervals. Bertram Brookes, for instance, was so obsessed with the mathematical implications of Bradford's law that he invoked a new "calculus of individuality" capable of competing with the standard calculus of classes in the foundation of a new statistical theory for the social sciences (e.g., Brookes, 1979). In general, though, bibliometricians appealed to the central limit theorem to reinstate the validity of standard statistical procedures whereby meaningful analyses and comparisons can also be performed with skewed datasets as long as average values are not used to represent individual observations (Glänzel, 2010).

Ironically, both the conditions in Price's manifesto quoted at the beginning of this section—the "not very mathematical fashion" and the "George-excluding gaze"—were dismissed by the subsequent evolution of the field, thanks also to the extension of Price's exemplary solutions to concrete bibliometric puzzles. An increasing number of individual "Georges" around the world were going to be evaluated on the basis of unsophisticated indicators of research performance, and increasingly sophisticated mathematical models were going to be introduced to explain the structure of informetric datasets. In the latter case, especially, the evolution has been steady but quite irregular, because almost any available mathematical model from other sciences has found its way into some form of bibliometrics-related exercise. Typically, the application of such models to a subset or the whole of the science system has taken place in the form of isolated experiments, without any reference to a common theoretical framework (Scharnhorst, Börner, & Van den Besselaar, 2012). Yet, a discernible pattern, since the late 1970s, is the emergence and increasing importance of the time dimension: dynamic models of information flows bear the hidden promise of allowing predictions on the future course of development or, more modestly, of fostering a precise formulation of otherwise ambiguous notions of scientific change. Given the complexity of the creative process leading to innovation in science and technology, however, real predictions in the sense of forecasting probabilities are impossible in bibliometrics. The social conditions affecting the production and dissemination of new scientific knowledge do not fit the stylized constructs employed to set up a model equation. This partly explains why mathematical models stand aloof from the applicative realm of science management, where overly simplistic rankings and indicators of research performance rule the stage in spite of their being considered an aberration on purely mathematical grounds.

Conclusions

Bibliometricians build indicators of research impact based on bibliographic citations and promote colorful maps to capture the structure of science or to translate in empirical terms traditionally elusive constructs, such as the Kuhnian notion of "paradigm." On the other hand, the bibliometric paradigm itself (or paradigms, as would seem more appropriate) would hardly fit a bibliometric map. The reason is simple: the history of bibliometrics cannot be reduced to a Whiggish chronicle of who-got-it-right-first. Rather, it is the unfolding of different and only partially overlapping histories bearing on practical, philosophical, mathematical, and political dimensions that only occasionally make it to the epidermal layer of formal bibliographic connections. It is also the story of the perennial interplay between the belief in the possibility to measure the unmeasurable and the setback inherent in such a long shot. At one extreme there are and will always be the unconditional believers, those who claim that bibliometrics is scientific just because (or only if) mathematics is in charge. A plethora of such statements on the inner superiority of mathematical knowledge is available throughout history, from Lord Kelvin to Joseph Schumpeter, bibliometricians being only the latest in a long line. At the opposite end, the nonbelievers will defend the methodological specificity of any science dealing with human artifacts. The fracture has been there all along: in the open space between the two extremes, room is made for political decisions taking advantage of either one or the other according to which party better suits the local network of power relationships at any given moment.

Something new is happening, though. The digital network revolution has the potential to expand and democratize the tools and the techniques for communicating, evaluating, and counting science. Plus, it can reveal textual fragments and bibliographic connections along with the contexts surrounding them. In principle, then, it is easier now than ever before for sociologists and information scientists to look into the local dynamics of research dissemination with fresh eyes. But a larger empirical base and a richer repertoire of analytical techniques will not improve the field of evaluative bibliometrics just by themselves. At least not until the bibliometric report is understood to be little more than a dry list of (more or less sophisticated) indicators detached from any context-sensitive picture of real scientific work. Fully embracing the ambiguities inherent in the history of science-metrics, then, might just be a humble first step toward next-generation research evaluation reflecting the consensus of a wide range of scholars and interested parties, not just bibliometricians.

Note

1. All translations from non-English sources have been done by the author.

References

Anderson, R. L. (2003). The debate over the *Geisteswissenschaften* in German philosophy. In T. Baldwin (Ed.), *The Cambridge history of philosophy*, 1870–1945 (pp. 221–246). Cambridge, UK: Cambridge University Press.

Baldi, S. (1998). Normative versus social constructivist processes in the allocation of citations: A network analytic model. *American Sociological Review*, *63*(6), 829–846.

Bensman, S. J. (1985). Journal collection management as a cumulative advantage process. *College & Research Libraries*, *46*(1), 13–29.

Bernal, J. D. (1967). *The social function of science*. Cambridge, MA: MIT Press. (Original work published 1939)

Börner, K. (2010). *Atlas of science: Visualizing what we know*. Cambridge, MA: MIT Press.

Bornmann, L., & Daniel, H. D. (2008). What do citation counts measure? A review of studies on citing behavior. *Journal of Documentation*, *64*(1), 45–80.

Bradford, S. C. (1934). Sources of information on specific subjects. *Engineering*, *137*(3550), 85–86.

Brookes, B. C. (1979). The Bradford law: A new calculus for the social sciences? *Journal of the American Society for Information Science, 30*(4), 233–234.

Callon, M., Law, J., & Rip, A. (Eds.). (1986). *Mapping the dynamics of science and technology: Sociology of science in the real world.* Basingstoke: Macmillan.

Candolle, A. P. de (1873). *Histoire des sciences et des savants depuis deux siècles.* Geneva: H. Georg.

Cattell, J. M. (Ed.). (1906). *American men of science: A biographical directory.* New York: Science Press.

Cole, F. J., & Eales, N. B. (1917). The history of comparative anatomy. Part I: A statistical analysis of the literature. *Science Progress, 11*(43), 578–596.

Comte, A. (1880). *A general view of positivism* (2nd ed.). London: Reeves & Turner.

Condorcet, J. A. N. de Caritat, Marquis de (1994). A general survey of science—concerning the application of calculus to the political and moral sciences. In I. McLean & F. Hewitt (Eds.), *Condorcet: Foundations of social choice and political theory* (pp. 93–110). Aldershot: Elgar. (Original work published 1793)

Cronin, B. (1984). *The citation process: The role and significance of citations in scientific communication.* London: Taylor Graham.

Durkheim, E. (1964). *The rules of sociological method.* New York: Free Press. (Original work published 1895)

Egghe, L. (2005). *Power laws in the information production process: Lotkaian informetrics.* Amsterdam: Elsevier Academic Press.

Elkana, Y., Lederberg, J., Merton, R. K., Thackray, A., & Zuckerman, H. (Eds.). (1978). *Toward a metric of science: The advent of science indicators.* New York: Wiley.

Galton, F. (1869). *Hereditary genius: An inquiry into its laws and consequences.* London: Macmillan.

Galton, F. (1874). *English men of science: Their nature and nurture.* London: Macmillan.

Garfield, E. (1955). Citation indexes for science: A new dimension in documentation through association of ideas. *Science, 122*(3159), 108–111.

Garfield, E. (1983). *Citation indexing: Its theory and application in science, technology, and humanities.* Philadelphia, PA: ISI Press.

Garfield, E., Sher, I. H., & Thorpie, R. J. (1964). *The use of citation data in writing the history of science.* Philadelphia, PA: Institute for Scientific Information.

Glänzel, W. (2010). On reliability and robustness of scientometrics indicators based on stochastic models. An evidence-based opinion paper. *Journal of Informetrics, 4*(3), 313–319.

Gross, P. L. K., & Gross, E. (1927). College libraries and chemical education. *Science, 66*(1713), 385–389.

Hacking, I. (1990). *The taming of chance*. Cambridge: Cambridge University Press.

Latour, B. (1987). *Science in action: How to follow scientists and engineers through society*. Cambridge, MA: Harvard University Press.

Leydesdorff, L. (2001). *The challenge of scientometrics: The development, measurement, and self-organization of scientific communications* (2nd ed.). Parkland, FL: Universal Publishers.

Lotka, A. J. (1926). Statistics—the frequency distribution of scientific productivity. *Journal of the Washington Academy of Sciences, 16*(12), 317–325.

Martin, B. R., & Irvine, J. (1983). Assessing basic research: Some partial indicators of scientific progress in radio astronomy. *Research Policy, 12*(2), 61–90.

Merton, R. K. (1973). The normative structure of science. In R. K. Merton (Ed.), *The sociology of science: Theoretical and empirical investigations* (pp. 267–278). Chicago: University of Chicago Press. (Original work published 1942)

Moed, H. F., Burger, W. J. M., Frankfort, J. G., & Van Raan, A. F. J. (1985). The use of bibliometric data for the measurement of university research performance. *Research Policy, 14*(3), 131–149.

Nacke, O. (1979). Informetrie: Ein neuer Name für eine neue Disziplin. *Nachrichten für Dokumentation, 30*(6), 219–226.

Nalimov, V. V., & Mulchenko, B. M. (1971). *Measurement of science: Study of the development of science as an information process*. Washington, DC: Foreign Technology Division.

Narin, F. (1976). *Evaluative bibliometrics: The use of publication and citation analysis in the evaluation of scientific activity*. Cherry Hill, NJ: Computer Horizons.

Otlet, P. (1934). *Traité de documentation: Le livre sur le livre*. Brussels: Editiones Mundaneum.

Price, D. J. de Solla (1963). *Little science, big science*. New York: Columbia University Press.

Price, D. J. de Solla (1965). Networks of scientific papers. *Science, 149*(3683), 510–515.

Price, D. J. de Solla (1976). A general theory of bibliometric and other cumulative advantage processes. *Journal of the American Society for Information Science, 27*(2), 292–306.

Pritchard, A. (1969). Statistical bibliography or bibliometrics? *Journal of Documentation, 25*(4), 348–349.

Quetelet, A. (1842). *A treatise on man and the development of his faculties*. Edinburgh: William and Robert Chambers.

Salsburg, D. (2001). *The lady tasting tea: How statistics revolutionized science in the twentieth century*. New York: W. H. Freeman.

Scharnhorst, A., Börner, K., & Van den Besselaar, P. (Eds.). (2012). *Models of science dynamics: Encounters between complexity theory and information sciences*. Berlin: Springer.

Schubert, A., & Braun, T. (1986). Relative indicators and relational charts for comparative assessment of publication output and citation impact. *Scientometrics, 9*(5–6), 281–291.

Small, H. G. (1973). Co-citation in the scientific literature: A new measure of the relationship between two documents. *Journal of the American Society for Information Science, 24*(4), 265–269.

Small, H. G. (1978). Cited documents as concept symbols. *Social Studies of Science, 8*(3), 327–340.

Stewart, J. A. (1983). Achievement and ascriptive processes in the recognition of scientific articles. *Social Forces, 62*(1), 166–189.

Thelwall, M., Vaughan, L., & Björneborn, L. (2005). Webometrics. *Annual Review of Information Science & Technology, 39*, 81–135.

van Raan, A. F. J. (1998). In matters of quantitative studies of science the fault of theorists is offering too little and asking too much. *Scientometrics, 43*(1), 129–139.

Wilson, P. W., & Fred, E. B. (1935). The growth curve of a scientific literature: Nitrogen fixation by plants. *Scientific Monthly, 41*(3), 240–250.

Wouters, P. (1999). The citation culture. Unpublished doctoral dissertation, University of Amsterdam. Retrieved from http://garfield.library.upenn.edu/wouters/wouters.pdf.

Zipf, G. K. (1936). *The psycho-biology of language: An introduction to dynamic philology*. London: Routledge.

II Critiques

3 The Citation: From Culture to Infrastructure

Paul Wouters

Introduction

In the professional lives of young researchers, performance assessment has become a matter of course. They are being taught to act strategically in choosing research topics and publication outlets (if they have that freedom). PhD projects are increasingly formatted on the template of a collection of journal articles or chapters in edited books, rather than on the scholarly monograph, even in fields where the monograph was the standard only a few decades ago. For many aspiring research leaders, citation-based indicators are natural. They may mention the value of their *h*-index in their curriculum vitae (CV), and depending on their field, many of them will be familiar with Thomson Reuters' (formally ISI's) Journal Impact Factor.

The shift of scholarly publishing to the Web is further amplifying scholars' performance awareness. Most new generations of researchers will have extensive experience with social media and the proliferation of annotations—"like" buttons in Facebook, musical league tables in LastFM, statistical analyses of job connections in LinkedIn, and a host of other web-based indicators. Online performance assessment in science mirrors the scorekeeping of popular games with which this generation may be familiar (e.g., Dungeons and Dragons, World of Warcraft) and, in many ways, is in keeping with technologically savvy Western and Asian lifestyles. Accordingly, it may not seem incongruous for young scholars to engage in web-based performance analysis to exploit the considerable potential of indicators drawn from the various forms of formal and informal communication among researchers and between academics and the public.

Online tracking tools extend a promise of moving beyond traditional citation-based performance analysis—this promise is perhaps best expressed in the "Altmetrics Manifesto" (Priem, Taraborelli, Groth, & Neylon, 2010). The manifesto calls for research and technology development for tomorrow's information filters. The manifesto notes that traditional forms of publication in the current system of journals and books are increasingly supplemented by other forms of science communication. These include the sharing of "raw science" (e.g., datasets, code, and experimental designs), new publication formats (e.g., the "nanopublication," basically a format for the publication of data elements; Groth, Gibson, & Velterop, 2010), and widespread self-publishing via blogging, microblogging, and comments or annotations on existing work (Priem et al., 2010). The literature on "altmetrics" outlines four basic arguments in favor of these new tracking tools: diversity, speed, openness, and informality (Wouters & Costas, 2012). A larger variety of publication formats can be accommodated and these can be monitored in real time, whereas it may take years for a document to attract a sizable number of citations. Many tracking tools are freely available on the Web, and a number of them are also open-source. They may also enable the measurement of dimensions of science and scholarship not covered by citation analysis, such as the societal impact of research.

The promise of these new sources of data and potential indicators is immense. In fact, the excitement is not dissimilar to the enthusiasm of the early band of scientometricians under the leadership of Derek de Solla Price and Robert Merton about the promise of the new citation database created by Eugene Garfield, Gordon Allen, and Joshua Lederberg in the 1950s (Wouters, 1999b). The citation database was a novel information tool that promised to help researchers to find scientific information much more quickly, and also to enable them to follow the way their own research was being used. Moreover, it enabled a new empirical sociology of science. Evaluation for science management was not high on the agenda. Yet it was this application that has become dominant, even though advanced bibliometrics also enables much more forward-looking applications such as the detection of research fronts and the identification of new niche areas in research.

How has this happened? This chapter aims to contribute to research and debate about performance measurement in a web-based context by juxtaposing the potential of the new tools for assessment purposes to the development of the "citation cultures" in the recent past (Wouters, 1999b). The longer-term goal of this research project is to establish the theoretical foundations for a sociological citation theory that is able to analyze the interaction between performance assessment and the primary process of knowledge creation. This goal is motivated by the fact that we urgently need more knowledge about this interaction. Most citation theories do not focus on this problem, but rather try to understand the interpretation of citations in terms of the citing behavior of researchers (Nicolaisen, 2007). This chapter takes a small step toward this goal by sketching a possible framework for such a theory in relation to both accountability regimes and communication configurations.

First, I examine the evidence of the proliferation of performance indicators. This is not meant to question the existence of a large number of such measures, but to point to the lack of robust knowledge we have about many aspects of the process of research evaluation. This is particularly true for the implications of these indicators for the process of knowledge creation.

Second, I discuss the problem of how research communities respond to the performance criteria applied to them and discuss the characteristics of strategic behavior.

Third, I put this behavior in the context of the ambivalent attitude of researchers toward citation counts and draw on the study of "folk citation theories." On this basis, I propose the notion of the *citation as infrastructure* as the key element for a sociological citation theory, introducing the notion of infrastructure to explain the transparent nature of citation representation of scientific and scholarly work. This builds the foundation for suggesting that future analyses of alternative metrics should take into account how web-based monitoring stimulates the evolution of new knowledge infrastructures that may unwittingly change the criteria for future scientific and scholarly performance. This is not to discourage the development of new metrics, but to guide their evolution with awareness of the way infrastructures interact with scholarly practices.

Proliferation of Performance Indicators

Since the 1990s, the volume and diversity of the use of citation analysis in scientific research management and assessment have been mushrooming. As befits the underground nature of mushroom growth, a substantial part of this development has happened under the radar of science policy analysts, perhaps because they are unwilling or unable to stay current with or are unaware of the latest developments in scientometrics. The literature on the development of research evaluation in science policy studies tends to focus on formal tools of assessment and national evaluation systems (e.g., Lane, Fealing, Marburger, & Shipp, 2011; see also Lane and colleagues, chapter 21, this volume). In most of these systems, citation analysis does not play an important role, even in the indicator-based ones such as the Australian evaluation system (Gläser, Lange, Laudel, & Schimank, 2010). In most formalized systems, indicators of funding success seem to be more important than citation-based indicators. And indeed, getting funding for the next series of projects, books, or research is perhaps one of the more pressing concerns of research group leaders in a number of fields.

Despite this, it would be mistaken to conclude that citation-based performance indicators play a marginal role. In only a couple of decades, citation-based indicators have settled themselves firmly in the daily routines of virtually all research groups in the most competitive areas, including, albeit somewhat more slowly, the humanities and social sciences. This is stimulated by the creation of new citation databases that attempt to cover a larger variety of research output, including national journals and books, sets of controlled research data, and web-based data. How can we understand this seemingly unstoppable rise in popularity of the citation and its derivatives? How can we characterize the incorporation of measurement in the way scientific quality is commonly understood and defined? What are the implications for the processes of knowledge production in a regime of permanent self-monitoring based on a necessarily limited number of objectified performance dimensions? Are indicator-based evaluation, assessment, and monitoring steering the direction in which research develops, and if so, how?

These questions have become the focus of an emerging body of work from several disciplines. The question of the meaning of citation has been the focus of citation theories in the field of library and information science from the early 1960s (Bornmann & Daniel, 2008; Elkana, Lederberg, Merton, Thackray, & Zuckerman, 1978; Leydesdorff & Wouters, 1999; Nicolaisen, 2007). The governance of science has been studied both from the perspective of science policy and management studies (Lane et al., 2011; Schimank, 2005) and the sociology of science (Whitley, 2000, 2011; Whitley & Gläser, 2007; Whitley, Gläser, & Engwall, 2010). The latter body of work has also started to analyze how researchers cope with the demand for accountability, and many new projects in this area will likely produce insight into the interaction between evaluation and knowledge production in the next few years. The different perspectives are, however, rarely combined (Gläser, 2010; Gläser & Laudel, 2001; Gläser, Laudel, Hinze, & Butler, 2002). This is partly because it is a relatively new research agenda. It may also be attributed to the intimate nature of the interaction between evaluation (often confidential) and knowledge creation (often in the realm of the daily life of researchers). As a result, we know some of the answers to these questions, but not all (Gläser et al., 2002; Lamont, 2009).

A recent review of international evaluation practices by the Canadian Expert Panel on Science Performance and Research Funding concluded that indicator-based assessments of scientific work are "increasingly common," although "a great deal of controversy remains about what and how to measure" (Colwell et al., 2012, p. 2). The evidence for this continuing penetration of indicators and evaluation is based on a limited number of case studies or amount of anecdotal evidence (Moed, 2007, p. 578), because a systematic worldwide survey of evaluation practices at the level of individual researchers and research groups does not exist.

It is safe to assume that we are indeed witnessing performance indicator growth at all levels and in almost all fields, provided we take into account that the types of application vary widely. The journals *Nature* and *Science*, perhaps the best trend watchers in the system, regularly report on new developments at the indicator front, often noting in passing that we are being confronted with a deluge of indicators and measurements. In 2010, *Nature* reported on "a profusion of measures": "Within the past decade, the development of ever more sophisticated measures has accelerated rapidly, fuelled by the ready availability of online databases such as the Web of Science from Thomson Reuters, Scopus from Elsevier and Google Scholar" (Van Noorden, 2010, p. 864). In a similar vein, Bollen was quoted as saying that we are going through "a Cambrian explosion of metrics" (Van Noorden, 2010, p. 864). According to Van Noorden (2010, p. 864), it has become "all but impossible even to count today's metrics." This has changed the position of citation metrics, which is no longer the preserve of a small bibliometrics community, but is widely used and debated "by a broad public of administrators, analysts, editors, librarians, and individual researchers" (Pringle, 2008, p. 86). The global university rankings have further encouraged the use of indicator-based assessments and rankings in universities across the globe (Hazelkorn, 2011).

Indicator proliferation has been confirmed by studies of management and labor relations in the higher education system. Roger Burrows (2012) places "the moment of metrics" in the United Kingdom at some point between the Research Assessment Exercise (RAE) carried out in 1996 and the one conducted in 2001 (p. 359). Indeed, although performance-based funding systems first emerged in 1984 in the United Kingdom, their international appeal is a recent phenomenon. By late 2010, 14 countries had adopted a system in which research funding is explicitly determined by research performance (Hicks, 2012). In addition, performance indicators play an increasing role in systems where funding is influenced in a more indirect way, such as the Standard Evaluation Protocol in The Netherlands. In the 1990s, the use of citation indicators by disciplinary evaluation committees in the Dutch system was already "common practice" (Meulen, 1997).

Academic accounts of the postwar university confirm that performance evaluation with material consequences is relatively new (Halsey, 1995; Slaughter & Leslie, 1997): "Measure, as we would now recognise it, simply did not exist in the post-war university or polytechnic" (De Angelis & Harvie, 2009, p. 10). Academics can "no longer avoid the consequences of the developing systems of measure to which they are becoming increasingly subject" (Burrows, 2012, p. 359). In a recent global survey of researchers' attitudes with respect to the Journal Impact Factor, almost 90% of the participants reported that this indicator is "important or very important" for the evaluation of scientific performance in their country (Buela-Casal & Zych, 2012, p. 289). The explosion of interest in the Hirsch Index (*h*-index) after its introduction in 2005 (Hirsch, 2005) is itself an indicator of indicator proliferation. Anne-Wil Harzing has developed a web service to calculate a range of citation indexes on the basis of Google Scholar, among others the *h*-index (Harzing, 2010), which seems to be relatively popular, not least in the social sciences, where the Web of Science coverage is less effective (Klandermans, 2009).

The Journal Impact Factor and the *h*-index are widely seen as the two most popular bibliometric indicators and perhaps also the most abused ones. They are also a demonstration of how this proliferation is relatively independent of formal protocols. For example, the Dutch national protocol does not recommend using either of the two measures, but rather favors a more systematic bibliometric assessment of research groups as an option in the context of a peer review–based judgment. Yet, both measures are used on a regular basis by deans, research group leaders, university provosts, and publishers and this remains undocumented. Hence, we do not really know how individual evaluators are being influenced by their attempts to use easily available yet potentially misleading indicators. We know even less about how these individual acts of evaluation accumulate to steer research agendas, funding decisions, and career trajectories. What we do know is that we need to look beyond the official politics of evaluation and try to understand the micropolitics of citation indicators.

Strategic Behavior

To understand the implications of this proliferating application of performance indicators, we first need to understand the way researchers and scholars respond to assessments and indicators. This is particularly relevant since scientists have played an active role in these measurement systems. Because the scientific system is a highly competitive social structure, it is in the interest of high-scoring researchers to actively display their scores, thereby fueling the further development of the system of indicators. It has now become routine for researchers in the medical sciences to put their *h*-index prominently on their CV and personal website, as noted by Acuna, Allesina, and Konrad (2012, p. 201): "The typical research CV contains information on the number of publications, those in high-profile journals, the h-index and collaborators." The virtues of the *h*-index have also been extolled by editors—for example, Clackson (2009, p. 311), of *ACS Chemical Biology*: "The brilliance of the h-index is that it provides a single, easy to compute, quantitative measure of your cumulative impact. You want your impact to go up! So, it follows directly and easily that all decisions in your career should be considered in terms of their potential to boost your h-index."

According to Butler (2007), any system used to assess research, whether peer review– or indicator-based, that involves money or reputation, will tend to affect researchers' behavior. Two different types of strategic behavior can be discerned. The first is goal displacement: high scores in the measures become the goal rather than a means of measuring whether an objective (or performance level) has been attained (Colwell et al., 2012, p. 27). The second is a more profound change in the research process itself in response to the assessment criteria, which may be more difficult to recognize (Butler, 2007, p. 572).

Goal displacement has indeed been shown by studies that have looked at the effects of funding and evaluation regimes on the production of scientific articles. Colwell et al. (2012) conclude that linking funding explicitly to research output (in terms of the number of publications) may lead researchers to produce a higher quantity of publications at the expense of their quality (p. 26). An analysis of the experiences in Australia in the years when funding was based on aggregate publication counts with little attention to the impact or quality of that output showed that journal publications increased at the cost of declining impact (Butler, 2003, p. 143). Strategic response by the research community was also shown in a longitudinal bibliometric study of UK science covering almost 20 years. In 1992, total publication counts were requested and British scientific production substantially increased. When, in 1996, a shift in criteria was announced from "quantity" to "quality," UK authors gradually increased their number of papers in journals with a relatively high impact factor (Moed, 2007, p. 579). Another study of the impact of the UK RAE concluded that the cumulative research productivity of individuals increased over time, but the effects differed across departments and individuals, revealing a rather precise strategic response:

> Individuals at higher-ranked programmes tended to respond by increasing their research output in higher-quality journals, while individuals at other programmes tended to increase their publications in other outlets. The productivity response occurred primarily among individuals whose pre-RAE output was below the requisite number of publications required to be included in the RAE. (Moore, Newman, Sloane, & Steely, 2002, abstract)

A survey among journal editors conducted at the end of the 1990s also showed that the RAE influenced where authors published. However, the common fear that the RAE would lead to "salami publishing"—the practice in which a project is published in as many separate articles as possible—was not confirmed (Georghiou et al., 2000).

There are indications that more fundamental changes in the research process induced by assessments and performance indicators may also take place. Sometimes the evidence is contradictory. For example, a decrease in the quality of submitted articles was not detected in Britain (Georghiou et al., 2000). In China, however, financial bonuses based on publications are suspected of stimulating fraudulent practices

(Colwell et al., 2012, p. 27). An analysis of researchers' responses to funding criteria in Australia has addressed the extent to which researchers are forced to focus their tasks and publication forms (Laudel & Gläser, 2006). The inherent tension here is between the ubiquitous urge for evaluations and the complex nature of communication in research and scholarship. In 10 of the 21 disciplines, the four publication types used in the evaluation were not identical to the four publication types that researchers found most important. This can lead to the abandonment of particular types of work. As a historian in the study put it:

> I mean, the way we are funded now by the government, by the faculty, by the university, we are severely discouraged from writing book reviews, we are severely discouraged from writing reference articles, encyclopedia articles. I mean, if somebody asked me to do that now, I always say no. (Laudel & Gläser, 2006, p. 294)

This study concludes that the arts and humanities are bound to suffer most since journal-oriented disciplines serve as archetypes for evaluations.

Task reduction has also been reported in the social sciences. A worldwide survey among demographers in developed and developing countries reports that traditional tasks of scholars and members of a scientific community, such as writing referee reports or translating research outcomes for the public or policymakers, are negatively affected by "the drive toward individual productivity" (Van Dalen & Henkens, 2012, p. 1291). A study of the RAE exercise in 1996 similarly reported that publication in professional journals "is actively discouraged" by some managers (McNay, 1998, p. 22). In science and technology studies, doctoral students have increasingly been advised to choose a dissertation topic based on journal articles rather than a monograph because the former would make them more visible in the labor market, given the existence of journal-based citation indexes (personal communications with doctoral students).

A related concern is the potential influence of disciplinary assessments on interdisciplinary research. Demographers show no tendency to focus on monodisciplinary research in terms of reading or publishing activity (Van Dalen & Henkens, 2012). In economics and many departments in business studies, however, publication productivity has been strongly shaped by the ubiquitous use of journal rankings. These lists are not based on citation, but on a qualitative consensus in mainstream economics and business studies about the top journals. A comparative analysis of the effect of these rankings in business and innovation studies found them to be biased against interdisciplinary work (Rafols, Leydesdorff, O'Hare, Nightingale, & Stirling, 2012). This study concludes that citation indicators may be more suitable than peer review for interdisciplinary work because criteria of excellence are essentially based on disciplinary standards. A survey of the impact of the 1996 UK RAE also reported evidence of negative effects for interdisciplinary work. Almost half of those in management positions felt the RAE "had hindered" interdisciplinary work (McNay, 1998, p. 20). The Australian Research Council decided to drop the established journal rankings in its assessment system when research managers began to set publication targets on the basis of the two top categories on the list (Colwell et al., 2012, p. 56).

In addition to possible reduction of both task complexity and interdisciplinary work, assessments have also been thought to affect institutional arrangements. According to Colwell et al. (2012), the RAE has unintentionally created a "transfer market" for faculty members in the United Kingdom: "The RAE has led some universities to focus on hiring younger staff with research potential, while others have taken a more conservative approach and focused on hiring well-established researchers" (Colwell et al., 2012, pp. 27–28). In Australia, universities responded to formula-based funding by more or less mirroring this practice (Gläser et al., 2002, p. 17). It is plausible that this mirroring effect is a more general phenomenon, due to the way the university organization tends to transfer responsibility. This type of institutional strategic behavior may have long-lasting effects on the position of universities and research institutes and so on research agenda setting.

In summary, the evidence indicates that performance-informed funding (with or without a formal link between performance and funding) does indeed increase the pressure on researchers and institutions to meet the performance criteria, irrespective of whether the latter are based on peer review or on citations. This is clearly an intended effect. After all, performance indictors became prominent in the governance of science in order to change the dynamics of scientific production and bring it more into line with current science and innovation policy priorities (Whitley & Gläser, 2007). The research community responds strategically, and this may in turn create unintended effects, either through the mechanism of goal displacement or through more structural changes in research priorities, publication activities, or research capacity and organization.

We should also note that the evidence is fragmented, incomplete, and sometimes contradictory. The most visible types of strategic behavior may obscure more fundamental shifts in knowledge production. Performance-based systems have been studied most frequently, but it is questionable whether a direct link between funding and performance is necessary for these effects to occur. The effect is not primarily based on the amount of funding that is shifted due to performance differences, but on the effect it has on researchers' reputations (Hicks, 2012). Systems where performance is publicly reported but not directly linked to funding may therefore lead to comparable or identical effects.

The great unknown is the long-term effect of current approaches to evaluation-based funding (EBF) on the nature of the knowledge created, as noted by Gläser et al. (2002, p. 17):

> While it has been established with some clarity that researchers indeed adapt their practices of knowledge production to the availability of resources, no ethnographic observation so far has provided sufficient data about how researchers adapt to EBF practices and how this adaptation changes the knowledge that is produced.

This remains an accurate observation.

Ambivalent Attitudes

Strategic responses by researchers are not only shaped by performance criteria, but also by the theories researchers hold about the research system and about science in general, as well as by the positions researchers maintain in disciplinary and institutional matrices (Whitley, 2010). We can expect that researchers are knowledgeable about their own discipline and institution, but this does not mean that they are in a position to systematically assess their theories about science. Hence, these theories and conjectures can best be analyzed as "folk theories"—theories that are empirically grounded, yet not scientifically created. Researchers often have an ambivalent attitude with respect to performance and citation indicators (Hargens & Schuman, 1990). On the one hand, researchers play an active role in making their performance scores visible, thereby contributing to the proliferation of these measures; on the other hand, they tend to protect their position as experts by pointing to the limitations of number-driven evaluations by outsiders.

Hence, citations are simultaneously reified in competitive struggles and criticized for not reflecting actual scientific contribution (Aksnes & Rip, 2009, p. 895). As Aksnes and Rip (2009, p. 899) note:

> To further their career and/or obtain credit to mobilise resources, scientists will apply themselves to publish and improve their chances of getting citations. And when these do not materialize, [they] find excuses related to visibility dynamics leading to discrepancies between citation counts and (self-assessed) contribution to science.

In other words, scientists' ambivalence about citation indicators is expressed in asymmetrical or contradictory interpretations of citation scores. Hargens and Schuman (1990) showed that researchers tend to protect their beliefs about the value of their work in their interpretation of citation scores. However, the sophistication and complexity of scientists' interpretation of citation should not be underestimated. Aksnes and Rip (2009, p. 904) observed that their respondents appeared to be "quite knowledgeable of citations." For example, scientists were able to mention a large number of reasons why citation counts do not correspond with scientific contribution: "This was based on their own experience, as well as their referring to a shared, and somewhat cynical repertoire about citations" (Aksnes & Rip, 2009, p. 904). Researchers also recognize "overcitedness," and these cases are often understood in terms of timeliness and relevance to larger audiences: "In other words, scientists have a sophisticated understanding of the citation process and its outcomes, and can explicate such understanding when there are no immediate stakes to be defended, as in a questionnaire asking them about earlier papers" (Aksnes & Rip, 2009, p. 904).

Scholars' ambivalence with respect to assessment and performance indicators cannot be explained by simple interest-based commonsense arguments, but rather by the interaction between individual and structural factors. According to Hargens and Schuman (1990), use and evaluation of citation counts vary by the level of consensus in a scholar's discipline, specialties' orientations regarding the value of empirical data, and the prestige of one's department (p. 219). In their study, quantitative sociologists were more positive toward citation counts than sociologists whose style was qualitative. Highly cited researchers tended to be more positive than less frequently cited ones. At the same time, researchers in a field with consensus on quality standards were less reliant on citation scores. Thus, Hargens and Schuman (1990) found that biochemists were less likely to consult the Science Citation Index for evaluation purposes than sociologists. Hence, researchers tended to mobilize citations mainly when uncertainties about quality prevent consensus in the relevant scientific community (Porter, 1995). Otherwise, peer judgment sufficed.

The degree to which this is still the case is unknown. Bibliometrics is becoming a normal part of the analytic toolset for the management of large research institutes, academic hospitals, and universities. For example, academic hospitals in The Netherlands have been monitoring their scientific production with the help of advanced bibliometrics and research information systems for many years, irrespective of the level of consensus about theory and method in the relevant specialties (Van Kammen, van Lier, & Gunning-Scherpers, 2009). In Scandinavia and Germany, libraries are increasingly positioned as the site for in-house bibliometric expertise (Åström & Hansson, 2012; Ball & Tunger, 2006). This trend is also visible in The Netherlands and the United Kingdom. This development is not restricted to Europe. The U.S. National Science Foundation has turned full circle by creating "the science of science policy," thereby reinstating the older tradition of "the science of science" (Lane et al., 2011).

In Asia, statistics and bibliometrics are even more popular. Scoring higher in the international university league tables and attracting more highly cited scientists are important elements of many Asian science policies. The new Australian program Excellence for Research in Australia (ERA) (started in 2009) is also based on the combination of metrics and peer deliberation (ARC, 2012). This new stage in the development of bibliometrics coincides with a general trend toward "informed peer review" (Butler, 2007; Colwell et al., 2012; Moed, 2007; Weingart, 2005), in which peer review experts and procedures provide the overall framework for evaluation, but statistical information and citation indicators have a specific, and often obligatory, role to play. As Colwell et al. (2012, p. 30) explain: "The implication is not that quantitative indicators should be eliminated from research assessments, but rather they should be used to inform expert deliberations."

The Citation as Institution

The citation has become firmly established as a social institution in a variety of forms. The extremely fast uptake of the *h*-index as a representation of an individual's scientific impact or quality may indicate that the institutionalization of the citation was already well advanced in 2005, albeit not in all fields of research. Journal editors and researchers alike are aware of and influenced by the Journal Impact Factor. Sonuga-Barke, an editor in chief, expressed his feelings in anything but weak terms:

> For many researchers the JIF is considered the only marker of a journal's quality and value and universities and other research institutions can perpetuate this oversimplistic view by basing promotions on a member of staff's record of publishing in high JIF journals. Some universities go so far as to instruct their staff to only publish in [a] journal with an impact factor above a particular score. (Sonuga-Barke, 2012, p. 915).

Both indicators have been naturalized as instantiations of quality irrespective of the methodological critiques by professional scientometricians, and even their inventors, Jorge Hirsch and Eugene Garfield, have not been able to limit the range of applications of the indicators.

Institutionalization is a process that can overpower the professional community from which the indicator first emerged. It may even create feelings of powerlessness among leading academics: "The index has become reified; it has taken on a life of its own; a number that has become a rhetorical device with which the neoliberal academy has come to enact 'academic value'" (Burrows, 2012, p. 361). Note that this is written in a country where the academic elite lobbied successfully against basing the national research assessment system (the new Research Excellence Framework) on metrics (Whitley & Gläser, 2007).

The institutionalization of the citation is the culmination of a decades-long process starting with the creation of the Science Citation Index. The impact of this emergence of a new social institution in science and scholarship is often underestimated because its development has been so strongly entwined with two more visible institutions: scientific communication with its codified bibliographies and referencing norms on the one hand, and the growth of accountability regimes in science and scholarship on the other hand (Shore, 2008; Stensaker & Harvey, 2011; Strathern & Mitchell, 2000; Woolgar, 2002). The first institution has provided a seemingly clear-cut justification to use citations as proxies for influence or quality, and gave Robert Merton the materials for his influential sociology of the citation as an instance of one of the norms of science (Elkana et al., 1978). The second institution has been the main context of the development and refinement of assessment and evaluation criteria, and seen from this perspective, bibliometric information simply seems one of the many possible sources of evidence and data. If we wish to understand the recent history of the citation and gauge its possible futures as a component of accountability regimes, we need to recognize the citation itself as a novel social institution with its own dynamics and analyze its international network of technologies, companies, and academic groups. We also need to analyze the interaction between the citation and both communication and accountability regimes.

For example, in many recent policy documents about research assessment it is advised to take an "informed peer review" approach (Colwell et al., 2012; KNAW, 2010, 2011; Phillips, 2012). This means that peer experts should be dominant in quality assessment but they should also use available bibliometric evidence. The idea is that this will help limit the bias in and increase the quality of the peer review process (Moed, 2007). But how should one combine the qualitative and quantitative information? This is often left to the discretion of the peers in question. Methodologically, the concept of informed peer review is not well developed. Often, it is not much more than the advice to combine qualitative and quantitative evidence in a prudent manner—and this is more a moral than a methodological standard. Therefore, informed peer review may lead to a variety of practices, and it is an open question whether the combined use of bibliometrics and peer judgment will indeed enhance the quality of the evaluation. For example, particular disciplinary biases may be reinforced rather than weakened, especially when subdisciplines have diverging publication or citation cultures.

Taking a more distanced analytic approach may help here. Since the peers are acting scientists (by definition), they will be subjected to the same pressures all researchers and scholars experience. Hence, they will tend to reason strategically, protect their beliefs about scientific quality, and have an ambivalent attitude with respect to citation analysis. Since they will be subject to evaluation themselves, they can be expected to take into account how their decisions may act as precedents for their own evaluations. Thus, one can expect that peers will tend to organize, use, and interpret citation data as if the data were being applied to themselves. Highly cited researchers will look more favorably on citations than peers hardly visible in the citation counts. Disciplines will also vary: some will use advanced bibliometric analysis, while others may depend on Google Scholar or other easily available indexes. Therefore, we can expect an alignment between the quality framework peers have before they entered the evaluation and the interpretation of the citation data.

We cannot blame the evaluators for this, since they cannot draw on an established theory of citation. If bibliometrics had developed an unambiguous theory that provided firm guidelines regarding how to interpret citation data in an evaluative context, the matter would be completely different. So far, however, such a theory has not yet emerged, although important building blocks for such a theory have been proposed (Bornmann & Daniel, 2008; Nicolaisen, 2007). This lack of methodological standards threatens to make "anything go" within the framework of informed peer review, even if peers wish to use state-of-the-art knowledge from the bibliometric community. The bibliometric community therefore increasingly feels the need to develop open quality standards for evaluative bibliometrics, an important agenda point for many scientometric research centers. These standards will be based on key concepts in citation theory.

In the long run this will require a paradigmatic shift in the field, and here the sociology of evaluation becomes relevant. So far, citation theory has mostly been developed in terms of science communication as an institution. However, the dynamics of the application of citations for evaluation, and the dynamics of the field of evaluative bibliometrics, are not determined by the institution of science communication, but by the institution of accountability.

In other words, many scientometricians have tried to ground the meaning of the citation in the way researchers cite, which makes sense in the framework of communication (Nicolaisen, 2007). They thereby often implicitly assume a transparent translation from the communication regime into the accountability regime. This assumption is flawed since it ignores the different dynamics of the latter regime, in which aggregated and normalized citation counts circulate and create a novel representation of scientific impact and quality, rather than the individual reference as the source of the citation (Wouters, 1997). A citation theoretical approach that aims to provide theoretical and methodological guidance to the use of citations in research assessments needs to acknowledge the accountability regime as the main framework and analyze how the communication system interacts with regimes of accountability. Solving this puzzle is of eminent practical relevance for the evaluation practice of the near future. This is confirmed by a recent Canadian panel on performance and funding, which expressed a distinct unease about the use of indicators in research evaluation:

> Past experiences with science assessment initiatives have sometimes yielded unintended, and undesirable, impacts. In addition, poorly constructed or misused indicators have created skepticism among many scientists and researchers about the value and utility of these measures. As a result, the issues surrounding national science assessment initiatives have increasingly become contentious. (Colwell et al., 2012, p. 3)

The Citation as Infrastructure

> The life-world of the university is now increasingly enacted through ever more complex data assemblages drawing upon all manner of emissions emanating from routine academic practices such as recruiting students, teaching, marking, giving feedback, applying for research funding, publishing and citing the work of others. Some of these emissions are digital by-products of routine transactions (such as journal citations), others have to be collected by means of surveys or other formal data capture techniques (such as the National Student Survey (NSS)) and others still require the formation of a whole expensive bureaucratic edifice designed to assess the "quality" of research. (Burrows, 2012, p. 359)

This observation of the effect of, in the words of Whitley (2010, p. 33), "a strong research evaluation system," indicates that it is not enough to signify the assemblage of the citation in its many forms as a social institution. We also need to specify what kind of social institution it is and how its dynamics are shaped. Burrows (2012) points to two properties of the citation-as-institution: its character as a database and its effect on "routine transactions." This resonates well with how historians of science have analyzed large technical systems: "To be modern is to live within and by means of infrastructures: basic systems and services that are reliable, standardized, and widely accessible, at least within a community" (Edwards, 2010, p. 8). Infrastructures should not be identified with network technologies per se, such as the railroad network. Rather, they are the taken-for-granted context that enables our life and work. Infrastructures are multilayered and complex and cannot be constructed top-down. Instead they evolve bottom-up. An important feature of infrastructures is their invisibility: they operate in the background and become only visible on breakdown. They are shaped by the standards and conventions of particular communities and are the embodiment of these standards, in turn shaping the conventions of a larger number of communities (Star & Ruhleder, 1996). All these characteristics apply to the citation network if we define it as the assemblage of databases, publishers, consultancies, bibliometric centers, and users of citation indexes in their various forms.

Therefore, if we speak of the application of citation indicators in research evaluation, we are actually considering how two social institutions—the citation infrastructure and the regimes of accountability that have developed in science and scholarship—interact. Analyzing this interaction requires a deep understanding of both institutions. Current studies of the governance of science have not yet paid sufficient attention to this interaction, partly because indicators often play a minor role in these case studies (Whitley & Gläser, 2007; Whitley et al., 2010).

Moreover, we need to understand how the communication system of science is generating traces that can be harvested as the raw materials for indicators in the citation infrastructure. There is urgency in this need, as the number of traces and indicators produced in science communication increases. We may be on the verge of the evolution of an increasing complexity of knowledge infrastructures, which may either frustrate or bring forward the development of scientific and scholarly knowledge. It is therefore timely to develop a better understanding of the citation infrastructure and its influence on epistemic cultures during the past 50 years, since these will tend to provide the templates for the evolution of new knowledge infrastructures.

References

Acuna, D. E., Allesina, S., & Konrad, P. (2012). Predicting scientific success. *Nature, 489*, 201–202.

Aksnes, D. W., & Rip, A. (2009). Researchers' perceptions of citations. *Research Policy, 38*(6), 895–905. doi:10.1016/j.respol.2009.02.001

ARC. (2012). *The Excellence in Research for Australia (ERA) initiative*. Australian Research Council, Commonwealth of Australia. Retrieved from http://www.arc.gov.au/era.

Åström, F., & Hansson, J. (2012). How implementation of bibliometric practice affects the role of academic libraries. *Journal of Librarianship and Information Science*. doi:10.1177/0961000612456867.

Ball, R., & Tunger, D. (2006). Bibliometric analysis—a new business area for information. *Scientometrics, 66*(3), 561–577.

Bornmann, L., & Daniel, H. (2008). What do citation counts measure? A review of studies on citing behavior. *Journal of Documentation, 64*(1), 45–80. doi:10.1108/00220410810844150.

Buela-Casal, G., & Zych, I. (2012). What do the scientists think about the impact factor? *Scientometrics, 92*(2), 281–292. doi:10.1007/s11192-012-0676-y.

Burrows, R. (2012). Living with the h-index? Metric assemblages in the contemporary academy. *Sociological Review, 2*, 355–372. doi:10.1111/j.1467-954X.2012.02077.x.

Butler, L. (2003). Explaining Australia's increased share of ISI publications—the effects of a funding formula based on publication counts. *Research Policy, 32*(1), 143–155. doi:10.1016/S0048-7333(02)00007-0.

Butler, L. (2007). Assessing university research: A plea for a balanced approach. *Science & Public Policy, 34*(8), 565–574. doi:10.3152/030234207X254404.

Clackson, T. (2009). My h-index turns 40: My midlife crisis of impact. *ACS Chemical Biology, 4*(5), 311–313.

Colwell, R., Blouw, M., Butler, L., Cozzens, S. E., Feller, I., & Gingras, Y. . . . Woodward, R. (2012). *Informing research choices: Indicators and judgment.* Ottawa: Expert Panel on Science Performance and Research Funding.

De Angelis, M., & Harvie, D. (2009). "Cognitive capitalism" and the rat-race: How capital measures immaterial labour in British universities. *Historical Materialism, 17*(3), 3–30. doi:10.1163/146544609X12469428108420.

Edwards, P. N. (2010). *A vast machine: Computer models, climate data, and the politics of global warming.* Cambridge, MA: MIT Press.

Elkana, Y., Lederberg, J., Merton, R. K., Thackray, A., & Zuckerman, H. (1978). *Toward a metric of science: The advent of science indicators.* New York: Wiley.

Georghiou, L., Howells, J., Rigby, J., Glynn, S., Butler, J., & Cameron, H. . . . Reeve, N. (2000). *Impact of the Research Assessment Exercise and the future of quality assurance in the light of changes in the research landscape.* Manchester: Policy Research in Engineering, Science and Technology, University of Manchester.

Gläser, J. (2010). Concluding reflections: From governance to authority relations? In R. Whitley, J. Gläser, & L. Engwall (Eds.), *Reconfiguring knowledge production: Changing authority relationships in the sciences and their consequences for intellectual innovation* (pp. 357–369). Oxford: Oxford University Press.

Gläser, J., Lange, S., Laudel, G., & Schimank, U. (2010). The limits of universality: How field-specific epistemic conditions affect authority relations and their consequences. In R. Whitley, J. Gläser, & L. Engwall (Eds.), *Reconfiguring knowledge production: Changing authority relationships in the sciences and their consequences for intellectual innovation* (pp. 291–324). Oxford: Oxford University Press.

Gläser, J., & Laudel, G. (2001). Integrating scientometric indicators into sociological studies: Methodical and methodological problems. *Scientometrics, 52*(3), 411–434.

Gläser, J., Laudel, G., Hinze, S., & Butler, L. (2002). *Impact of evaluation-based funding on the production of scientific knowledge: What to worry about, and how to find out.* Karlsruhe, Germany: Fraunhofer ISI.

Groth, P., Gibson, A., & Velterop, J. (2010). The anatomy of a nanopublication. *Information Services & Use, 30,* 51–56. doi:10.3233/ISU-2010-0613.

Halsey, A. H. (1995). *Decline of donnish dominion: The British academic professions in the twentieth century*. Oxford: Clarendon Press.

Hargens, L. L., & Schuman, H. (1990). Citation counts and social comparisons: Scientists' use and evaluation of citation index data. *Social Science Research, 19*(3), 205–221. doi:10.1016/0049-089X(90)90006-5.

Harzing, A.-W. (2010). *The publish or perish book: Your guide to effective and responsible citation analysis*. Melbourne, Australia: Tarma Software Research.

Hazelkorn, E. (2011). *Rankings and the reshaping of higher education: The battle for world-class excellence*. New York: Palgrave Macmillan.

Hicks, D. (2012). Performance-based university research funding systems. *Research Policy, 41*(2), 251–261. doi:10.1016/j.respol.2011.09.007.

Hirsch, J. E. (2005). An index to quantify an individual's scientific research output. *Proceedings of the National Academy of Sciences of the United States of America, 102*(46), 16569–16572. doi:10.1073/pnas.0507655102.

Klandermans, P. (2009). *Het sturen van wetenschap: Sociale wetenschappen in bedrijf.* Amsterdam: Vrije Universiteit, Faculteit der Sociale Wetenschappen.

KNAW. (2010). *Quality assessment in the design and engineering disciplines*. Retrieved from http://www.knaw.nl/Pages/DEF/27/160.bGFuZz1FTkc.html.

KNAW. (2011). *Quality indicators for research in the humanities*. Retrieved from http://www.knaw.nl/Content/Internet_KNAW/publicaties/pdf/20111024.pdf.

Lamont, M. (2009). *How professors think: Inside the curious world of academic judgment*. Cambridge, MA: Harvard University Press.

Lane, J., Fealing, K., Marburger, J., & Shipp, S. (Eds.). (2011). *The science of science policy: A handbook (Innovation and technology in the world)*. Stanford, CA: Stanford Business Books.

Laudel, G., & Gläser, J. (2006). Tensions between evaluations and communication practices. *Journal of Higher Education Policy and Management, 28*(3), 289–295.

Leydesdorff, L., & Wouters, P. (1999). Between texts and contexts: Advances in theories of citation? (A rejoinder). *Scientometrics, 44*(2), 169–182.

Mcnay, I. (1998). The Research Assessment Exercise (RAE) and after: "You never know how it will all turn out. *Perspectives: Policy and Practice in Higher Education, 2*(1), 19–22. doi:10.1080/713847899.

Meulen, B. J. R. van der. (1997). The use of S&T indicators in science policy: Dutch experiences and theoretical perspectives from policy analysis. *Scientometrics, 38*(1), 87–101.

Moed, H. F. (2007). The future of research evaluation rests with an intelligent combination of advanced metrics and transparent peer review. *Science & Public Policy, 34*(8), 575–583. doi:10.3152/030234207X255179.

Moore, W. J., Newman, R. J., Sloane, P. J., & Steely, J. D. (2002). Productivity effects of research assessment exercises. Retrieved from http://www.bus.lsu.edu/economics/papers/pap02_15.pdf.

Nicolaisen, J. (2007). Citation analysis. *Annual Review of Information Science & Technology, 41*, 609–642.

Phillips, M. (2012). *Research universities and research assessment.* League of European Research Universities? http://www.ub.edu/farmacia/recerca/LERU/LERU_PP_2012_May_Research_Assesment.pdf.

Porter, T. M. (1995). *Trust in numbers: The pursuit of objectivity in science and public life*. Princeton, NJ: Princeton University Press. Retrieved from http://books.google.nl/books?id=oK0QpgVfIN0C.

Priem, J., Taraborelli, D., Groth, P., & Neylon, C. (2010). Altmetrics: A manifesto. Retrieved from http://altmetrics.org/manifesto.

Pringle, J. (2008). Trends in the use of ISI citation databases for evaluation. *Learned Publishing, 21*(2), 85–91. doi:10.1087/095315108X288901.

Rafols, I., Leydesdorff, L., O'Hare, A., Nightingale, P., & Stirling, A. (2012). How journal rankings can suppress interdisciplinary research: A comparison between Innovation Studies and Business & Management. *Research Policy, 41*(7), 1262–1282. doi:10.1016/j.respol.2012.03.015.

Schimank, U. (2005). "New public management" and the academic profession: Reflections on the German situation. *Minerva, 43*(4), 361–376. doi:10.1007/s11024-005-2472-9.

Shore, C. (2008). Audit culture and illiberal governance: Universities and the politics of accountability. *Anthropological Theory, 8*(3), 278–298. doi:10.1177/1463499608093815.

Slaughter, S., & Leslie, L. (1997). *Academic capitalism: Politics, policies, and the entrepreneurial university*. Baltimore: Johns Hopkins University Press.

Sonuga-Barke, E. J. S. (2012). Editorial: "Holy grail" or "siren's song"?: The dangers for the field of child psychology and psychiatry of over-focusing on the Journal Impact Factor. *Journal of Child Psychology and Psychiatry, and Allied Disciplines, 53*(9), 915–917. doi:10.1111/j.1469-7610.2012.02612.x.

Star, S. L., & Ruhleder, K. (1996). Steps toward an ecology of infrastructure: Design and access for large information spaces. *Information Systems Research,* 7(1), 111–134.

Stensaker, B., & Harvey, L. (Eds.). (2011). *Accountability in higher education: Global perspectives on trust and power*. New York: Routledge.

Strathern, M., & Mitchell, J. P. (2000). *Audit cultures: Anthropological studies in accountability, ethics and the academy*. New York: Routledge.

Van Dalen, H. P., & Henkens, K. (2012). Intended and unintended consequences of a publish-or-perish culture: A worldwide survey. *Journal of the American Society for Information Science and Technology, 63*(7), 1282–1293. doi:10.1002/asi.22636.

Van Kammen, J. (2009). Assessing scientific quality in a multidisciplinary academic medical centre. *Netherlands Heart Journal* 17: 500.

Van Noorden, R. (2010). Metrics: A profusion of measures. *Nature, 465*(7300), 864–866. Retrieved from http://www.nature.com/news/2010/100616/full/465864a.html.

Weingart, P. (2005). Impact of bibliometrics upon the science system: Inadvertent consequences? *Scientometrics, 62*(1), 117–131. doi:10.1007/s11192-005-0007-7.

Whitley, R. (1984). *The intellectual and social organization of the sciences*. Oxford: Clarendon Press.

Whitley, R. (2000). *The intellectual and social organization of the sciences* (2nd ed.). Oxford: Oxford University Press.

Whitley, R. (2010). Reconfiguring the public sciences: The impact of governance changes on authority and innovation in public science systems. In R. Whitley, J. Gläser, & L. Engwall (Eds.), *Reconfiguring knowledge production: Changing authority relationships in the sciences and their consequences for intellectual innovation* (pp. 3–47). Oxford: Oxford University Press.

Whitley, R. (2011). Changing governance and authority relations in the public sciences. *Minerva, 49*(4), 359–385. doi:10.1007/s11024-011-9182-2.

Whitley, R., & Gläser, J. (Eds.). (2007). *The changing governance of the sciences: The advent of research evaluation systems*. Dordrecht, the Netherlands: Springer.

Whitley, R., Gläser, J., & Engwall, L. (Eds.). (2010). *Reconfiguring knowledge production: Changing authority relationships in the sciences and their consequences for intellectual innovation*. Oxford: Oxford University Press.

Woolgar, S. (2002). *The boundaries of accountability: A technographic perspective*. Paper presented at the European Association for the Study of Science and Technology Conference, York, UK, July 31–August 3, 2002.

Wouters, P. (1997). Citation cycles and peer review cycles. *Scientometrics, 38*(1), 39–55.

Wouters, P. (1999a). *The citation culture*. Amsterdam: University of Amsterdam.

Wouters, P. (1999b). The creation of the Science Citation Index. In M. Bowden, T. Halin, & R. Williams (Ed.), *Proceedings of the 1998 Conference on the History and Heritage of Science Information Systems*, 127–136. Medford, NJ: Information Today Inc.

Wouters, P., & Costas, R. (2012). Users, narcissism and control—tracking the impact of scholarly publications in the 21st century. Utrecht: SURF.

4 "The Data—It Is *Me!*" ("Les données—c'est *Moi!*")

Ronald E. Day

Citation Analysis and the Social Sciences

The literature on bibliometrics and citation analysis is vast. To start with citation analysis alone, there seem to be many claims for its reason for being. As a start, we may list the following:

- Showing the influence of a work or scholar in a discipline
- Showing the influence of a work or scholar on another work or scholar
- Showing the citing behavior of scientists (broadly understood)
- Showing networks of influence (for reasons of sociological studies or policy decisions)
- Showing core or peripheral subjects, influences, and authors in disciplines
- Showing the major and minor producers of works in a discipline and their citation relationships
- Showing research disciplines, their subdomains, and their relations to one another

If one were to ask how citation analysis shows the above, one would likely get many answers, ranging from the counting of traditional bibliometric forms (books, journals, chapters, etc.) to the counting and formation of indices out of documentary fragments, such as titles, author names, and vocabulary within the documents, to the counting of metadata. As diverse as these methods are, they share the quality of resulting in quantitative indicators of something. That is, their scientific practice is ultimately reducible to that of describing by quantitative analysis some sort of "citing behavior" of persons or groups. Indeed, citation analysis is distinct from bibliometrics more generally in that it shows relations in the citation data, which then suggest citing behaviors.

The scientific claims of citation analysis, like other sciences, lie in showing, preferably causal but also correlational, regularities. Such regularities may be stated as "laws," but more commonly they are established as "empirical" by studies that confirm or contest previous findings, or even more commonly, by studies that simply pick up on the claims of previous studies.

As with other social sciences, the major epistemological and ontological questions that occur are whether anything real is actually being shown in the studies or whether what is represented are simply grammatical or logical properties of the terms that have been operationalized. Such operationalization may occur either upon the "objects" in the particular research itself by the unstated a priori and grammatical assumptions of a study or in the taxonomies and their ontological types that form the epistemological and practical foundations for a research project, its area of specialization, or its discipline. Social science investigations, founded on descriptions of human behavior that are themselves characterized by the cultural tools of language in norms of social deployment, are notorious for demonstrating not independent real entities, but the grammatical or logical properties of their own central terms, which are often borrowed, moreover, from folk knowledge or popular metaphors (e.g., understanding language according to the conduit metaphor or the mind as a computational information processing device; for this last, see, e.g., Ekbia, 2008). This results in the generally positive nature of their findings, insofar as the research merely unpacks and confirms the grammatical and logical properties of the nominal entities that it takes up as a priori "objects" for "empirical" or mathematical study. However, even when this is true, it still may not negatively affect, and indeed, may be a very positive boost, to the overall political success and power of a research field or agenda.

For the social sciences, truth claims are often established not simply through the correspondence between research studies, but through their intersection with practical activities. The extension of quantitative research findings to broader social, cultural, and political values (for example, in policy formation) is evidence of their usefulness, and so it is also often taken as evidence of research that shows real entities and relations. Social science research projects and their tools may have a social role no matter whether their a priori epistemological and ontological assumptions are flawed or not. A priori questions are often seen to lie outside the doing and writing of scientific research and engineering, other than, in their acceptance, continuing a founding discourse, method, and/or technique/technology. In other words, the a priori is disposed of as *critical*.

Citation Analysis and Social Computing

Today, citation indexing is rightfully seen as a forerunner to many types of digitally automated social representations, and not just in the domains of "scientific communication." It has reached full bloom in social computing, where indices of social relations are not just used for descriptive analyses, but are recursively fed back into the searches themselves (Thomas, 2012). As in traditional citation analysis, much rides on the assumptions that these indices rest on, not only for "scholarly" or "scientific" descriptions or "visualizations" (which are loaded with epistemological and cultural assumptions), but also for social and psychological descriptions. Recursively constructed computational indices (Thomas, 2012) (as well as algorithmic ranking) can narrow the intentional potentialities of selves to the logical possibilities of socially recognized persons through the strengthening of previous searches and the searches of others. The self, in other words, limits its own possibilities to those roles and rules that others assume a person should perform because of the rewards and punishments (both explicit and implicit) that recursive indexes give to past behavior. This interpellation of the "I" by the "me" constitutes an interpellation of the self as a site of hypothetical potential expressions by a social psychological understanding of recognized persons according to socially normative rules and roles. While cultural psychologies blur the line between personal "self-psychology" and social "person psychology" in many varied instances, a central concern of Western ethical and political theory in modernity has been the increasing emptying out of the former by the latter in areas of human agency and accountability, so it worth pursuing this issue in an area where so much of human life is now lived, namely, online.

With social computing, the issues of human behavior are similar, but broader, than in classic scholarly citation analysis: What does vocabulary tell us about the identity and actions of persons that we assume lie behind proper names, particularly authorial names, in many different social and cultural spheres? With recursive social computing algorithms the fears of astute bibliometricians and citation analysts come to fruition, namely, that the epistemic problems of citation analysis are compounded by recursive inclusion into further searching and communication online. What in citation analysis start as behavioral explanations for the purpose of citation analysis end up as algorithms that control identity construction and intention in information searching and communication through group psychological and sociological assumptions. These algorithms are relatively hidden to the user, and they function (a) not only in constructing searches through recursive and recommendation systems, but (b) when the psychological and sociological explanations are internalized by actors in their search and research expression and abilities, and (c) when these then are used for policy and social engineering evaluations of performance and for representing the work or even the identity of persons and social institutions.

For example, certain terms and names will give me more documents or more social contacts, and then these will be reinforced by social computing algorithms. The search engines here are devices for the psychological "introjection" and the ideological "interpellation" (Althusser, 2001) of the subject by an "objective" subject ("this is who I am in the field!"; "this is what I like!"). And then, based on search results and social networks, I am institutionally or more generally socially evaluated accordingly as being central, marginal, or not at all belonging to a discipline or other social group. Further, the symbolic capital then piles up when such indices are "visualized," providing heuristics for representations of what are the initial assumptions of the algorithms. Whether this translates to the social capital that we want or not, we have to adapt ourselves in some way (toward or against) the representations that we were and now are for the system and its users.

What we find is that social computing algorithms, such as PageRank (link analysis algorithms) and recommender systems, strengthen the "truthfulness" of bibliometric "laws" (such as Lotka's law), simply because they automate the group behavioral assumptions inherent in such "laws" and then feed this back into user behavior. Their recursivity, in particular, gives us the technological component of a "cybernetic system" (Sosteric, 1999, para. 16). The laws describe in mathematical form publishing, research, or linguistic behavior. People do citations, and then citations "do" or shape people when the latter's expressive agency (the individual as "self" of personal psychology) and social representations (reputation, etc.; the individual as a socially recognized "person," in a technical sense of social positioning) are shaped and dependent on systems of citations.

Citation and social computing indices encode these psychological frames of human behavior and then feed them back into sociocultural systems by means of techniques and technologies that are not easily understood by users, but that deliver "pragmatic," ready-to-hand, devices that act as heuristics of judgments and social positioning. Indeed, the academic "citation rat race" has expanded to blogging, tweeting, etc., because of the exportation of citation indices and analytic modeling. These "make sense" because they reflect sociological and psychological assumptions and politics. The explanations are "pragmatic" and "make sense" because they logically and grammatically follow from the social and personal group psychologies and sociologies (in general, the political economies) that are encoded in the algorithms. That a small number of authors publish a greater number of works is a sociological fact, not a bibliometric one. It belongs to the logic and distributions (the "grammars") of social power in particular types of sociocultural systems. Feeding this back into the production system in terms of social rewards or in terms of favored search terms leads to exponentially increasing the powers of the sociological systems and does little for the more marginal or unrepresented authors and works that were present (or not) for counting in the first place.

The expansion of the logic of citation indices to what are online social indices represents a largely unheralded and unrecognized triumph of documentary/information science, from the obscurity of post–World War II Anglo-American documentation-based "library and information science" (and its limited claims to be modeling science) to the social engineering of much of the modern world. The new social computing systems bear and prescribe their own modes of classification (Bowker, 2005) at the implicit or "unconscious" levels of social and cultural norms and personal and social psychology. Like citation indices, social computing indices first of all index, or point to, the sociological, psychological, and generally the political inscriptions of that which they are indexing.

Empirical findings never simply show themselves. Citation analytics, either explicitly or implicitly, as a social science *must* indicate social explanations of various types of regular behaviors. Once again, the epistemic problem of social science operationalization—which becomes political and psychological when citation analyses are highly valued in restricted (e.g., academic) or general (e.g., social) economies—is what happens when these explanations are the very basis for the metrics to begin with. Then, the "objects" of study and their empirical measurements (and the tools and algorithms that aid this) may be nothing other than devices in the restating of social, cultural, and political norms. What they would assert is the certainty of ideology.

Citation Economies

Citation scholars have argued that citation behavior originates in personal choices, in rhetorical functions, and in sociological functions for citations. Citation indices encompass all these intentions to various degrees.

Rhetorical rules for citation are formal at the level of style, but different styles for citation lead to different styles of argument. (For example, MLA citation practices in literature often demand detailed page citations, whereas APA citation practices in the social sciences allow for a breezy reference to a whole work.) And even within a citation style, there is room for some ambiguity as to the rhetorical element that is being exactly referred to (Cronin, 1994). These differences result in differences in values and practices for the evidential function of citation.

Rhetorically, citations provide evidence on behalf of rational persuasion. If there is any scholarly communication going on, it is one of the enlisting of resources for an argument. Personal choices and interest certainly come to the fore in this as well, but these are part of social functions within a domain of cultural forms and social norms, of which rhetorical functions play a part. Rhetoric is a means of persuasion by the use of cultural forms in social situations, including those delineated by genre and discipline practices.

To study citation analysis, however, is to study not just the practices of using citations as rhetorical tools, but also the practices of positioning one's self within social norms, cultural forms, and ultimately, political economies that regulate these in time and across epistemic borders, within which persuasion functions. To become well positioned, one must master the tools of that positioning. But the tools also shape the individual deploying them, particularly over time. For example, like it or not, one becomes "disciplined" in a certain discipline over time, leaving behind to some extent what one was before. As it is said, "Culture seduces us all."

Citation indices, like all documentary indices, including those used in social computing, act as amalgamating tools for rhetorical or, more broadly, "expressive" performances. They do so through algorithms that, particularly with commercial providers, are not transparent in how they function or how and how much of a user's activity they track and store. Nonetheless, as the citation economy expands to nontraditional scholarly or even nonscholarly domains and "smaller" rhetorical units ("altmetrics") for academics and nonacademics, *political economy forces into being a "need"* (along with "information needs" and "communication needs") for individuals to try and maintain their social status through these tools.

Such is a tricky act, indeed. Beyond more traditional issues, like quality versus quantity, prestige versus popularity (Ding & Cronin, 2011), scholarly versus non-scholarly, and traditional citation practices versus web-page visits and such, individuals must deal with their own self-citations vis-à-vis their own "unforgotten" (Blanchette & Johnson, 2002) histories, and their own "packaging" by others' linking and citation practices (e.g., RateMyProfessor and other crowd-sourced evaluations, such as those discussed by the SURF report [Wouters & Costas, 2012], as well as the algorithms' interpretation of similar [e.g., Facebook or Amazon's recommendation systems] individuals and works).

Not only this, but the very notion of a citation has been elevated to represent something more or less than a traditional document. Citation analysis was one of the first full-blown attempts to move beyond documents themselves (and indeed authors themselves) and to deal instead with their expressive elements (words) and social contexts for expressions (journals, conferences, etc.) as *documentary qualities* that were then inferred to be properties of documents, persons (as authors and as scholars and scientists), and some sort of intellectual community to which these elements supposedly point.

Blaise Cronin, in this volume and elsewhere (for example, Cronin & Shaw, 2002), has discussed the multiple issues of inferring social capital from symbolic capital through citation analysis and other digital analytics. Social computing systems help to shape the former into the latter by shaping communication and information over time and across social and cultural groups. In a sense, the chief computational mediation of recursive social computing systems is to not only "cleanse" the data, but to cleanse the searching subjects. While there have been many claims to go beyond this by looking at searches as cases rather than as instances of formal logic, norms tend to be established over time by feedback loops between persons and texts.

Social capital in Marx's works (for example, in the popularly called "fragment on machines" in Marx's *Grundrisse* [Marx, 1973]) refers to the "general intellect" of the social whole, as found through the people's upbringing, education, innovation, and their social relationships. For Marx, general intellect precedes capital and is what capital funnels and exploits in the pursuit of profits. The "general intellect" of the Internet is both general and made up of microdiscourses, shaped by user vocabulary and social relations. While professional microdiscourses can be managed to some degree by working within established review channels, the "general intellect" of the Internet can be both scholarly and non-scholarly, and of course, even scandalous. From an academic administrator's viewpoint, the combination of scholarly and popular exposure and evaluation provides a basis for evaluating a faculty in terms of both academic and popular virtues. From the scholar's viewpoint, increased metrics require increased expenditures of energy and time into those populations and mechanisms for increased exposure, whether they be grant applications or student evaluations or popular institutions and means or traditional publications—or increasingly likely, all the above—in order to enroll these resources toward increasing one's symbolic and social capital. Increased evaluations and increased markets for value within a symbolic economy demand a self-commodification of the scholar, which has been seen in many areas of social life in late neoliberalism, not least in the "star system" of academe during the past few decades.

It becomes necessary within such a political economy for an academic to act like the entrepreneur that the neoliberal university desires—advancing his or her career by attempting to grow and tend various forms of "evidence" for impact (see Burrows' wonderful analysis of the UK universities under such metrics [Burrows, 2012]). But, like all evidence, such documentary evidence also enrolls the actor himself or herself within the systems of evidence. Within the citation rat race and citation mongering, it becomes unclear what the role of truth is or how one can find a position for critique that itself is not a commodity or at least seen as a commodity and self-commodification. Indeed, given past examples of popular or "mass" rankings, the expansion of the evaluation system to the popular sphere and the expansion of algorithms rooted in group psychology to the production of knowledge lead one to wonder what the delegation of knowledge to opinion, as well, may look like.

We have an inkling of this last in the expansion of scholarly citation in Google's PageRank. As Thomas (2011) notes, ISI indices provide evaluative measures based on traditional systems of peer review, while Google page rankings are established through more general social systems. In sum, the academic is left with the prospect of having to manage a much bigger "image" among various indices, particularly as academic evaluation becomes more cursory and less careful in the use of metrics.

On the other hand, the resources for building up a popular (as well as an academic) image are more available than ever. First, the tools for production are now more available and easier to use than before: software packages; Wikipedia and commentaries on classic intellectual works; and the rise of "surface reading" as opposed to "deep reading" among students (Hayles, 2012) and even among more advanced scholars and scientists. Alongside the pressures of time because of multitasked and leisured online lives, these allow for quick production and shallow scholarship, especially within communities of practice where everyone is living by these same tools and rules. Second, the means of circulating one's "self" through different media types and through various communities allows everyone to have the proverbial "15 seconds of fame," at the least (Cronin, 1999). Such events are not new to the modern research university, whose topics and methods of research have long been directed by funding agencies and publishers and, more recently, by popular knowledge. But they represent new stages, aided by social technology and an attention economy, in the speeding up of academic production. In sum, the Internet has greatly afforded the growth of what constitutes knowledge and truth and the agents of such, for good or bad.

Further, we need to note the historicity of recursive indexing (Thomas, 2012). Such recursivity constitutes a subsumption or gathering up (Hegel's *Aufhebung*), wherein one's past actions are not "forgotten" in a database (Blanchette & Johnson, 2002), but instead, form the basis for future searches and rankings. The indices represent historical subsumptions of personal and broadly understood "textual" agencies and powers that shape users through ranges of possible "needs" and shape documents out of ranges of possible "information." In this manner, *the identities of users and documents are functions of a dialectic of social and technical devices*, with normative cultural forms (particularly language) and social norms for deployment on the one hand, and indexing algorithms, indices, databases, and information and communication recommendations on the other hand, each being mediated by the other through the logic of algorithms that formalize group behavioral assumptions.

Citation analysis was the beginning of the informational reformation of selfhood and of textuality. In the documentation era and perspective, users and documents were joined by professionally constructed vocabularies. But now, just as neoliberalism has reversed the modernist notions of the "public" and "private" spheres, so, too, does social computing shape the self as a person of logical possibilities (rather than hypothetical potentials) and present information (i.e., "aboutness") as the source of documentary evidence. Where the self was, so the commodity shall be.

Consequently, citation analysis must be seen as the great predecessor of today's social computing and the reconstruction of self and personhood. A neoliberal age of digital mediation in social relationships and the presence of selves and persons, mediated by nontransparent and corporately owned algorithms and the databases and indexing that accompany them. In contrast, simple documentation organizations and their missions of document retrieval, as well as privacy, as Marx wrote, "melt into air" (Marx & Engels, 1848/1969, chap. 1, para. 18).

Three Problematics of Reference or Representation

I would identify three central social-epistemic problematics in citation analysis: (a) the issue of defining and cutting the referential pie of citations and determining what those parts and the pies themselves should be called; (b) the issue of social positioning within symbolic and social capital orders; and (c) the issue that underlies the first two and many other issues with citation analysis: the general issue of representation or "reference" in citation analysis.

The first problem permeates citation analysis at every level in the general form of: What do citations show? This must be inferred from the data, but this inference, as is the case with all data gathering and interpretation, means that citation studies are guided by theory—that is, by founding epistemic as well as methodological assumptions, which are then folded into the techniques and technologies of research, including in the rhetoric of their "reporting" structures.

"Maps of science" (De Bellis, 2009) are a nice example of this: How do we know that certain citations belong to biology, to chemistry, to computer science? The answer could be that the metadata or word frequency or coassociation or documentary relationships may tell us this. But each of these is problematic in terms of stating a domain, not because the data may be flawed, but because *disciplinary domains are not real entities, but rather, they are nominal entities.* Hence, each of the above indicators may be contrary to the others in providing evidence as to what domain an article belongs to.

Moreover, traditional domain names may not indicate current research or newer trends because science, like all research, is situated in cultural and social traditions and institutions. "Maps of science" are not maps of an object called "science" (there is no such object as a real entity) nor are they even maps of behavioral practices of scientists, but rather, they are maps of bibliometric indicators *that are then taken as* representations of assumed sociological domains and behaviors, *not as inferred from the data, but as a framing assumption in the very metaphorically borrowed concept of "maps" or "atlases."* "Science," and its "parts" and "geographic relationships," may be taken as empirically real, when in reality it is numerical data literally shaped within a (nominal) metaphor. If this is the case, the reality of such maps depends on a technically produced reification of nominal entities. This reification is done for various pragmatic reasons.

Maps of science are representations of representations within a convenient metaphorical vehicle ("atlas," "maps"). There is no epistemic reason for choosing the metaphor of maps and their representational forms as a vehicle for showing bibliometric data nor is this even logically cohesive (science is a nominal entity, not a real entity like land masses are). The choice is a pragmatic choice of choosing an easily available and understandable visualization frame for data. It is this pragmatic choice, and in this sense an opportunistic choice, rather than any logical relationship between "science" and "maps" that drives the privileging of this metaphor as the frame for the engineering practice of mapping science. Keeping in mind the general problem of teleological nominal operationalization in the social sciences that was discussed earlier, one must also bear in mind the particular dangers of unacknowledged tropes in fostering such operationalization, including in the truth claims of maps of science.

The second issue, that of social positioning via symbolic capital, follows from this positing of nominal entities as real entities by the erasure or marginalizing of basic a priori critiques in the practice of "science" or engineering. Social positioning within science takes place not only through sociological forms, but also within the rhetoric of producing truth statements of empirical research and fact, which for Latour (1987) are "enrolled" within the sociological practice of science toward the advancement of the research and the researcher.

As is seen throughout science, but as becomes quite evident in the exposure of subsequent fraud, the researcher's social power is intimately tied to his or her ability to stabilize conflicting or competing taxonomies, methods, techniques, and truth claims toward discovering the "real" in phenomena. The power of turning nominal entities and their relationships into the facts of real objects and successful policy or other "practical" applications is a very important part of establishing the scientist's legitimacy. Indeed, it is what separates scientists from nonscientists and why positivism remains in the sciences. Claims of representing reality are necessary for a scientist's work *not* to be considered "mere" rhetoric or "not science."

The title of this chapter, "The Data—It Is *Me!*," points to the necessary, and sometimes sufficient, condition for the data to be taken seriously and for research to continue. Despite whether this "me" is that of a single author or refers to a research team, through various institutional channels for any given project social, technical, institutional, and rhetorical devices must be deployed for winnowing out and asserting that *this* research project is more likely to give us a true picture of phenomena. For this to occur, a priori and critical questions—ranging from that of questioning the purpose of the field and the research, to its founding taxonomic and other theoretical presuppositions, to fundamental refutations of positivism (that is, that "the data speaks for itself")—must be suppressed, or more commonly, are kept to the margins of productive work, and are excluded from the scientific report or publication by means of rhetorical norms of writing, publication, and review. In these ways, nominal entities are established as real, at least for a field of research. But this also includes the author of a document, whose name and further thoughts and works become symbolic capital to which social capital becomes attached.

This brings us to the issue of the representations of the citations themselves, and how they themselves constitute "citations" or references to other rhetorical and sociological events, not least to citation and social computing indices. As Small (1978) suggests, citations are symbols. Symbols may be "representations" in the sense that they mimetically or metaphorically represent an object, or in the sense that they metonymically refer to some order, larger or smaller than the object within which the object belongs. Small's notion of "concept symbols" may include either. Far from originating in scientific communication, as Walsh (2012) has suggested, citations or documentary fragments lie at the heart of classical rhetoric and religious iconography. References or "citations" here are not only to individual works (as with modern bibliography), but also to entire orders of works, or in the case of medieval iconography, to entire symbolic orders or "worlds" (the city of God, the life of Jesus, etc.), which appear in medieval icons (Walsh, 2012).

If we follow the mid-20th-century French documentalist Suzanne Briet's (2006) 1951 discussion of the relationship of indexical signs (*indice*) to documents, Briet discusses two types of indexicality: "initial" and secondary. Initial occur in regard to scientific taxonomies or bibliographic ontologies, where a new object is brought into a structure of naming through its belonging to or differing from already-named entities. Taxonomy is a necessary first stage in scientific or bibliographic activities. Secondary indexicality for Briet consists in this named entity's position and circulation in various social discourses: as in her example, how a newly discovered antelope is then discussed in scientific forums, in classrooms, in the newspapers, etc. (Briet, 2006).

In the context of the present argument I suggest that there are two different senses to indices and two different functions to indices that Briet (2006) was discussing. The first is that of the sign's reference to an entity—to use a philosophical vocabulary, its extensional reference—and second, the sign's intensional reference to its own circulation in a discursive space (a disciplinary field or social domain). In a previous work (Day, 2001), following the concept of the quasi-object, I called the objects that simultaneously do both these indexing functions—mimetic, and reflexively iconic—"informational objects." ("Informational objects" both reference or "inform" us about the external objects or events that they reference, and they inform us about their social positions and the economies in which they circulate.)

Citation indices, and broader social computing indices, too, have extensional and intensional reference. Citation indices index citations using taxonomic categories such as author, title, etc., but they trace as well cultural and social categories of importance within which they circulate and are seen as valuable. These categories of importance are not simply "external" to their functions, but such indices would not exist at all unless they literally indicated such values. They are indicative of social and cultural values insofar as they index and create metrics of such values. In short, as behavioral measures, citation metrics represent metrics of value—not just metrical values, but social and cultural values. *The "practical" values of citation indices do not, and cannot, lie in their metrical activities alone. Rather, their social usefulness comes from the behavioral inferences that we make from them, but the possibilities for these are embodied in their very indexing and metrical functions. Citation and social indices represent the amalgamation and embodiment of such factors in technological and technical devices.*

Indexical algorithms, including citation indices, must be constructed so as to already include their social and cultural needs in order for their data to be meaningful and useful. This is to say that the techniques and technologies of citation indices, like all social indices, are not just mechanical but are, a priori, constructive of both documentary objects and social subjects of need. In principle, not from an epistemological but from an engineering perspective, there cannot be simply "data itself," and certainly not "data" "speaking" for itself. Data both exists and "speaks" for cultural forms and social orders in its very creation.

Data is what is given—it is "empirical." In French, it is *donné*. And like any *donnée*, any gift, it comes with conditions. These conditions are the products of the social and cultural practices to which it addresses itself as a need. Citation indices not only fulfill needs by their indexing and algorithms, but they themselves represent a fulfillment of needs. This fulfillment of needs a priori conditions the types of metrics and the relations that they will show, which are part of the "scientific" process that is characterized by method, techniques, and measures. It is this a priori that is excluded from the scientific reporting that is the rhetorical and documentary form of the scientific article, conference presentation, citation analysis, etc. This a priori shows the conditions for the "reproduction of the production" (i.e., the "ideological" Althusser [2001] conditions) of social and cultural values through the scientific "report." Through the reporting structure of a certain practice of "science" that features the "thesis of the precedence of method" (Heidegger, 1977b, p. 382), the a priori is taken as the unwritten introduction or set of assumptions for the report. "Science," relegated to the calculation of "facts" by methods and techniques and by the exclusion of critical inquiry into its a priori, either outright in the procedures and report or through nonexperimental and nonreplicable practices, is thus little different from engineering.

But, it is then this a priori that haunts the studies as a whole and appears again and again in the call for a "theory" (e.g., a "theory of citation," and thus, by extension, a theory of citation indices and social indices, and of citation analyses). What is *given* as the data in the indices of citation analysis and social computing—that is, what is truly "empirical" in them—is the technological arrangement of social and cultural values, over and over again—that is, a reproduction of social life lived within the sphere of the technical and technological. The social and cultural that are reflected and reinforced, *that are enfolded*, in the algorithms and indexes are that of a certain technics of modernity, namely, a calculability of the social and cultural and the reproduction of what is calculated. Recursively, inscription within such measures strengthens these measures and their privileged social and cultural categories over time. This reflection and reinforcement follow the classic forms of modern technical reproduction in the political sphere, despite claims to the contrary. *"Digital experiences" are modern experiences, and they are therefore not "beyond" modernist critiques (such as critiques of ideology, and the analyses of critical theory and, in general, the tradition of critique since Kant), but rather, are very much at the center of such.*

The "gaming" of citation systems, and whatever similar that can be done in larger social computing systems, is not simply an issue of a person understanding the logic of the algorithms involved, but also understanding the values of social norms and cultural forms.

Citation indices have a metonymical relationship with sociocultural values and their economies of production. They both index these latter through documentary fragments and they reorder them through documentary orders and rankings. Through this use of documentary fragments, both documents and ultimately persons evolve and appear as properties of social and cultural values in political economies. Citation and social computing indices begin with social and cultural categories and create documents and persons (users, as well as authors) out of algorithms of information and communication "needs" (Thomas, 2012), which reflect these social and cultural categories in political circulation.

Conclusion

What do the metrics of citation analyses represent? They are said to represent the objects and relations of bibliographic or documentary data. These, in turn, are said to represent social behaviors of citation. Citation indices are databases and mechanisms for automated manipulation or analysis of such data, for the purposes of representing social and individual behaviors on a "macro" scale.

This is how they are said to function. But what do they socially and culturally mean? This is to ask, what do they *index* and point to, not just extensionally, but by their own practical importance and by the fact of their very existence and functions?

Obviously, as many authors note, they index the importance of *metrics* for determining value. Their social and epistemological enigma for many authors—the question of how they continue to hold such sociocultural value despite lingering and significant epistemological questions regarding not only citation behaviors, but also, what such metrics do or can show—exists because of this very role for metrics in the play of power. They are products of technological modernity, not only in their construction, but, more importantly, in their social and cultural values.

But this metrics is of a particular type. It is an informational metrics, in the sense that the metrics construct not only the identities of documents and persons out of the "information" or "aboutness" of documentary fragments through metrical relations, but they also extend this power outward to "online existence" or what some have called "digital experience," as a whole, which is increasingly communicatively based.

This is rather stunning, not least from a historical perspective. From a little recognized field, 20th-century documentation, and following this more recently, a very small, largely Anglo-American, discipline, "library and information science," an entire episteme, indeed, an entire metaphysics of identity and power has arisen. It displaces previous evaluation systems, pervious systems of textuality (i.e., displacing texts by documents, and then documents by information), previous modes of document representation (classification, cataloging, and other professionally driven, "whole-document" representation systems), previous psychologies and metaphysics of identity (the self or "I" as hypothetical and situational powers of expression by the "me," as a product of recognition by others and by my past expressions—that is, as a product of my accumulated and indexed evidence). Behavior is seen as explicit and logically understandable (good-bye the unconscious and the right to neurosis; good-bye Romantic reserve and existential will; basically, good-bye Western subjective metaphysics), and identity is seen as traceable and forever "there" (good-bye being forgotten [Blanchette & Johnson, 2002]). In social computing, even the authority of previous "knowledge-based" citation indices and their sociological basis (peer review) are upended (good-bye ISI and its successors, hello Google Scholar and broader webometrics).

In short, citation indices and analyses open us to an era of the individual being a function of algorithmically determined documentary relationships and identities through valorized fragments of evidentiary "aboutness" (i.e., "information"). These fragments reflect categorical norms of social and cultural values, taking place in political economies that organize these values and give them powers of expression and of exclusion. Metrical citation systems valorize these fragments further, strengthening the normative value of their social and cultural categories, both in the present and toward the future (Thomas, 2012). If the fragments do not start off as human language, they end up as human language, at least in analysis, social and cultural regulation, and political governance. That is, *they mediate language through computational algorithms*, at a secondary, as well as a primary, level of social indexicality or positioning.

This overturning of the metaphysics of the "individual" or "subject" in Western culture by a group-defined subject, a subject arrived at through other subjects including the subject's own past selves, is a momentous event and constitutes a new social, cultural, and political (informational) episteme, precisely in its very consolidation and empowerment of the previous modernist (documentation) episteme. It enfolds professional documentation structures and tools into everyday, as well as specialized, infrastructures for social relations and expression. It opens up to different ways of seeing "individuals" across a variety of fields. It bridges Western and Eastern cultures and politics in sometimes facile and uneasy manners for both. It redefines expression as communication, in a way that was foreseen by the recursive theory of cybernetics, but over a vastly larger scale.

Traditional citation analysis has sort of hit the historical lottery. It may not be the winning that some of its most insightful founders, practitioners, and critics would fully want, but it has hit it big time. What remains largely out of sight, though, is what often lies at the heart of the practice of science and lies outside of scientific practices and reporting (i.e., what lies outside the practices and reporting of proper methods, techniques, and metrics): the social and cultural forms and values that drive this. Or to put it succinctly: the social engineering of science as engineering. There has been little social, cultural, and political accounting of citation analysis and citation indices outside of a very restricted professional practice.

But, some would argue that this "practical" view is the nature of information, as the representation of what is. This is the triumph of the scientific worldview, of an empiricism based on data, based on facts, based on (a certain modern meaning of) *information*. The data shows us and leads us. While many would say that this is nonsense, that this is no longer nor has ever been the belief of scientists, it is unclear whether this is still not the regulative ideal that allows the *practice* of science. It may also allow certain practiced ambiguities of the word *science* (the natural sciences, physical sciences, social sciences, computational sciences, science/engineering, and so forth) and may justify the funding and valorization of science, at least *as a social practice*. The belief that what science, and citation analysis among this, shows, is what is. And what is, is me—that is, me among other things (of course). The data—that which is given—is me. And so, increasingly, I am, this me. The data—c'est *Moi*!

What does this mean, that the only social value one has is as a commodity—a *product*, a value and a ranking, in a market—and that all human relationships and communication should be viewed as markets of commodities, contracts, and beyond all, exchange and competition? Or is "the data—c'est *Moi!*" just plain narcissism, made ever more seductive by the publicizing, visualization, and spectacle production of metrics and computational information companies and such produced spectacles of "the new" and "knowledge" as the TED lectures—what Liu called the "laws of cool" (Liu, 2004)?

References

Althusser, L. (2001). Ideology and ideological state apparatus (notes towards an investigation). *Lenin and philosophy and other essays* (pp. 85–126). New York: Monthly Review Press.

Blanchette, J. F., & Johnson, D. G. (2002). Data retention and the panoptic society: The social benefits of forgetfulness. *Information Society*, *18*, 33–45.

Bowker, G. (2005). *Memory practices in the sciences*. Cambridge, MA: MIT Press.

Briet, S. (2006). *What is documentation?: English translation of the classic French text*. Lanham, MD: Scarecrow Press.

Burrows, R. (2012). Living with the h-index? Metric assemblages in the contemporary academy. *Sociological Review*, *60*(2), 355–372.

Cronin, B. (1994). Tiered citation and measures of document similarity. *Journal of the American Society for Information Science*, *45*(7), 537–538.

Cronin, B. (1999). The Warholian moment and other proto-indicators of scholarly salience. *Journal of the American Society for Information Science and Technology, 50*(10), 953–955.

Cronin, B., & Shaw, D. (2002). Banking (on) different forms of symbolic capital. *Journal of the American Society for Information Science and Technology, 53*(14), 1267–1270.

Day, R. E. (2001). *The modern invention of information: Discourse, history, and power.* Carbondale: Southern Illinois University Press.

De Bellis, N. (2009). *Bibliometrics and citation analysis: From Science Citation Index to cybermetrics.* Lanham, MD: Scarecrow Press.

Ding, Y., & Cronin, B. (2011). Popular and/or prestigious? Measures of scholarly esteem. *Information Processing & Management, 47*(1), 80–96.

Ekbia, H. (2008). *Artificial dreams: The quest for non-biological intelligence.* Cambridge, UK: Cambridge University Press.

Hayles, K. (2012). *How we think: Digital media and contemporary technogenesis.* Chicago: University of Chicago Press.

Heidegger, M. (1977a). The age of the world picture. In D. F. Krell (Ed.), *Martin Heidegger: Basic writings from* Being and Time *(1927) to the* Task of Thinking *(1964)* (pp. 115–154). New York: Harper & Row.

Heidegger, M. (1977b). The end of philosophy and the task of thinking. In D. F. Krell (Ed.), *Martin Heidegger: Basic writings from* Being and Time *(1927) to the* Task of Thinking *(1964)* (pp. 373–392). New York: Harper & Row.

Latour, B. (1987). *Science in action: How to follow scientists and engineers through society.* Cambridge, MA: Harvard University Press.

Liu, A. (2004). *The laws of cool: Knowledge work and the culture of information.* Chicago: University of Chicago Press.

Marx, K. (1973). *Grundrisse.* Harmondsworth, England: Penguin.

Marx, K., & Engels, F. (1848/1969). *Manifesto of the Communist Party.* Marx/Engels Selected Works, Vol. 1. Moscow: Progress Publishers. Retrieved from http://www.marxists.org/archive/marx/works/1848/communist-manifesto/ch01.htm#a1.

Small, H. G. (1978). Cited documents as concept symbols. *Social Studies of Science, 8*(3), 327–340.

Sosteric, M. (1999). Endowing mediocrity: Neoliberalism, information technology, and the decline of radical pedagogy. Retrieved from http://www.radicalpedagogy.org/Radical_Pedagogy/Endowing_Mediocrity__Neoliberalism,_Information_Technology,_and_the_Decline_of_Radical_Pedagogy.html.

Thomas, N. (2011). *Social computing as social rationality.* Unpublished doctoral dissertation, McGill University, Montréal.

Thomas, N. (2012). Algorithmic subjectivity and the need to be in-formed. Paper presented at the conference of the Canadian Communication Association, Kitchener-Waterloo, Ontario, May 2012. http://www.tem.fl.ulaval.ca/en/waterloo-2012.

Walsh, J. A. (2012), "Images of God and friends of God": The holy icon as document. *Journal of the American Society for Information Science, 63*(1), 185–194.

Wouters, P., & Costas, R. (2012). Users, narcissism and control—tracking the impact of scholarly publications in the 21st century. SURFfoundation. http://www.surf.nl/nl/publicaties/Documents/Users%20narcissism%20and%20control.pdf.

5 The Ethics of Evaluative Bibliometrics

Jonathan Furner

Introduction

The purpose of this chapter is to present a theoretical framework for the study of the ethical aspects of evaluative bibliometrics. The practice of evaluative bibliometrics involves the use of quantitative methods to analyze the decisions made by authors and readers of documents, and the use of the results of that analysis to inform decision making in the processes by which authors are rewarded for their work. Expressions of partial or complete theoretical frameworks for the study of bibliometric practice abound in the literature, but few provide foundations appropriate for study of its ethical dimension. This chapter is intended to fill that gap.

In the first two sections to follow this introduction, evaluative bibliometrics is situated in the context of its overlapping parent fields of bibliometrics and research evaluation, and a version is presented of a theoretical framework for bibliometrics that, as is typical, omits ethical categories. A justification is then provided for the decision to focus on the ethical dimension. The existence of such a dimension is demonstrated, its content and scope defined, and its significance evaluated. The next section provides the chapter's primary contribution: a framework for the study of bibliometric ethics. The values and principles of participants in evaluative studies are reviewed, and the lack of community-wide consensus on principles of distributive justice is highlighted as a core concern. The chapter concludes with a remark on the potential for applying a framework of the kind developed here to the study of other types of uses of bibliometric techniques.

Bibliometrics and Evaluation

Bibliometrics has been defined as "the study of the quantitative aspects of the production, dissemination, and use of recorded information" (Tague-Sutcliffe, 1992, p. 1;

quoted in Bar-Ilan, 2010, p. 2755). More informally, we might say that bibliometrics is about what *people* (authors, readers, etc.) do with *documents* (books, journal articles, web pages, tweets, etc.), for what reasons, and with what effects. It involves the observation, classification, and counting of document-related actions (writing, submitting, reviewing, editing, publishing, viewing, buying, reading, citing, etc.), and the ranking and mapping of classes of such actions, in order to produce representations of patterns and trends in document-related behavior. These representations, in the form of descriptions and indicators of various numerical and graphical kinds, can in turn be used to (a) *reward* people for their past activity as authors or readers; (b) *recommend* particular documents, or classes of document, for future use; or (c) simply improve our *understanding* of the processes underlying the structures and dynamics of networks of documents and related entities.

An assumption at the core of this conception of the nature and scope of the field of bibliometrics is that any document-related action of the kinds listed above is the outcome of a decision to select, at time *t*, one particular document (or class of document) rather than any other as the object of the action. In other words, the action is treated as an expression of a preference ordering over the universal set of documents. Analysis of multiple such preference orderings allows us to produce (a) composite *rankings* of documents (or of classes of documents), which may then be used as the basis for rewarding the authors of highly ranked documents, and/or recommending highly ranked documents to information seekers, and (b) *maps* or graphs showing the relationships among documents (or among classes of documents), which may then be used as the basis for recommending strongly related documents to information seekers, and/or representing or describing the structure of document networks.

Evaluative bibliometrics (see, e.g., Narin, 1976) is the branch of the field that focuses on (a) the specification of techniques for the production of rankings, and (b) the use of such rankings as the bases for distributing resources or credit among the individuals responsible for ranked documents, or among the institutions with which authors are affiliated. University administrators use the techniques of evaluative bibliometrics in faculty tenure cases, in the course of identifying authors deemed most worthy of promotion; government agencies use evaluative bibliometrics in the allocation of research funding, in the course of identifying departments, programs, and projects deemed most worthy of support (see Lane and colleagues, chapter 21, this volume); librarians use evaluative bibliometrics in collection development, in the course of identifying journals deemed most worthy of purchase or licensing for access by library users (see Haustein, chapter 17, this volume).

Viewed as a set of techniques, evaluative bibliometrics is just one of several options available to would-be evaluators of research and/or researchers. The distinct but overlapping subfield of *research evaluation* (see, e.g., Whitley & Gläser, 2007) is dedicated to the study and application of such sets of procedures for the systematic determination of the value[1] of research projects and programs, of their outputs and outcomes, and of those who lead and participate in them. Research evaluation is itself a branch of the field of *evaluation* (see, e.g., Scriven, 1991), whose practitioners inquire into the general process of determining the value of agents, objects, events, etc. (i.e., "evaluands"), of any given kinds. An important result of work in the latter field is an outline of a general procedure for evaluation that involves the following tasks:

- Specification of the *variables* (aka properties, states, conditions, qualities, attributes, criteria, dimensions) whose values[2] are to be used to characterize evaluands
- Specification of the methods to be used of *operationalizing* the chosen variables so that measurements may be taken easily and reliably
- Specification of the methods to be used of *normalizing* values of the chosen variables so that measurements taken under different conditions (e.g., over different time periods) are comparable
- Optionally, specification of the methods to be used of *weighting* the chosen variables so that measurements may be combined in a single, overall metric

Justifications of particular choices of variables may make claims for the *intrinsic value* (sometimes known as merit) of selected variables, and/or for their instrumental or *extrinsic value* (sometimes known as worth or "goodness-for"). Justifications of the latter type may include additional specification of the purposes, goals, or *functions* of evaluands for the members of one or more groups of *stakeholders*.

Evaluations are themselves undertaken for a variety of purposes. The instigators of evaluative studies may be primarily interested simply in knowing how the evaluands in a given population compare with one another. They may also wish to use the results of an evaluation as warrant or grounds for choosing among evaluands, or for allocating varying quantities of resources or rewards to different evaluands. Alternatively, they may wish to determine how the value of evaluands might be improved, or to encourage evaluands to consider the fact of evaluation (or the prospect of reward) as a motivation or incentive to achieve their goals more successfully. Finally, administrators may consider it their duty or responsibility to undertake an evaluation in order to meet professional standards of accountability.

Together with the effects typically intended by administrators—better decision making, fairer allocation of resources, improved performance and/or reputation—evaluative studies can also have *unintended side effects* of various kinds. Evaluands might see their involvement in the evaluation, or their expectation as to its outcome, as an incentive to change their behavior with results that run counter to those desired by administrators. Methods of evaluation may themselves be treated as evaluands, and their intrinsic and extrinsic value determined in a process of *metaevaluation* in which undesired effects are set against desired ones.

Two types of analysis, distinguished by the variable on which subjects are assessed, are dominant in evaluative bibliometrics. *Publication analysis* is based on counts of the occasions on which the documents produced by each author (or by each organization, each subject area, each country, etc.) have been published; *usage analysis* is based on counts of the occasions on which the documents produced by each author (or by each organization, each subject area, each country, etc.) have been used.[3] Citation analysis is a specialized form of usage analysis in which it is assumed that counts of citations serve as reliable evidence of the amount of use to which citing authors have put cited documents. Usage analysis is itself sometimes conceived of as a form of *impact analysis*, on the assumption that counts of usage events (i.e., citations, links, loans, holdings, downloads, views, etc.) serve as reliable indicators of the amount of impact that documents have had on a given population of users. Similarly, publication analysis is sometimes conceived of as a form of *productivity analysis*, on the assumption that counts of publications serve as reliable indicators of the rate at which their authors are productive.

Our assessment of the validity of analysis of each of these kinds rests on our attitudes toward each of a chain of successively more basic premises: the claim that values of the chosen variable of evaluation (rate of productiveness, amount of impact, etc.) are positively correlated with measurements of the level of *quality* (i.e., the goodness) of research, and thus that rankings derived from publication and/or citation counts can be used as surrogates for measures of quality; the belief that the quality of research is the most appropriate basis on which to assess the extent to which researchers are *deserving* of reward; and the belief that desert[4] is the most appropriate basis on which to distribute reward. In summary, arguments in justification of the validity of using bibliometric techniques in research evaluation need to demonstrate (a) that publications are evidence of productivity and citations are evidence of impact; (b) that productivity and impact are evidence of quality; (c) that quality is the appropriate basis for the assessment of desert; and (d) that desert is the appropriate basis for the distribution of reward.

A Conceptual Framework for Bibliometrics

As one would expect, the literature of bibliometrics is vibrant and multifaceted, replete with contributions to many different debates on methodological and other foundational issues, as well as with reports of the findings of studies in which bibliometric techniques have been applied (see, e.g., Bar-Ilan, 2008; Borgman & Furner, 2002).[5] A framework for classifying the most significant foundational issues might include the following categories, among others. Taken in combination, contributions in these categories allow for detailed description and explanation of the nature and scope of the field, of subfields such as evaluative bibliometrics, of the distinctions between bibliometrics and related areas of inquiry, and of its disciplinary affiliations:

- *Purposes*: specification of the general kinds of questions, problems, and issues, and of the particular instances of those kinds, that bibliometricians seek to answer, resolve, or understand
- *Uses*: specification of the kinds of contexts and environments, and the kinds of ways in which the outcomes of bibliometric research may be applied
- *Ontology*: clarification of the commitments that bibliometricians have to the existence, in reality, of entities in various fundamental categories
- *Epistemology*: clarification of the processes by which bibliometricians believe it is possible to acquire knowledge of the subject matter of bibliometrics
- *Methodology*: generally, specification of the methods by which valid and reliable data may be collected, and of the methods by which relevant and appropriate analysis of data may be carried out
- *Metamethodology*: explanation and evaluation of the general approaches that may be taken, and the particular methods that may be used, to address the foundational issues listed above
- *Paradigms:* at the most general level, identification of the paradigms within which bibliometricians may (consciously or unconsciously) operate

Turning to focus on methodology in particular, we find contributions of the following kinds, most of which are generic to fields that involve the development and application of statistical techniques:

- Specification of the kinds of *phenomena* (objects, properties, actions, agents, etc.) about which data may be collected and analyzed for bibliometric purposes, and of the *levels* or units at which phenomena may usefully be aggregated and analyzed
- Specification of the kinds of *data* that may serve as evidence of the influences on and/or effects of human document-related activity
- Specification of the kinds of *observation* required to produce data that are valid and reliable indicators of the existence of structures and operation of processes

- Specification of the methods by which *descriptions* of sets of bibliometric data are produced in the form of summary statistics (aka metrics, indicators), ranked lists, and graphical visualizations
- Specification of the methods by which mathematical *functions* are generated as putative descriptions of the regularities found in distributions of the probabilities of occurrence of observed phenomena
- Specification of the methods by which we may calculate the *goodness of fit*, to the data collected, of the functions proposed
- Specification of the methods by which *models and theories* may be produced as explanations of regularities
- Specification of the methods by which we may *evaluate* the utility, coherence, and/or correspondence with reality, of the models and theories proposed as explanations of regularities
- Specification of the kinds of *technologies and tools* that may be used to support efficient and effective data collection and analysis

Lastly, consideration of aspects most germane to evaluative bibliometrics leads to the following list of the kinds of choices among available alternatives that must be made and justified by analysts working on any given evaluative study:

- Selection of the *unit type(s)* of evaluands to be studied: e.g., documents, authors, journals, departments, institutions, nations, fields
- Selection of a method of identifying the particular *population(s)* of evaluands to be studied: e.g., institutional membership, database coverage
- Selection of the *variable(s)* whose values are to be used to characterize evaluands: e.g., productivity, impact on science/scholarship, impact on society, research quality, equality, diversity
- Selection of a method of *operationalizing* the chosen variables so that measurements may be taken: e.g., counting publications, counting citations
- Selection of a method of *normalizing* values of the chosen variables so that measurements are comparable: e.g., by time period, by frequency of citable documents
- Selection of a method of *weighting* the chosen variables so that measurements may be combined in a single, overall metric
- Selection of a method of *ranking* normalized values of operationalized variables for the evaluands in the chosen population

Neither the general process, nor the specific outcome presented above, of constructing a theoretical framework for evaluative bibliometrics along these lines could be construed as a novel contribution. The level of detail is necessary, however, to demonstrate a significant omission: the *ethical* dimension, which (I claim) cuts across many of the categories listed. Treating evaluative bibliometrics as a discrete set of techniques for the evaluation of the agents and products of authorship, we may engage in a form of metaevaluation in which we determine the intrinsic and extrinsic value of the kinds of methodological choices made in each of the given categories. Extrinsic value is assessable relative to the goals of stakeholders, but how might we go about measuring intrinsic value (of choices and/or the goals of choosers)? This is where a foray into the field of ethics is helpful.

Ethics, Values, and Principles

Ethics (see, e.g., Shafer-Landau, 2010) is the area of inquiry, normally treated as a branch of philosophy, in which answers are sought to questions like "What is the right thing to do?" and in which methods and results of thinking and reasoning about such questions are studied and evaluated. Well-established subfields of ethics include *normative ethics*, which is productive of specifications of criteria for distinguishing between right and wrong actions, and of theories that provide justifications for those specifications; *metaethics*, which is productive of methods of classifying ethical theories; and *applied ethics*, which is productive of demonstrations of the consequences of applying criteria of particular kinds as guides to action in situations of particular kinds.

Professional ethics is that subbranch of applied ethics concerned with the ethical aspects of work in the various professions.[6] A tool found to be useful by the leaders of many professional associations is the *code of ethics*, which can take any of a variety of forms (and a variety of titles) but the primary purpose of which is typically intended to be to ensure that members of the given profession have the opportunity (by studying the code) to develop an awareness and understanding of the kinds of practices generally considered by their peers to be justifiable by ethical principles. Secondary purposes of codes of ethics include (a) the communication of the values of the profession to nonmembers, that is, to the consumers of the goods and/or clients of the services provided by members of the association, as well as to policymakers, journalists, and members of the public; and (b) the establishing of a means of holding members of the profession to account for actions perceived not to be justifiable by ethical principles.

The forms taken by codes of ethics do vary, but one structure commonly adopted involves a distinction being made between statements of the profession's values, and statements of principles. *Values* are those kinds of states, conditions, properties, etc.—variously attributable to agents, objects, events, or other phenomena, as individuals or in aggregations—that (it is claimed) are *good*. *Principles* are specifications of the kinds of conditions that must be satisfied, the kinds of states that must prevail, the kinds of properties that must be instantiated, for any given action to be deemed *right*.

When a person is said to "hold" a certain value, then the claim is that that person believes that a certain kind of state, property, etc., is good. Definitions of goodness proliferate, as do typologies of kinds of goodness, but one feature commonly (if only implicitly) attributed to goodness is its quantifiability, in one or both of two senses: all other things being equal, the more we have of a good thing, the better; and again, all other things being equal, the more things we have that are good, the better.

Different ethical theories propose different kinds of justification for action-guiding principles, and different conceptions of the ways principles relate to values. According to theories in a family known as consequentialism, for example, principles are justified to the extent that the actions they recommend tend to produce effects characterized by greater quantities of values. The rightness of actions, in other words, is determined by the goodness of their *consequences*. Such theories suggest that, if we are interested in the possibility of a better world, it is rational for us to act in whatever way is productive of higher frequencies of occurrence of those states, properties, etc., that we identify as values. On a view of this kind, principles may be treated as specifications (ranging from the very general to the very specific) of the kinds of actions that (it is claimed) have higher probabilities than do alternatives of producing greater quantities of values.

Other theories propose justifications for principles that pay less attention to the goodness of the consequences of the actions recommended by those principles, and more to the goodness of the *reasons* that agents have for acting in those ways. On some views of this kind, values may be treated as virtues attributable to agents, and principles as specifications of the kinds of actions that tend to be characteristic of virtuous agents.

A Conceptual Framework for Bibliometric Ethics

However we decide to theorize the relationship between values and principles, it appears that one productive way we might structure an inquiry into the ethics of evaluative bibliometrics would be to focus on the following tasks:

1. Identification of relevant *subgroups*, each distinguished by their members' shared goals, of the population of agents responsible for actions taken in the course of bibliometric evaluations

2. Identification of the kinds of *actions* taken by the members of each subgroup in the course of bibliometric evaluations

3. Identification of the *values* held by the members of each subgroup
4. Identification of the *principles* for which the members of each subgroup advocate
5. Identification of *holes* in the ethical systems analyzed, where guiding principles would be useful and yet are absent
6. Identification of *violations*—that is, activities indicative of the values and/or principles of one subgroup lacking correspondence, or coming into conflict, with those of another

Subgroups

The three subgroups of agents at the heart of any bibliometric evaluation are as follows:

- *Analysts*: i.e., the bibliometricians responsible for collecting and analyzing data on the document-related activities of specific groups of subjects and reporting on their findings
- *Users*: i.e., the administrators and policymakers responsible for commissioning bibliometric studies, and for using the results of such studies to inform decision making in the distribution of resources
- *Subjects*: i.e., the researchers responsible for the document-related activities observed by the analysts

Actions

The main kinds of tasks involving choices among alternatives to be made by bibliometricians were summarized above, in the section on "A Conceptual Framework for Bibliometrics." These tasks include selection of the following:

- The unit type of evaluands
- A method of identifying the population of evaluands
- The variables used to characterize evaluands
- A method of operationalizing the variables
- A method of normalizing values
- A method of weighting the variables

The main kinds of decisions to be made by users of the results of bibliometric evaluations are:

- The level at which a given researcher, project, program, department, institution, etc., is to be funded or otherwise supported

- The formula or principle according to which available resources are to be distributed among the population of potential recipients

The main kinds of document-related choices to be made by researchers are:

- The frequency with which the researcher writes documents
- The coauthors with whom the researcher collaborates on a given document
- The order in which coauthors are listed on the document
- The topic(s) that the document is to cover
- The other documents that the document is to cite
- The venue(s) (e.g., the journal) to which the document is to be submitted

Values

The preeminent professional association for bibliometricians, the International Society for Scientometrics and Informetrics (ISSI), does not currently maintain a code of ethics for reference by its members.[7] Candidates for the values and principles that are promoted by bibliometricians, policymakers, and researchers may instead be sought in the codes of ethics developed by standards-making bodies in closely related fields, such as evaluation, statistics, and publishing. For the present chapter, several such codes were mined with the aim of producing the lists of values and principles that follow:

- *Declaration on Professional Ethics* (International Statistical Institute [ISI], 2010)
- *Norms for Evaluation in the UN System* (United Nations Evaluation Group [UNEG], 2005a), *UNEG Ethical Guidelines for Evaluation* (UNEG, 2008), and *Standards for Evaluation in the UN System* (UNEG, 2005b)
- "Responsible Research Publication: International Standards for Authors" (Wager & Kleinert, 2011), and "Responsible Research Publication: International Standards for Editors" (Kleinert & Wager, 2011)
- *The European Code of Conduct for Research Integrity* (European Science Foundation [ESF], 2011)

Analysts The kinds of values that typically appear in statements purporting to summarize the values held by professional statisticians and evaluators may be classified into three broad groups, according to their status as characteristics of the products, methods, or agents of evaluative work.

Valued characteristics of the *products* (i.e., the outputs) of such work, such as rankings, include the following pair, each of which may be interpreted as a family of subproperties of varying significance:

- Quality (i.e., credibility, trustworthiness) of data: e.g.,
  - Accuracy
  - Completeness
  - Consistency
  - Absence of bias
- Fitness for purpose (i.e., utility, usefulness): e.g.,
  - Relevance
  - Timeliness
  - Accessibility
  - Clarity and transparency

The motivation for making the binary distinction drawn here is to highlight the difference between (a) final or intrinsic values, and (b) instrumental or extrinsic values. The usefulness of a given ranking can be determined only by external reference to the use to which it is put, whereas the credibility of a ranking can, at least in principle, be determined without reference to external purposes. In the normal absence of a "ground truth" against which the product of an evaluative study may be compared, however, levels of data quality may be estimated in practice by examining the propensity of the methods selected by evaluators to produce outputs that are trustworthy. A breakdown of the valued characteristics of analysts' *methods* (i.e., processes) might proceed along the following lines:

- Fitness for purpose (i.e., propensity to produce outputs that are trustworthy and useful): e.g.,
  - Validity (i.e., extent to which methods are capable in practice of providing answers to the research questions to which they are applied)
  - Reliability (i.e., extent to which methods are capable in practice of providing reproducible results)

Valued characteristics (i.e., virtues) of analysts as *agents* include the following (commonly grouped under the family name of *integrity*):

- Impartiality
- Honesty
- Respectfulness (e.g., of rights)
- Accountability

Users According to the codes examined, administrators and policymakers who make decisions informed by the results of bibliometric evaluations value characteristics of the *outcomes* of those decisions as follows:

- Cost-effectiveness (i.e., extent to which the benefits for the administrator of applying the results of the evaluation outweigh its costs for the administrator)
- Maximization of benefit-harm ratio (i.e., extent to which the combined benefits for members of all stakeholder groups, including evaluands, of applying the results of the evaluation outweigh the harms)

Valued characteristics of administrators' *methods* of applying the results of evaluative studies include the following:

- Fitness for purpose (i.e., propensity to produce outcomes that maximize welfare and cost-effectiveness): e.g.,
    - Fairness in distribution of reward (i.e., extent to which the resources distributed on the basis of the results of the evaluation are allocated in a manner demonstrated to be fair to recipients)
    - Transparency of purpose (i.e., extent to which administrators' goals, intentions, assumptions, and values are clarified)

The *virtues* of administrators and policymakers may be broken down in a similar way to that applied to analysts:

- Impartiality
- Honesty
- Respectfulness
- Accountability

Subjects Values reported to be held by members of the research community may similarly be categorized in relation to outputs, methods, and agents; and, as before, intrinsic or final values may be distinguished from extrinsic or instrumental values that are defined relative to some external goal or purpose.

Valued characteristics of researchers' *outputs* are:

- Quality (i.e., credibility, trustworthiness) of work: e.g.,
    - Accuracy
    - Consistency
    - Completeness
    - Absence of bias

- Fitness for purpose (i.e., utility, usefulness): e.g.,
  - Relevance
  - Timeliness
  - Accessibility
  - Clarity
  - Completeness of documentation
  - Impact

The last value mentioned—*impact*—may be treated roughly as equivalent to the "maximization of benefit-harm ratio" applied to administrators' outputs, above, since the particular kind of impact that is valued is positive impact. Impact on different groups may be valued to varying degrees, and a distinction is often drawn between impact on science or knowledge (i.e., impact within academia or the research sector) and impact on society.

Valued characteristics of researchers' *methods* include:

- Fitness for purpose (i.e., propensity to produce outputs that are trustworthy and useful): e.g.,
  - Validity (i.e., extent to which methods are capable in practice of providing answers to the research questions to which they are applied)
  - Reliability (i.e., extent to which methods are capable in practice of providing reproducible results)

Virtues of researchers as *agents* are:

- Impartiality: e.g.,
  - Impartiality in distribution of credit for prior work (i.e., extent to which all and only those works used by researchers are cited and/or acknowledged)
- Honesty: e.g.,
  - Honesty in submission (i.e., extent to which works submitted for publication are original, substantial, unique, genuine products of those claiming to be their authors)
- Respectfulness: e.g.,
  - Respectfulness of stakeholders' rights (i.e., extent to which the various rights of the members of all stakeholder groups are taken into account in the course of research)
- Accountability

Principles

Principles specifying the kinds of actions that have ethical warrant—that is, that are justifiable by reference to intentions or expected consequences that are intrinsically good—may be formulated by considering the kinds of decisions to be made in light of the values identified. The codes listed earlier provide some examples, of which a selection follows.

Analysts

Quality of Data

[Statisticians should] strive to collect and analyze data of the highest quality possible. (ISI, 2010, p. 5)

Clarity and Transparency

Evaluators should discuss, in a contextually appropriate way, those values, assumptions, theories, methods, results, and analyses that significantly affect the interpretation of the evaluative findings. (UNEG, 2005b, p. 17)

[Statisticians should be] transparent about the statistical methodologies used and make these methodologies public. . . . In order to promote and preserve the confidence of the public, statisticians should ensure that they accurately and correctly describe their results, including the explanatory power of their data. It is incumbent upon statisticians to alert potential users of the results to the limits of their reliability and applicability. . . . Adequate information should be provided to the public to permit the methods, procedures, techniques, and findings to be assessed independently. (ISI, 2010, pp. 5–7)

Validity and Reliability

Evaluation methodologies . . . should reflect the highest professional standards. . . . Evaluation processes [should] ensur[e] that evaluations are conducted in an objective, impartial, open and participatory [manner], based on empirically verified evidence that is valid and reliable, with results being made available. . . . The evaluation methodologies to be used for data collection, analysis and involvement of stakeholders should be appropriate to the subject to be evaluated, to ensure that the information collected is valid, reliable and sufficient to meet the evaluation objectives, and that the assessment is complete, fair and unbiased. . . . Evaluation methodologies should be sufficiently rigorous to assess the subject of evaluation and ensure a complete, fair and unbiased assessment. . . . Evaluation methods depend on the information sought, and the type of data being analysed. The data should come from a variety of sources to ensure its accuracy, validity and reliability, and that all affected people/stakeholders are considered. Methodology should explicitly address issues of gender and under-represented groups. (UNEG, 2005b, pp. 6, 13)

[Evaluators should carry out] thorough inquiries, systematically employing appropriate methods and techniques to the highest technical standards, validating information using multiple measures and sources to guard against bias, and ensuring errors are corrected. (UNEG, 2008, p. 8)

[Statisticians] are responsible for the fitness of data and of methods for the purpose at hand. . . . [They should] pursue promising new ideas and discard those demonstrated to be invalid . . . [and] work towards the logical coherence and empirical adequacy of . . . data and conclusions. (ISI, 2010, p. 5)

Impartiality

Evaluators must ensure the honesty and integrity of the entire evaluation process. [Evaluators] also have an overriding responsibility to ensure that evaluation activities are independent, impartial and accurate. (UNEG, 2005b, p. 10)

In carrying out his/her responsibilities, each statistician must be sensitive to the need to ensure that his/her actions are, first, consistent with the best interests of each group and, second, do not favor any group at the expense of any other. . . . [Statisticians should] use . . . statistical knowledge, data, and analyses for the Common Good to serve the society. . . . [Statisticians should] produce statistical results using . . . science and . . . not [be] influenced by pressure from politicians or funders. . . . [Statisticians should] strive to produce results that reflect the observed phenomena in an impartial manner. . . . Statisticians should pursue objectivity without fear or favor, only selecting and using methods designed to produce the most accurate results. . . . Available methods and procedures should be considered and an impartial assessment provided to the employer, client, or funder of the respective merits and limitations of alternatives, along with the proposed method. (ISI, 2010, pp. 4–6)

Respectfulness

Evaluations [should be] carried out with due respect and regard to those being evaluated. . . . Evaluators should be sensitive to beliefs, manners and customs and act with integrity and honesty in their relationships with all stakeholders. . . . In line with the UN Universal Declaration of Human Rights and other human rights conventions, evaluators should operate in accordance with international values. . . . Evaluators should be aware of differences in culture, local customs, religious beliefs and practices, personal interaction and gender roles, disability, age and ethnicity, and be mindful of the potential implications of these differences when planning, carrying out and reporting on evaluations. . . . Evaluators should protect the anonymity and confidentiality of individual information. . . . The rights and well-being of individuals should not be affected negatively in planning and carrying out an evaluation. (UNEG, 2005b, pp. 7, 10, 17)

Evaluations can have a negative effect on their objects or those who participate in them. Therefore evaluators shall seek to: minimize risks to, and burdens on, those participating in the evaluation; and seek to maximize the benefits and reduce any unnecessary harms that might occur from negative or critical evaluation, without compromising the integrity of the evaluation. (UNEG, 2008, p. 8)

[Statisticians should] respect the communities where data is collected and guard against harm coming to them by misuse of the results. . . . Findings should be communicated for the benefit of the widest possible community, yet attempt to ensure no harm to any population group. . . . In collaborating with colleagues and others in the same or other disciplines, it is necessary and important to ensure that the ethical principles of all participants are clear, understood, respected, and reflected in the undertaking. (ISI, 2010, pp. 5–7)

Users

Transparency of Purpose

Make clear from the outset how the evaluation report will be used and disseminated. (UNEG, 2008, p. 11)

Respectfulness of Stakeholders' Rights

Anticipate the different positions of various interest groups and minimize attempts to curtail the evaluation or bias or misapply the results. (UNEG, 2008, p. 11)

Subjects

Impartiality in Distribution of Credit for Prior Work

Authors should represent the work of others accurately in citations and quotations. . . . Relevant previous work and publications, both by other researchers and the authors' own, should be properly acknowledged and referenced. The primary literature should be cited where possible. . . . Data, text, figures or ideas originated by other researchers should be properly acknowledged and should not be presented as if they were the authors' own. Original wording taken directly from publications by other researchers should appear in quotation marks with the appropriate citations. (Wager & Kleinert, 2011, p. 3)

Important work and intellectual contributions of others that have influenced the reported research should be appropriately acknowledged. Related work should be correctly cited. References should be restricted to (paper or electronically) printed publications and publications "in print." (ESF, 2011, p. 14)

Honesty in Submission

Work should not be submitted concurrently to more than one publication unless the editors have agreed to co-publication. . . . Authors should inform editors if findings have been published previously or if multiple reports or multiple analyses of a single data set are under consideration for publication elsewhere. Authors should provide copies of related publications or work submitted to other journals. . . . Multiple publications arising from a single research project should be clearly identified as such and the primary publication should be referenced. Translations and adaptations for different audiences should be clearly identified as such, should acknowledge the original source, and should respect relevant copyright conventions and permission requirements. (Wager & Kleinert, 2011, pp. 3–4)

[Authors should not engage in] repeated publication [or] salami-slicing. . . . Publication of the same (or substantial parts of the same) work in different journals is acceptable only with the consent of the editors of the journals and where proper reference is made to the first publication. In the author's CV such related articles must be mentioned as one item. (ESF, 2011, pp. 6, 14)

The authorship of research publications should accurately reflect individuals' contributions to the work and its reporting. . . . The criteria for authorship and acknowledgement should be agreed at the start of the project. Ideally, authorship criteria within a particular field should be agreed, published and consistently applied by research institutions, professional and academic societies, and funders. . . . Researchers should ensure that only those individuals who meet authorship criteria (i.e. made a substantial contribution to the work) are rewarded with authorship and that deserving authors are not omitted. Institutions and journal editors should encourage practices that prevent guest, gift, and ghost authorship. (Wager & Kleinert, 2011, pp. 1, 4)[8]

All authors, unless otherwise specified, should be fully responsible for the content of publication. Guest authorship and ghost authorship are not acceptable. The criteria for establishing the sequence of authors should be agreed by all, ideally at the start of the project. Contributions by collaborators and assistants should be acknowledged, with their permission. (ESF, 2011, p. 7)

Editors should work to ensure that all published papers make a substantial new contribution to their field. Editors should discourage so-called "salami publications" (i.e., publication of the minimum publishable unit of research), avoid duplicate or redundant publication unless it is fully declared and acceptable to all (e.g., publication in a different language with cross-referencing), and encourage authors to place their work in the context of previous work (i.e., to state why this work was necessary/done, what this work adds or why a replication of previous work was required, and what readers should take away from it). (Kleinert & Wager, 2011, p. 5)

Editors should not attempt to inappropriately influence their journal's ranking by artificially increasing any journal metric. For example, it is inappropriate to demand that references to that journal's articles are included except for genuine scholarly reasons. In general, editors should ensure that papers are reviewed on purely scholarly grounds and that authors are not pressured to cite specific publications for non-scholarly reasons. (Kleinert & Wager, 2011, p. 3)

Omissions

Codes of ethics may themselves be evaluated along the lines developed above. Statements of normative principles are "fit for purpose" to the extent that they are accessible to, and considered relevant by, the members of their intended audience. Different methods, of course, would be needed if we wished to conduct a sociological study of the values that individuals actually claim to hold, and thus of the degree to which researchers, bibliometricians, and administrators are guided, in practice, by the norms long codified by their professional associations. Meanwhile, we can proceed by pointing to gaps in the codes, where guidance on certain specifics would be especially useful, yet is unfortunately absent.

The most significant omission is that of a principle of *distributive justice*. Existing statements of principles are largely silent on the issue of the right way to allocate rewards to researchers. The general question addressed by theories of distributive justice is this: On the basis of what principle should benefits and burdens of any kinds (including economic and cultural goods and services) be distributed among populations of recipients? Justice, or fairness, is the label conventionally given to the valued property of distributions of benefits. Different theories of distributive justice provide justifications for different principles by which that value may be maximized (see, e.g., Cozzens, 2007; Lamont & Favor, 2007). For example: Principles of *strict equality* define fairness as the extent to which every member of a population receives the same quantity of net benefits. No characteristics of individual recipients are relevant to distributions based on strict equality; principles of *relative equality* specify some particular characteristic of recipients (such as need, desert, or status) in accordance with which benefits should be distributed. Yet other principles allow for inequalities only to the extent that the least advantaged are better off than they would be under strict equality. *Libertarian* theories deny the primacy of equality as a value, and consider distributions to be fair to the extent that certain freedoms and rights of recipients are respected.

A prevailing, if frequently left unstated, assumption held by participants in the national and international governance of research is that resources should be distributed in accordance with *desert* (i.e., the extent to which recipients are deserving of reward). The absence of an attendant justification for this general principle is less problematic than is the (equally understandable) absence of guidance in dealing with measurement issues of four related kinds that are perennials for distributive-justice theorists and evaluation theorists alike:

- On what basis should we determine which *elementary characteristics* (e.g., merit, need; past performance, future potential; intrinsic quality, extrinsic impact) are to be included in the calculus of overall desert?
- On what basis should we determine how these various variables are to be *operationalized* in forms (e.g., publication counts, citation counts) that are measurable?
- On what basis should we determine how the values of these various variables are to be *normalized* (e.g., by time frame)?
- On what basis should we determine how these various variables are to be *weighted*?

In general, the need is for a rational principle for determining the right way of measuring amounts of desert. Existing codes of ethics lack advocacy of any such principle, and policymakers' and analysts' selections of relevant characteristics, and of methods of operationalization, normalization, and weighting, tend to be made on a largely ad hoc basis.

Violations

The technical and methodological problems faced by evaluative bibliometricians are numerous and widely discussed, and their effects on the validity and reliability of bibliometric methods are relatively well understood (see, e.g., Bornmann, Metz, Neuhaus, & Daniel, 2008; Moed, 2007; Pendlebury, 2009; Sivertsen, 1997). For example:

• The databases of publications and citations on which counts are based tend to contain errors that are not always distributed uniformly, and tend to lack unbiased coverage of all types of citing documents (e.g., books as well as journal articles), all languages, and all fields. Conclusion: The use of counts derived from such databases to compare authors whose oeuvres are well covered with those whose oeuvres are not is invalid.

• Citation counts are not distributed uniformly or normally across cited documents; rather, the distributions are heavily skewed, with the result that mean counts work poorly as descriptions. Conclusion: The use of metrics based on mean counts (e.g., impact factors) as proxies for individual counts is invalid.

• Documents in some disciplines tend to attract higher citation counts simply because those disciplines are large or highly productive. Conclusion: The use of citation counts to compare evaluands across disciplines is invalid.

• Many authors base their decisions to cite a given document on reasons other than whether they have used it or not. Conclusion: The use of citation counts as evidence of impact is invalid.

Should we remain in any doubt about the desirability of using methods whose validity has already been shown to be suspect, we are now in a position to ask serious questions about the intrinsic value of such methods, on the basis of our observation of the absence of justification for (or of systematic compliance with) principles of distributive justice. Is it *fair* to compare citation counts and impact factors without normalizing for disciplinary differences? If such normalization is required, what are the relevant dimensions of difference, and at what level of aggregation (discipline, field, area, individual researcher) should the normalization take place? Is it *fair* to treat productivity and impact as indicators of research quality? Is it *fair* to treat publication counts as evidence of productivity, and citation counts as evidence of impact?

Meanwhile, much is made in the codes of the supposed moral unacceptability of the various kinds of activities in which candidates for reward (rather than its distributors) engage with the aim of "gaming" the system, by corrupting indicators so that they can no longer be treated as valid measures of desert:

• The "salami-slicing" strategy, by which authors divide up their research results for separate publication in a series of "least publishable units"

- The "repeated-publication" strategy, by which authors submit very similar papers to multiple venues
- The "guest-author" strategy, by which those who did not contribute to a publication nevertheless claim authorship of it
- The "citation trawling" strategy, by which journal editors encourage (or even require) authors of submissions to cite their journals

(Nothing is said in the codes about the acceptability of the similarly motivated institutional practice of hiring, and paying large salaries to, highly cited scholars in order to boost institutional counts in advance of national research assessment exercises.)

It might be argued that such instances of "gaming" are quite rational reactions to the perception that one is being forced to participate in a system of evaluation that is unfair to begin with (see, e.g., Frey & Osterloh, 2011). Administrators typically have reasons of two good kinds for using bibliometric techniques in evaluations:

- It is *possible* to distribute reward on the basis (at least partially) of quantitative measurement of the frequency of occurrence and/or strength of document-related events (publications, citations, etc.).
- Distributing reward on the basis of quantitative measurement is more *cost-effective* than doing so on the basis of qualitative peer review, which requires hard work over long periods by experts on the topics of a wide range of publications.

A far greater challenge for administrators is to demonstrate the intrinsic *fairness* of the quantitative approach. With such a challenge in mind, the framework presented in this chapter is intended for use in identifying the issues requiring attention, in reaching an understanding of the reasons for bibliometricians' past disinclination to adopt a code of professional ethics, and ultimately in exerting appropriate levels of pressure, on groups and institutions with authority and influence in the field, to require their members routinely to provide justifications *on ethical grounds* of their decisions, actions, and practices.

Conclusion

In this chapter, I have focused on the ethical implications of using evaluative bibliometrics to inform decision making in the distribution of reward. The self-imposed limitation, whereby consideration of applications of bibliometric techniques to information retrieval (IR) was excluded, is quite arbitrary. Other uses of bibliometrics are no less fraught with ethical issues. One challenge for the designers of search engines that make recommendations of documents in accordance with counts of prior usage events (links, views, downloads, etc.) is to develop a convincing response to the charge that the *Matthew effect* (see, e.g., Rigney, 2010)—an ever-increasing inequality between higher-ranked and lower-ranked documents resulting from the tendency of higher-ranked documents to attract more usage—is the product of a mechanism that distributes rank unfairly. Any context in which a "rich-get-richer" phenomenon of cumulative advantage is observed would certainly seem to be a prime candidate for justice-theoretic analysis. It is hoped that the framework presented in this chapter may help to stimulate further work in this area.

Notes

1. *Value* is a term that, potentially confusingly, has at least three distinct senses: (1) an amount, quantity, or number serving as a measurement of the extent or degree to which (or the level or rate at which) any phenomenon exhibits a given property; (2) any kind of state, condition, property, etc., attributable to agents, objects, events, or other phenomena, as individuals or in aggregations, that is held by some agent (or group of agents) to be good; and (3) (as here) the amount or quantity of goodness intrinsic to, or potentially generated by, an evaluand. On different occasions in this chapter, different senses are intended; it is hoped that the context makes the intention clear in each case.

2. See previous note. Here *value* is used in sense 1. (The sense of *variable* here is close to that of sense 2.)

3. Some indicators are the products of analysis of a hybrid form. The *h*-index, for example, is a measure in which publication counts and citation counts are combined.

4. That is, the extent to which researchers are deserving, or worthy of receiving reward.

5. Core journals in which such contributions are published include the *Journal of Informetrics*, the *Journal of the American Society for Information Science and Technology*, *Research Evaluation*, and *Scientometrics*.

6. A useful resource in this context is the Center for the Study of Ethics in the Professions (CSEP) at the Illinois Institute of Technology (IIT). According to its website (http://ethics.iit.edu/about/history-mission-center), CSEP was established in 1976 "to promote research and teaching on practical moral problems in the professions." It is "the first interdisciplinary center for ethics to focus on the professions," and "one of the nation's leading centers for practical and professional ethics." CSEP maintains an online collection of over 850 codes of ethics.

7. According to its website (http://www.issi-society.info/mission.html), ISSI was established in 1993 with the aims "to encourage communication and exchange of professional information in the field of scientometrics and informetrics, to improve standards, theory, and practice in all areas of the discipline; to stimulate research, education, and training, and to enhance the public perception of the discipline."

8. "Guest authors are those who do not meet accepted authorship criteria but are listed because of their seniority, reputation or supposed influence; gift authors are those who do not meet accepted authorship criteria but are listed as a personal favour or in return for payment; ghost authors are those who meet authorship criteria but are not listed" (Wager & Kleinert, 2011, p. 4). Reliable measurements of the prevalence of guests, gifts, and ghosts are hard to come by. One might reasonably expect to see substantial disciplinary differences, both in the frequencies of occurrence of these quasi-authorial acts, and in administrators' and scholars' perceptions of the demerits of such acts. For example, there is anecdotal evidence to suggest that, in medicine and some related fields, ghost authorship is a practice that is both relatively common and generally perceived as benign.

References

Bar-Ilan, J. (2008). Informetrics at the beginning of the 21st century: A review. *Journal of Informetrics, 2*(1), 1–52.

Bar-Ilan, J. (2010). Informetrics. In M. J. Bates & M. N. Maack (Eds.), *Encyclopedia of library and information sciences* (3rd ed., pp. 2755–2764). Boca Raton, FL: CRC Press.

Borgman, C. L., & Furner, J. (2002). Scholarly communication and bibliometrics. *Annual Review of Information Science & Technology, 36*, 3–72.

Bornmann, L., Metz, R., Neuhaus, C., & Daniel, H.-D. (2008). Citation counts for research evaluation: Standards of good practice for analyzing bibliometric data and presenting and interpreting results. *Ethics in Science and Environmental Politics, 8*, 93–102.

Cozzens, S. E. (2007). Distributive justice in science and technology policy. *Science & Public Policy, 34*(2), 85–94.

European Science Foundation. (2011). *The European code of conduct for research integrity*. Strasbourg, France: European Science Foundation. Retrieved from http://www.esf.org/fileadmin/Public_documents/Publications/Code_Conduct_ResearchIntegrity.pdf.

Frey, B. S., & Osterloh, M. (2011). *Ranking games* (CREMA Working Paper No. 11). Basel, Switzerland: Center for Research in Economics, Management and the Arts.

International Statistical Institute. (2010). *Declaration on professional ethics*. The Hague, The Netherlands: International Statistical Institute. Retrieved from http://www.isi-web.org/images/about/Declaration-EN2010.pdf.

Kleinert, S., & Wager, E. (2011). Responsible research publication: International standards for editors: A position statement developed at the 2nd World Conference on Research Integrity, Singapore, July 22–24, 2010. In T. Mayer & N. Steneck (Eds.), *Promoting research integrity in a global environment* (pp. 317–328). Singapore: Imperial College Press / World Scientific Publishing. Retrieved from http://publicationethics.org/international-standards-editors-and-authors.

Lamont, J., & Favor, C. (2007). Distributive justice. In E. N. Zalta (Ed.), *Stanford encyclopedia of philosophy*. Stanford, CA: Stanford University. Retrieved from http://plato.stanford.edu/entries/justice-distributive.

Moed, H. F. (2007). The future of research evaluation rests with an intelligent combination of advanced metrics and transparent peer review. *Science & Public Policy, 34*(8), 575–583.

Narin, F. (1976). *Evaluative bibliometrics: The use of publication and citation analysis in the evaluation of scientific activity*. Cherry Hill, NJ: Computer Horizons.

Pendlebury, D. A. (2009). The use and misuse of journal metrics and other citation indicators. *Archivum Immunologiae et Therapiae Experimentalis, 57*, 1–11.

Rigney, D. (2010). *The Matthew effect: How advantage begets further advantage*. New York: Columbia University Press.

Scriven, M. (1991). *Evaluation thesaurus* (4th ed.). Newbury Park, CA: Sage.

Shafer-Landau, R. (2010). *The fundamentals of ethics*. New York: Oxford University Press.

Sivertsen, G. (1997). Ethical and political aspects of using and interpreting quantitative indicators. In M. S. Frankel & J. Cave (Eds.), *Evaluating science and scientists: An East-West dialogue on research evaluation in post-communist Europe* (pp. 212–220). Budapest, Hungary: Central European University Press.

Tague-Sutcliffe, J. (1992). An introduction to informetrics. *Information Processing & Management, 28*, 1–3.

United Nations Evaluation Group. (2005a). *Norms for evaluation in the UN system*. New York: United Nations Evaluation Group. Retrieved from http://www.uneval.org/normsandstandards.

United Nations Evaluation Group. (2005b). *Standards for evaluation in the UN system*. New York: United Nations Evaluation Group. Retrieved from http://www.uneval.org/normsandstandards.

United Nations Evaluation Group. (2008). *UNEG ethical guidelines for evaluation*. New York: United Nations Evaluation Group. Retrieved from http://www.unevaluation.org/ethicalguidelines.

Wager, E., & Kleinert, S. (2011). Responsible research publication: International standards for authors: A position statement developed at the 2nd World Conference on Research Integrity, Singapore, July 22–24, 2010. In T. Mayer & N. Steneck (Eds.), *Promoting research integrity in a global environment* (pp. 309–316). Singapore: Imperial College Press / World Scientific Publishing. Retrieved from http://publicationethics.org/international-standards-editors-and-authors.

Whitley, R., & Gläser, J. (Eds.). (2007). *The changing governance of the sciences: The advent of research evaluation systems*. Dordrecht, Netherlands: Springer.

6 Criteria for Evaluating Indicators

Yves Gingras

Introduction

Over the course of the last decade, much discussion has been devoted to the creation of new indicators, essentially in the context of policy-related demands for more thorough evaluation of research and higher education institutions. We have thus seen the multiplication of indicators—there are already many dozens integrated in various rankings of universities. Whatever the reasons for this recent excitement about evaluation and rankings, it raises a basic question that has curiously received scant attention in the literature, as if all the energies devoted to the manufacture of new indicators has left no time for serious reflection about what *exactly* these indicators are measuring, as if that was obvious and not worth explaining. This question is simple: Which criteria can tell us when a given indicator is valid and really measures what it is supposed to measure? Using concrete examples taken from well-known indicators, this chapter introduces three criteria that should be applied to each proposed indicator in order to evaluate its validity. However, before discussing them in detail, I briefly describe the recent transformation in the uses of bibliometric indicators, because it helps explain why it has become more urgent than ever to evaluate indicators before using them for assessment purposes.

Bibliometrics: From the Sociology of Science to Evaluations and Rankings

During the 1960s and 1970s, bibliometrics was a tool used by sociologists of science to analyze the dynamics of scientific disciplines. Whether to study knowledge communication (Crane, 1972), social stratification (Cole & Cole, 1973), citation (Cronin, 1984), or cocitation patterns (Small, 1973), relatively simple and transparent indicators were constructed using publications and the references they contain (citations) to understand the internal dynamic of scientific communities. Though evaluative bibliometrics slowly emerged at the end of the 1970s (Narin, 1976; Elkana, Lederberg, Merton, Thackray, & Zuckerman, 1979), it long remained a limited specialist concern before governments absorbed, in the 1990s, the "new management" ideology that focused on the evaluation of everything using indicators and benchmarks as "objective" measures of efficiency and return on investment. In higher education and research, the effect of that new approach to management became particularly visible in the last decade. It took the form of simple rankings of so-called best colleges and universities based on a series of supposedly commonsense indicators like "employer reputation," "academic reputation," or the presence of "international faculty and students," to name some of those used in the *U.S. News & World Report* ranking of universities (Gladwell, 2011). Rankings of research universities based on reputational measures and indicators of "quality" followed suit with the publication in 2003 of the so-called Shanghai Ranking of the "top 1,000" universities of the world. Published each year, this ranking now has "competitors" in the QS World University Ranking and the *Times Higher Education* World University Ranking, also published annually.

Confronted with the many competing rankings, a French senator has observed that each of them seems to have the unfortunate tendency of being biased in favor of different institutions: "The Shanghai Ranking favors American universities, while the Leiden ranking seems to have a certain bias towards Dutch universities," the Bourdin (2008, p. 53) report claims. The French senator could have added to this list the more recent École des Mines ranking, which sees French Grandes Écoles in a very flattering light while they are marginal in the Shanghai one, all this using "the number of alumni holding a post of chief executive officer or equivalent in one of the 500 leading international companies" as an indicator (Mines ParisTech, 2011, p. 3). Knowing the strong relations between French Grandes Écoles and France's large companies (many of them formerly state-owned), the chosen indicator inherently favors these institutions.

Until recently, the use of bibliometric indicators for evaluation was effectively limited to experts by the fact that access to Web of Science and Scopus is costly and they were the only sources of bibliometric data. With the rapid development of the Web, the Internet can now be used freely to make bibliometric (and, by extension, webometric) "evaluations" and rankings. Freely accessible but uncontrolled databases such as Google Scholar have undoubtedly contributed to a certain degree of anarchy in research evaluation, because users with some technical skill are now able to measure the "quality" or "visibility" of their research (or that of their "enemies") by fashioning their own ad hoc indicator, based, for instance, on the number of hits (to Google

Scholar or some other website containing publications) and now even the number of tweets (see Priem, chapter 14, this volume). The number of "spontaneous" or "raw" evaluation methods and impact indicators has exploded in recent years. This, in turn, has contributed to a chaotic state of affairs in academia, whereby no one really knows how to interpret these "measures." Rankings circulate as "black boxes," presented as indisputable "facts" about individuals or institutions and are supposed to help policy-makers and university administrators set research and academic priorities. This evaluation fever has resulted in rampant multiplication and misuse of faulty indicators. One has lost count of the letters to *Nature* and *Science* and the proliferation of blogs by scientists who claim to evaluate their colleagues or their institution. Scientists now flaunt their *h*-index (more on that later), while universities fret over their position in the Shanghai Ranking or other rankings.

Three Necessary Criteria for Evaluating Indicators

Despite the numerous publications now devoted to analyzing and criticizing these rankings, most critics have addressed the unintended consequences of their uses but few have really dug into the indicators themselves except to point out their limits, stopping short of questioning their epistemological foundation—that is, whether these indicators really have any definite meaning and thus measure what they are supposed to measure. For before assessing the inevitable limits of any indicator, one must first ensure it indicates what it is supposed to indicate. If not, then the chosen indicator is better characterized as irrelevant or misleading rather than "limited" and should be replaced by another, more appropriate, indicator. Using rankings based on faulty measurements could result in bolstering policies based on poorly analyzed problems. The lack of serious methodological reflection given to the question of the validity of each of the indicators used to concoct rankings has not stopped university managers from investing scarce resources in order to improve their position in their preferred ranking, though they, in fact, ignore what really is being measured. As we will see, most ranking schemes do not have any of the properties necessary for good indicators and it would be foolhardy to use them as a guide for policymaking. It should also be noted that, in both cases, *a single number* is used for ranking and evaluating the "quality" and "impact" of the research performed by organizations and individuals, despite the obviously multidimensional nature of research (Bollen, Van de Sompel, Hagberg, & Chute, 2009). The very existence (and persistence) of such biased indicators and rankings seems to be a consequence of the unwritten rule that *any number beats no number*.

It is striking that among the large number of essays for or against rankings, and in papers promoting new measures of scientific activity, few take time to discuss in detail the conditions under which an indicator can be taken as valid (Salmi & Saroyan, 2007; Hazelkorn, 2007). The "Berlin Principles on Rankings of Higher Education Institutions," for example, contains sixteen principles to which rankings should adhere, but only one of these specifies in general terms that indicators should be chosen "according to their relevance and validity" and that "the choice of data should be grounded in recognition of the ability of each measure to represent quality and academic and institutional strengths" (IREG, 2011, p. 22). Finally the producers of the indicator should be "clear about why measures were included and what they are meant to represent" (IREG, 2011, p. 22). Given the actual properties of existing rankings, it is clear that this principle is rarely, if ever, applied seriously since most measures used are not shown to be fit for purpose by comparing the dynamic of the indicator with that of the concept to be measured. Faced with a proliferation of indicators that tend to be used to justify policy changes in higher education and research, one must go beyond the vague appeal to "relevance" and "validity," terms whose precise meaning is rarely made explicit. Other criteria for evaluating indicators are also invoked that relate to the quality of data sources (e.g., timeliness) or the transparency of the construction of the indicator (Council of Canadian Academies, 2012, p. 64). Here, however, I concentrate on criteria directly related to the internal validity of the indicator evaluated through its adequacy to the reality behind the concept it is supposed to measure. I thus propose three criteria as necessary conditions for validity. These define the essential properties a well-constructed indicator should possess in order to be considered valid:

1. Adequacy of the indicator for the object it measures
2. Sensitivity to the intrinsic inertia of the object measured
3. Homogeneity of the dimensions of the indicator

Using different examples, I now discuss each of them in more detail.

Adequacy of the Indicator for the Object It Measures

By definition, an *indicator* is a *variable* that can be *measured* and that aims at faithfully representing a given *concept* referring to the property of an object that one wants to measure (Lazarsfeld, 1958). Typical examples of such concepts and indicators are inflation, which measures changes in the price of goods over time, and gross domestic product (GDP), which measures the national economic production of a country. The indicator is not the concept itself but a proxy, used as a way of measuring how the reality behind the concept changes over time and/or place. Thus, the properties of the indicator should always be checked against the properties the concept is assumed to have, based on intuition and prior knowledge of the object that has the properties we want to measure or on independent empirical measures of that concept. The indicator should thus be strongly correlated with what we presume to be the inherent characteristics of the concept we want to measure using that specific indicator.

The first property of a good indicator is that it must be fit for purpose—that is, correspond to the object (or concept) being evaluated. Are the results produced by the indicator of the correct order of magnitude, given what we already know about the object? Do they correspond to our intuition of the concept? For instance, the level of investment in R&D is a good first measure of the intensity of research activity in a given country. It refers to an investment and this cannot be considered a measure of quality, for example. Likewise, the total number of scientific papers published by a given country provides a good measure of the volume of its public research output but not of its industrial research, the results of which one does not expect to see published, though there are papers coming from industry. Those two classic indicators are based on the basic intuition that the more money you put into a research system, the greater the likelihood that it will produce more output. This intuition is supported by the fact that we typically find a relation between the size of countries, as measured by their GDP and their total number of papers. The money is translated into human resources and equipment and it is intuitive that, though the relation is not linear, more money should give rise to more papers since it makes possible the enrollment of more students, the hiring of more professors, and the purchase of better equipment. Therefore, we can say that we have a good grasp of the meaning of these indicators and they behave as one would expect.

Whereas indicators of production are relatively well understood, things become more complicated when we move to abstract concepts like "quality" or "impact" of research as opposed to sheer quantity. To measure the scientific impact of a given author or institution, one could conduct some type of survey, asking many "experts" what they think of the "quality" or "scientific impact" of the work of a given individual or institution and computing an average, using a qualitative scale. This could serve as an indicator of "quality" though the subjective aspects of such evaluations and the danger of circularity—the "best" are the ones most "experts" tend to say are the best—are difficult to control. Despite its limitations, such a definition of the indicator is clear and can make sense, provided the set of experts chosen corresponds to the domain being evaluated. Alternatively, and much less difficult and subjective, one could compute the number of citations received by a paper and use it as an *indicator* of scientific "quality" and "impact," here taken as synonymous since both use the same indicator. One could also decide not to try to measure "quality" with citations and use the term "visibility": frequently cited papers are more visible than infrequently cited ones. If one wants to keep "quality," then it is not sufficient to make the claim in a tautological manner; one must first test the connection between the concept (quality) and the indicator (citations) by finding a relationship between citations and an *independent* measure of "quality," already accepted as a valid measure. In this regard, sociological and bibliometric studies since the 1970s have consistently shown that there is a correlation between how often an author is cited and how renowned he or she is, as measured by other indicators of eminence like important prizes and awards or academic nominations to scientific academies (Cole & Cole, 1973). This relation is also consistent with our understanding of the role of recognition in scientific communities (Merton, 1973). Indeed, the myth of great, as yet uncited, scientists is just that—a myth. "A Structure for DNA," the famous 1953 paper by Watson and Crick on the structure of the DNA, for example, quickly became one of the most cited papers published that year in *Nature*, contrary to what has been often asserted by critics of bibliomeric evaluations (Lawrence, 2007; Gingras, 2010). The same is true for the 1905 papers of Albert Einstein. The crux of the matter is not to check whether they received 100 or 104 citations, but rather to note that this number is much higher than that of the average at the time, which was indeed the case for Einstein as early as 1907 (Gingras, 2008). An important caveat, however, is that indicators such as number of published papers and citations obtained have mostly been validated in the case of the natural sciences, so one cannot mechanically and blindly transfer their use to the social sciences or the humanities. One has to take into account differences between these disciplines—for example, the fact that books are much more common than articles as a means of disseminating new results and that they are more often cited over a long period of time (Larivière, Archambault, Gingras, & Vignola-Gagné, 2006; Archambault, Vignola-Gagné, Côté, Larivière, & Gingras, 2006).

An example of an indicator that is not fit for purpose is the one based on the number of Nobel Prizes associated with a university. Though it is obvious that the Nobel Prize in itself can be taken as an excellent measure of the quality or scientific impact of the work of a researcher, the problem comes from the fact that in evaluating a *university* at a given time, the index also counts prizes awarded decades earlier. Time being an important variable in the measure of an attribute, it should be obvious that taking into account the Nobels associated with a university more than twenty years previously sheds no light on the quality of that institution today. Though that should be obvious, this indicator remains a component of the Shanghai Ranking, which I discuss below. One could also ask what a ranking of universities based on "presence" on the Web really measures beyond the fact that it is "present" on the Web, or when one defines the "quality" of a university by "counting all the external links that the University webdomain receives from third parties" (Ranking Web of Universities, 2012, para. 22).

Sensitivity to the Intrinsic Inertia of the Object Measured

A major intrinsic characteristic of an object is its *inertia*—that is, its resistance to change, which thus affects the swiftness at which change can happen. Thus, a good indicator is one that varies in a manner consistent with the inertia of the object being measured, since different objects change with more or less difficulty (and rapidity) over time. Consider a digital thermometer (instead of an older mercury-based thermometer). If, in a given room (without any drafts), we first measure a temperature of 20 degrees and then, a minute later, 12 degrees and a minute after that, 30 degrees, common sense would lead the observer to conclude that the instrument is defective, not that the temperature of the room varies wildly. We know very well that the temperature of a room cannot change that much in the space of three minutes with no door or window opened. Now take the case of universities. It is well known that large academic institutions are like supertankers and cannot change course quickly (and thankfully so, because this allows them to avoid responding swiftly to shortsighted or frivolous so-called social demands[1]). Therefore, an annual ranking where an institution moves, *in a single year*, from, say, 12th to 18th or 12th to 9th, should strongly suggest that the indicator "showing" that movement is defective, not that the "quality" of the institution has plummeted or risen significantly during the year. Given the inevitable variance in the data from one year to the next, it is clear that most annual changes in positions are random and devoid of any real significance. For this reason, it does not really make sense to "measure" (or "evaluate") them every year. In the United States, for instance, the National Research Council produces a ranking of all doctoral programs at all American universities in every discipline, but it does so only once every 10 years. Why choose such a low frequency? Simply because, in addition to the high costs associated with conducting such an evaluation properly, the probability of a given academic program being "excellent" in 2008 but "mediocre" in 2009 is very small, if one excludes a sudden wave of retirements. This timescale thus respects the fact that universities are relatively inertial institutions. It also suggests that evaluating large research groups every two or three years makes little sense (and constitutes a waste of resources), and that doing so every six to eight years might be a more realistic and more economical way of observing any *real* changes.

In light of this analysis, one can conclude that annual rankings of universities, be they based on surveys, bibliometrics, or webometrics, have no foundation in methodology and can only be explained by marketing strategies. They serve no serious scientific purpose and can even have negative unintended consequences if used by academic managers who think they should adjust the priorities of their institutions to fit the latest wave of indicators.

Homogeneity of the Dimensions of the Indicator

A third crucial property that any valid indicator should possess is being homogeneous in its composition. With respect to research, homogeneous indicators of research output (say, at the level of a country) can be constructed using the number of articles published in leading scientific journals. Indicators based on papers can help construct a descriptive *cartography* of research activities using input and output measures. In addition, one can obtain a *productivity* index from the input/output ratio. However, if one were to somehow combine number of papers with a citation measure as does, for example, the "*h*-index," one would then obtain a heterogeneous indicator. The problem is the same with all indicators that combine, arbitrarily, different kinds of indicators like "academic reputation" and "international faculty and students." The fundamental methodological problem with such composite heterogeneous indicators is that when they vary, it is impossible to have a clear idea of what the change really means since it could be due to different factors related to each of the heterogeneous parts of the composite indicator.[2] Combining different indicators into a single number is like transforming a multidimensional space into a zero-dimension point, thus losing nearly all the information contained on the different axes. One should keep each indicator separate and represent them on a spiderweb diagram, for example, in order to make visible the various components of the concept being measured.

These three validity criteria are sufficient for detecting any invalid indicator. I now turn to two widely used ones: the Shanghai Ranking and the *h*-index. I limit the discussion to these, since they are prime examples of well-known indicators constructed at different scales or levels of aggregation: the Shanghai Ranking evaluates institutions, while the *h*-index focuses on individuals. The reader may already suspect that these two measurements do not constitute valid indicators, and scientists and policymakers promoting these as "objective" and "international" measures should think twice before using them as the basis of evaluations and for decision making.

The Shanghai Ranking of Universities

The Shanghai Ranking is computed by summing six different measures. The first four, worth 20% each, involve the number of

1. Faculty members who have received a Nobel Prize or the Fields Medal (for mathematics)
2. Researchers at the institution who are on the "most cited" list compiled by Thomson Reuters
3. The institution's papers that are published in *Nature* and *Science*
4. Articles found in the Web of Science

Two additional measures, worth 10% each, round out the indicator:

5. The number of *alumni* who have received a Nobel Prize or the Fields Medal
6. An adjustment of the five preceding indicators based on the size of the institution

Clearly, the final index is a composite of several heterogeneous measures, since the number of publications in *Science* and *Nature* is not commensurable with the number of Nobel Prizes. Thus, it does not meet criterion 3. What is even more surprising is that the results on which the final rankings are based turn out to be very difficult to reproduce (Florian, 2007). One could also question the adequacy (criterion 1) of an indicator such as "the number of articles in *Science* and *Nature*," given that both these journals do not cover all disciplines and, moreover, are heavily biased in favor of the United States: in 2004, for example, 72% of the articles in *Science* and 67% in *Nature* had at least one author with an address from the United States. Most importantly, given what we know about the inertia of universities, we should certainly cast doubt on an index that causes universities—in this case the Free University of Berlin and Humboldt University—to move (up or down) almost 100 places, simply by being associated with (or not) Einstein's 1922 Nobel Prize! One might also wonder, in the first place, whether the quality of a university really depends on the research conducted on its campus many decades earlier (Enserink, 2007). Consequently, it does not meet criterion 2.

To show how the rankings can go against what we otherwise generally know about universities, let us take the case of Canadian universities compared with French ones. In the 2009 Shanghai Ranking of the "best" universities in the social sciences, we observe that among the "top 100" one finds 8 Canadian universities but *zero* French universities. Now, does anyone *really* believe that French social science is that much worse than Canadian social science? Of course not, and this bizarre result is simply an artifact of the data used, which are strongly biased against the European social sciences because papers written in French and German journals are underrepresented (Archambault, Vignola-Gagné, Côté, Larivière, & Gingras, 2006). So, using this ranking as a guide can only produce bad decisions. I show below that even in the sciences, the ranking is not consistent with the quality of French science. Clearly, most of the indicators used to construct the Shanghai Ranking do not meet our basic criteria.

The *h*-index

We turn now to another fashionable indicator, which is widely used by researchers themselves, not just by university managers. The *h*-index was developed by Jorge E. Hirsch (2005), a physicist at the University of California at San Diego. It is defined as the number of articles (*n*) having received at least *n* citations over a given time period. For instance, a researcher having published 20 articles, 10 of which have been cited at least 10 times, would have an *h*-index equal to 10. The author wanted to go beyond the mere number of papers, given that those cannot be taken as a measure of "quality" since one may publish many useless papers simply to respond to the "publish or perish" injunction. The *h*-index is supposed to be a measure of the overall "quality" of an individual's research and is a composite measure of productivity (number of articles published) and "visibility," "quality," or "impact" (number of citations).

Being a heterogeneous indicator, it fails criterion 3. Interestingly, its heterogeneous combination of quantity and "quality" confuses the author himself about what his *h*-index really measures, for the title of Hirsh's paper was "An Index to *Quantify* an Individual's Scientific Research *Output*" (my italics). It is obvious that this index is neither a measure of output (quantity) nor a measure of quality but an admixture. Despite its obvious flaws, and probably because it feeds into the narcissism of scientists, the *h*-index has rapidly become widely used in the scientific community. It has even been incorporated into several databases, thus *reifying* it despite its major technical flaws. The index has become easily accessible and cheap to use without a need to understand its properties. According to Hirsch, his index should contribute to "a more democratic assessment of people's research" (Rovner, 2008, para. 41). Given that its aim was to counteract the emphasis on the sheer number of papers by combining it with a count of citations, it is ironic that the *h*-index is in fact highly correlated with the number of published papers and thus essentially determined by it (Van Leeuwen, 2008). This indicator does not meet criterion 1 either.

Being an invalid indicator, the *h*-index could (in fact, will) have perverse effects if used in evaluations. Though a detailed analysis has shown that the *h*-index is inconsistent as a measure of quality (Waltman & Van Eck, 2011), a simple example should be sufficient to illustrate the fact that it is inconsistent with our intuitive understanding of the relation between the quality of papers and the citations they receive. Take the hypothetical case of two scientists: the first is a young researcher who has published only 3 articles, but each has been cited 60 times (over a given time period); the second researcher (of the same age) is more prolific and is the author of 10 articles, each having received 11 citations (over the same time period). The latter case results in an *h*-index of 10, while the former would have an *h*-index of 3. Now, given what we know about the meaning of citations in science, should we conclude that the second scientist is three times "better" than the first? Clearly not: from a citation point of view, which provides a rough but direct measure of scientific impact, the first researcher is almost twice as good as the second. This example shows a fundamental defect of the indicator: it is not, as it should be, a *strictly increasing function* of the measure of the concept. It is obvious that if the value of the concept to be measured ("temperature," for example) goes up, then the value of the indicator used to measure it (the number indicated by the thermometer) should also go up. A good indicator that sees its values go up (or down) should be interpreted in all cases as meaning an increase (or decrease) of the thing being measured: on all thermometers 10 degrees is hotter than 3 degrees and one cannot accept an apparatus where the constructor claims that 10 is in fact *colder* than 3. A thermometer with such a scale would not sell.

In the case of the *h*-index, we have situations where a lower index often hides a better-quality researcher. Therefore, this indicator is not fit for purpose because it does not correspond to our intuitive notion of "quality" or "scientific impact," for which the normalized measure of citation is more appropriate. Given that a good indicator must retain its intuitive relation to the concept it seeks to measure, one should not expect that the problems of the *h*-index can be corrected by inventing even less intuitive and more complicated indexes like squaring it or dividing it by something else. Such operations only make an invalid indicator even less transparent.

Why Are Invalid Indicators Used?

Given the above analysis, it is surprising that so many university presidents and managers lose all critical sense and take such rankings at face value. Only a psychosociological analysis of high-ranking administrators could potentially explain the appeal of a grading system that has no scientific basis. Undoubtedly, the rhetoric surrounding the globalization of the university market has exacerbated sensitivity around evaluation issues, because universities are keen to attract foreign students in order to compensate for demographic decline or insufficient government subsidies or to jump on the internationalization bandwagon, albeit at the cost of neglecting their local duties. As the president of University of Toronto said: "Canadian universities have been complicit, *en masse*, in supporting a ranking system that has little scientific merit because it reduces everything to a meaningless, average score." And he added: "My institution has found Maclean's useful for one thing only: Marketing. None of us really believes that the ranking has much intellectual rigor."[3] Overinterpretation of rankings can also lead to ridicule, as in the case of the sudden presence among the "top" research-intensive universities in the 2010 QS World University Ranking of Alexandria University in Egypt. The university boasted of its new status on its website, and the editor of the *Times Higher Education* magazine wrote to the university saying that "any institution that makes it into this table is truly world class."[4] None of them asked the obvious question: How could an institution move *in a single year* from unknown to world class? Fortunately, more informed people raised eyebrows. The year after, the QS ranking put Alexandria in position "601+" and the data for the year 2010 disappeared, leaving a blank instead of the original position "147."[5] This extreme example should be enough to caution university managers to look inside the black box of ranking instead of taking that box as a welcome gift.

The reasons bad rankings still get used are also political. As the case of France clearly shows, ranking systems can be used as a means of justifying university reforms. Given the priorities of the Sarkozy government following his election in 2007, we can assume that if French universities had been highly ranked in the Shanghai Ranking, it would have been much more difficult for Sarkozy to justify his policies, and the government would have had a very different perspective on such a ranking, probably dismissing it as not representative of the French university system they wanted to reform. In such a case, we might have seen the opposite of what materialized (where the French government used the ranking to justify reforms) and the ranking would have been used by those *opposed* to the reform as proof that French universities are in fact excellent and should be left alone. Finally, an important contributory factor to the rise of rankings is the role played by the communication and marketing divisions of universities, which tend to see their institutions simply as a product to be sold using the standard rhetoric of marketing (Gingras, 2009b).

It is also important to emphasize that, contrary to the belief that only managers seek to promote indicator-based evaluations, the rapid proliferation of the *h*-index within certain scientific disciplines has been largely a grassroots phenomenon. Indeed, scientists themselves are often the ones who succumb to the anarchic uses of "raw" bibliometrics. As members of committees and boards with the power to affect policy decisions, they frequently push for the generalized use of such indicators, even though they have not tested their properties. This only confirms that for scientists, the "enemy" is more often one's colleague than a distant bureaucrat.

Examples of Valid Indicators

Keeping the above critical perspectives in mind, it is obviously possible to construct aggregate indicators of research that provide a good sense of the relative position of universities or countries in a national or global research system. The most common one is, of course, the total number of articles published in the main journals indexed by the Web of Science and Scopus databases. This provides a useful ranking of outputs, which can then be used with personnel or R&D investment to construct a productivity measure. In fact, despite differences in bibliographic coverage, different databases (for example, Web of Science and Scopus) yield essentially the same rankings, at least for the top 25 countries (Archambault, Campbell, Gingras, & Larivière, 2009). In addition, there is a strong correlation between the number of articles produced per country and its R&D expenditures. Canada, for instance, ranks 8th in total number of articles published in 2007. We could go one step further and compute an index for the visibility of articles, as measured by the number of citations received by articles in the two years following their publication (though we could equally well choose three or five years, for instance), then normalizing in order to take into account the varying rates of citation of different scientific fields. According to this measure, which, it should be noted, is both homogeneous and distinct from the number of publications, Canada moves up to the 4th position (in 2007). In other words, although China has surpassed Canada in terms of the number of publications, it still lags far behind in terms of citations received, thus suggesting a real lag in terms of quality of papers. Given the diversity of research fields, it is most effective to construct indicators for different research areas, since universities and countries are not equally active or visible in all sectors. In terms of relative citations, Canada for example is 5th in biology but 7th in mathematics in the G-8 of most publishing countries, compared to France, for instance, which is 2nd in mathematics and 4th in biology (Gingras, 2009a). We could continue in the same way by differentiating within these sectors, but the point here is simply to suggest that constructing homogeneous indicators is possible and allows us to interpret them more easily and to see whether they change abruptly or smoothly. Again, small fluctuations in a country's position from one year to the next are due to the natural variance of the data and should not lead to overinterpretation and even less to policy decisions. However, a regular decline in the value of an indicator (the number of articles or the relative citations) over a longer time period—say five years—should give rise to more rigorous attempts at interpretation. For instance, the noticeable rise of China in terms of overall production—from 8th in 2000 to 4th in 2005—points to a strategy of increasing publications in international journals, but one that has not yet translated into an important rise in the number of citations, because China ranks last among the G-8 on this indicator.

Conclusion

What, then, should be done to counter the evaluation anarchy that has been gaining ground over the last decade? First, institutions should learn to evaluate the validity of indicators before using them for policy decisions. They will thus reduce the danger of adversely affecting decisions or policies by using bad indicators. Only valid indicators can lead to informed decisions based on local and national priorities. Second, we must educate and convince scientists themselves of the dangers of misusing bibliometrics and indicators based on dubious combinations of data. Although the bibliometric community often likes to play with indicators for their own sake, it should be more critical in evaluating new indicators and resist succumbing to their false concreteness.

Finally, one should remember that evaluations can affect people and institutions. As a consequence, a basic ethical principle should be that before creating "new" indicators based on anything that happens to be measurable,[6] we should first make sure that the indicators being promoted have at least the three basic properties I have proposed here in order to make sure they are fit for purpose. In such a way, it would become possible to perform evaluations that could be truly useful in helping decision making. Despite inevitable policy debates, we would at least have constructed robust and relevant indicators and controlled the research evaluation epidemic by eradicating the insidious virus that gave birth to it: makeshift indicators.

Notes

1. It is absurd, for instance, to state that "education should be tied to the extremely volatile labour market," since education operates on timescales incompatible with those of the job market ("Les cahiers de la compétitivité," *Le Monde*, May 21, 2008, p. 1). This problem highlights the importance of having basic "training" that transcends this volatility.

2. Summing many indicators having different measures is what makes the indicator heterogeneous. In the case of the price index, for example, we add different objects (eggs, meat, etc.) but the unit is the price, which is homogeneous. It would not make sense to sum up the number of units of the different objects themselves because they are heterogeneous. So a *composite* index is a different thing from a *heterogeneous* one.

3. *Maclean's* is a Canadian magazine that annually publishes a ranking of Canadian universities. The citation comes from the newspaper *The Ottawa Citizen*, April 23, 2006.

4. "Questionable Science behind Academic Rankings," *New York Times*, November 15, 2010, retrieved from http://www.nytimes.com/2010/11/15/education/15iht-educLede15.html?pagewanted=all&_r=0.

5. http://www.topuniversities.com/institution/alexandria-university

6. With the many appeals to "altmetrics" for research activity, it may be worth recalling that there are in fact only a few variables that can be combined: human resources, dollars, papers, citations. The latter could be the traditional citations in papers or the new forms related to Internet use like downloads, page views, and now tweets. Although we do not have any clear understanding of the meaning of these new "measures," they are still limited in number and still turn on the same basic units: input, output, outcome, and impact.

References

Archambault, É., Campbell, D., Gingras, Y., & Larivière, V. (2009). Comparing bibliometric statistics obtained from the Web of Science and Scopus. *Journal of the American Society for Information Science and Technology, 60*(7), 1320–1326.

Archambault, É., Vignola-Gagné, É., Côté, V., Larivière, V., & Gingras, Y. (2006). Benchmarking scientific output in the social sciences and humanities: The limits of existing databases. *Scientometrics, 68*(3), 329–342.

Bollen, J., Van de Sompel, H., Hagberg, A., & Chute, R. (2009). A principal component analysis of 39 scientific impact measures. *PLoS ONE, 4*(6), e6022.

Bourdin, J. (2008). *Rapport d'information fait au nom de la délégation du Sénat pour la Planification sur le défi des classements dans l'enseignement supérieur*. Appendix to the minutes of the July 2 meeting. Retrieved from http://www.senat.fr/rap/r07-442/r07-4421.pdf.

Cole, J. R., & Cole, S. (1973). *Social stratification in science*. Chicago: University of Chicago Press.

Council of Canadian Academies. (2012). *Informing research choices: Indicators and judgment. Report of the Expert Panel on Science Performance and Research Funding:* Ottawa: Council of Canadian Academies.

Crane, D. (1972). *Invisible colleges: Diffusion of knowledge in scientific communities*. Chicago: University of Chicago Press.

Cronin, B. (1984). *The citation process: The role and significance of citations in scientific communication*. London: Taylor Graham.

Elkana, Y., Lederberg, J., Merton, R. K., Thackray, A., & Zuckerman, H. (Eds.). (1979). *Toward a metric of science: The advent of science indicators*. New York: Wiley.

Enserink, M. (2007). Who ranks the university rankers? *Science, 317*(5841), 1026–1028.

Florian, R. V. (2007). Irreproducibility of the results of the Shanghai academic ranking of world universities. *Scientometrics, 72*(1), 25–32.

Gingras, Y. (2008). The collective construction of scientific memory: The Einstein-Poincaré connection and its discontents, 1905–2005. *History of Science, 46*(151), 75–114.

Gingras, Y. (2009a). Le classement de Shanghai n'est pas scientifique. *La Recherche*, (430), 46–50.

Gingras, Y. (2009b). Marketing can corrupt universities. *University Affairs*, January. Retrieved from http://www.universityaffairs.ca/marketing-can-corrupt-universities.aspx.

Gingras, Y. (2010). Revisiting the "quiet debut" of the double helix: A bibliometric and methodological note on the "impact" of scientific publications. *Journal of the History of Biology, 43*(1), 159–181.

Gladwell, M. (2011, February 14). The order of things: The problem with college rankings. *New Yorker*, 68–75.

Hazelkorn, E. (2007). The impact of league tables and ranking systems in higher education decision making. *Higher Education Management and Policy, 19*(2), 87–110.

Hirsch, J. E. (2005). An index to quantify an individual's scientific research output. *Proceedings of the National Academy of Sciences of the United States of America, 102*(46), 16569–16572.

IREG. (2011, November). IREG ranking audit manual. Brussels: IREG. Retrieved from http://www.ireg-observatory.org/pdf/ranking_audith_audit.pdf.

Larivière, V., Archambault, É., Gingras, Y., & Vignola-Gagné, É. (2006). The place of serials in referencing practices: Comparing natural sciences and engineering with social sciences and humanities. *Journal of the American Society for Information Science and Technology, 57*(8), 997–1004.

Lawrence, P. (2007). The mismeasure of science. *Current Biology, 17*(15), R583–R585.

Lazarsfeld, P. F. (1958). Evidence and inference in social research. *Daedalus, 87*(4), 99–130.

Merton, R. K. (1973). *Sociology of science*. Chicago: University of Chicago Press.

Mines Paris Tech. (2011). *International Professional Ranking of Higher Education Institutions*. Retrieved from http://www.mines-paristech.fr/Actualites/PR/Ranking2011EN-Fortune2010.html#1.

Narin, F. (1976). *Evaluative bibliometrics: The use of publication and citation analysis in the evaluation of scientific activity*. Parsippany, NJ: Computer Horizons.

Ranking Web of Universities. (2012). Methodology. Retrieved from http://www.webometrics.info/en/Methodology.

Rovner, S. L. (2008). The import of impact: New types of journal metrics grow more influential in the scientific community. *C&EN, 86*(20), 39–42. Retrieved from http://pubs.acs.org/cen/science/86/8621sci1.html.

Salmi, J., & Saroyan, A. (2007). League tables as policy instruments: Uses and misuses. *Higher Education Management and Policy, 19*(2), 31–68.

Small, H. G. (1973). Co-citation in the scientific literature: A new measure of the relationship between two documents. *Journal of the American Society for Information Science, 24*(4), 265–269.

Van Leeuwen, T. N. (2008). Testing the validity of the Hirsch-index for research assessment purposes. *Research Evaluation, 17*(2), 157–160.

Waltman, L., & Van Eck, N. J. (2011). The inconsistency of the *h*-index. Retrieved from http://arxiv.org/ftp/arxiv/papers/11081108.3901.pdf.

III Methods and Tools

7 Obliteration by Incorporation

Katherine W. McCain

Introduction

After a scholarly work is published (or presented at a conference), its post-publication visibility can be charted via its citation history, and its standing in its field evaluated, in part, by the total number of citations received. At one extreme, the work may never be cited, even by its author(s); at the other, it may become a "citation classic" (Garfield, 1977), garnering a large number of citations annually for many years after initial publication. Between the two extremes of uncitedness (Burrell, 2012; Schwartz, 1997; Stern, 1990) and classic status, researchers have described a variety of citation history profiles. Most works have a limited citation lifespan—being replaced by more recent works in the natural course of scholarship or subsumed in a scholarly review. Aversa (1985) and McCain and Turner (1989) distinguished between two citation profiles, one with counts peaking early in the profile (years 2–3 post-publication) and declining swiftly thereafter, while others exhibited a later peak in years 6–7 followed by a slower decline. Other researchers (see the review in Costas, Van Leeuwen, & Van Raan, 2010) have described three categories of documents—"normal" documents (a peak in years 3–4 and an exponential decline thereafter), "flash-in-the-pan" documents (a peak soon after publication followed by a sharp decline), and "delayed" documents (a late or very late peak in citation counts; see Garfield, 1980). This last category has also been characterized as "sleeping beauties" (Van Raan, 2004) that are "awakened" by a much later document (the "prince"; Braun, Glänzel, & Schubert, 2010) or that represent "premature discovery" (Stent, 1972).

Overall, higher counts for cited works and later profile peaks in citation histories are considered to be useful sources of evidence for the greater visibility and utility of the published work as perceived by citing scholars (see the discussion in De Bellis, 2009, pp. 243–284; Moed, 2005, pp. 193–219). Some researchers (e.g., Edge, 1979; MacRoberts & MacRoberts, 1989) have expressed strong reservations about any reliance on citation counts and profiles in assessment of individual scholars, research groups, institutions, or countries. Their primary concern appears to be that well-received publications and elite authors are more likely than others to garner more than their share of additional citations over time. In her response to a MacRoberts and MacRoberts (1987) article that warned of significant bias and incompleteness in citation analyses, Zuckerman (1987) noted that citation data had been found to be well correlated with other sources of evidence of utility and influence. She also indicated that there was a variety of potentially offsetting sources of error in measurement when using citation data that needed to be better understood through empirical investigation. In addition to possible overcitation because of existing visibility (the MacRoberts' concern) and "citation cartels" (an explicit agreement to cite specific authors or journals; see Frank, 1999), Zuckerman (1987) pointed to overcitation of students and young colleagues by their mentors and to two sources of "undercitation": (1) the citation of "intermediary publications" that in turn cite the work of interest (depriving the older work of additional citations), and (2) "obliteration by incorporation"—in which the valued contribution of the cited work becomes decoupled from its bibliographic identity and is discussed or invoked without attribution. These two conditions that lead to undercitation—obliteration by incorporation and the effect of intermediary publications on a work's citation history—are the subject of this chapter.

The notion of "obliteration by incorporation" (OBI) was first suggested by Robert Merton and is most succinctly defined by him in his 1988 paper on the Matthew effect: "the obliteration of the source of ideas, methods, or findings by their incorporation in currently accepted knowledge" (Merton, 1988, p. 622). Earlier, in his preface to Garfield's monograph on citation analysis (Garfield, 1979), Merton extended his explanation of OBI and speculated on its effect on the usefulness of explicit citations in both evaluation and historical study:

> In the course of this hypothesized process [of OBI], the number of explicit references to the original work declines in the papers and books making use of it. Users and consequently transmitters of that knowledge are so thoroughly familiar with its origins that they assume this to be true of their readers as well. Preferring not to insult their readers' knowledgeability, they no longer refer to the original source. And since many of us tend to attribute a significant idea or formulation to the author who introduced us to it, the altogether innocent transmitter sometimes becomes identified as the originator. In the successive transmission of ideas, repeated use may erase all but the immediately antecedent versions, thus producing an historical palimpsest in which the source of those ideas is obliterated. To the extent that such obliteration does occur—itself an empirical question that is only beginning to be examined—explicit citations may not adequately reflect the lineage of scientific work. As intellectual influence becomes deeper, it becomes less readily visible. That influence may operate through acceptance of a theoretical framework, with its basic assump-

tions, or through standardized procedures of inquiry. In short, it may be canonical knowledge that is most subject to obliteration of source. (Merton, 1979, p. ix)

Merton's notion of OBI was popularized by Garfield in a *Current Contents* essay (Garfield, 1975) as the "obliteration phenomenon" and the notion has been discussed, or at least mentioned, with some frequency in the bibliometrics literature since. Indeed, OBI appears to be an accepted fact by scholars discussing various aspects of citation analysis—citing Garfield's essay or Merton's writings or, in at least one case, subjecting OBI to its own obliteration by using the phrase without attribution (Hargens & Felmlee, 1984). Scholars outside information science/informetrics have also mentioned OBI as having a potential effect on the visibility of formerly well-cited works (e.g., in applied linguistics see Hyland, 1999; in statistics see Stigler, 1994).

In his various discussions of the processes affecting citation visibility, Merton (1965) also introduced a related notion—the "palimpsestic syndrome." He defined this as "the covering over of earlier versions of an idea by ascribing it to a comparatively recent author in whose work the idea was first encountered" (p. xxiii); he gives his attribution of the phrase "on the shoulders of giants" to Newton as an example. Later, he put even more emphasis on the role of the nonoriginating author in obliterating the earlier source:

[The] palimpsestic syndrome: assigning a striking idea or formulation to the author who first introduced us to it when in fact that author had simply adopted or revived *a formulation that he (and others versed in the same tradition) knew to have been created by another.* (Merton, 1972, p. 31; emphasis mine)

In the foreword to Garfield's book (see above), Merton combines this notion of (deliberate?) misattribution of intellectual credit with the more general process of OBI. Obliteration by incorporation is the disappearance of citations to the older work, although the ideas live on; the palimpsestic syndrome is the attribution of the idea to an author who did not originate it, but made the work accessible or visible, thus acquiring the citations that should otherwise go to the original author.

Merton's palimpsestic syndrome describes the situation where the later work, rather than the earlier, is specifically recognized as the source of the idea, reminiscent of Stigler's law of eponomy (Stigler, 1980): "No scientific discovery is named after its original discoverer" (p. 147). But intentional (or unintentional) misattribution is only one of the processes that can lead to a loss in citations and OBI. Scholarly reviews, texts, and monographs—publications that review, summarize, synthesize, and cite prior work—are part of the dissemination process of scholarly research (see, e.g., Björk, 2007; Garvey & Griffith, 1967; Price, 1965). Citing one of these kinds of publications can be an effective way to point the reader to a fuller explanation or discussion of the topic at hand while retaining connections to the original work. The intermediate work serves as a "cognitive conduit" (Zuckerman, 1987) and can be tallied as an "indirect citation" to the original source (Rousseau, 1987; Thomas, 1992). In these cases as well, citations to the original highly cited work or eminent author will also be reallocated to the later work.

Obliteration by incorporation will make itself visible as a decline and disappearance of citations to key works and authors as the ideas become part of the knowledge base of scholarship—discussed, mentioned, or assumed to be known, but not tagged with a reference to their intellectual origin. However, the careful reader will have noted that, in his preface to Garfield's book, Merton refers to "this *hypothesized* process [of OBI]" (my emphasis). Hypothesized because, while the notion itself seems self-evident or supported by anecdote and personal experience, he cites only one empirical study as having focused on charting a citation history that could demonstrate OBI. In this study, Messeri (1978) looked at the concept of "seafloor spreading," reporting annual citation counts to a small set of core papers identified by citation and subject indexing. He reported a growth in the seafloor-spreading literature along with a decline in citations to the key papers—noting a "striking increase in uncitedness" (in this case articles on the topic that did not cite a core paper).

As late as 1987 (Zuckerman, 1987), the Messeri paper apparently stood alone as an empirical investigation specifically focusing on OBI. Since then, a small literature on OBI and the processes of palimpsestic replacement and indirect citing has developed. The remainder of this chapter reviews this literature, focusing on the findings and on the challenges and trade-offs in designing a careful and informative OBI study. It ends with a brief consideration of possible "reasons" underlying the OBI/indirect citation processes, drawing on the literature that has studied citation behavior and motives for citing/non-citing.

Citations as Concept Symbols

The bibliometrics literature is replete with discussions of the roles and functions of citations in scholarly publications, discussions of the development of classification schemes to sort out the citation context, and, related to this, discussions of the motives and behaviors of scholars as citers (see the reviews in Borgman & Furner, 2002; Cole, 2000; Nicolaisen, 2008; Small, 1982; White & McCain, 1989; Wouters, 1999). For studies of OBI and palimpsestic replacement/indirect citations, I think the most fruitful perspective is that of Henry Small, who proposed that citation was an act of symbol usage, that the cited work was a "concept symbol," and that the text surrounding the embedded citation or footnote number identified the concept for which the document was cited (Small, 1978; see also the discussion in Cozzens, 1989; Schneider, 2006; White, 2004). Small quantified the consistency with which the same or very similar turns of phrase were used in the citation context to label the cited concept as "percent uniformity" (PU). Journal articles tend to be cited for a single concept, showing a high PU in citing texts (Small, 1978, 1998; Small & Greenlee, 1980), although McCain (2012c) found multiple frequently invoked citation concepts relating to Price's paper on citation networks (Price, 1965). This last case was more similar to the tally of cited concepts in books (see Furner, 2003; Garfield, 1985; McCain & Salvucci, 2006).

The concepts examined in these studies have generally been represented in the citation contexts as eponyms or short noun phrases. In citation context analyses of highly cited articles in chemistry, Small (1978) reported a high PU for eponymic phrases, names of methods, and catchphrases. Small and Greenlee (1980) found high-PU catchphrases in a recombinant DNA document cocitation cluster but do not report any eponymic phrases. McCain and Salvucci (2006) did not distinguish between use of the eponym "Brooks' law" and the phrase "mythical man-month" in their citation context analysis of citations to *The Mythical Man-Month* (Brooks, 1975)—the concept represented by these two phrases accounted for just under 25% of all citation contexts (but articles discussing Brooks' topics while not citing the book were not studied, so these results are not representative of OBI).

By taking the view that citations have a rhetorical function in the text—to link an idea (expressed in eponyms or catchphrases, formulas, or possibly images) with the source of the idea—we can explore the citation history of important, highly cited works in the context of their use. Through citation-in-context analysis of available texts, we can observe, over time, the extent to which the concept has become decoupled from a source and the eponym or catchphrase stands alone, without an accompanying explicit or indirect reference—OBI in action.

Empirical Work on OBI and Palimpsestic Replacement

Since the early work of Messeri, Small, and Greenlee, a small, scattered literature has developed that looks at OBI, with or without the added consideration of indirect citations. This literature can be viewed from two perspectives—the text representation of the concept symbol (eponym vs. catchphrase) and the level of data analysis (database record vs. full text).

Most articles focus solely or primarily on eponyms and their link or disconnection from the eponymized author's relevant, cited publications. Thomas (1992) studied the mentions, in a variety of text sources (including journal articles, textbooks, vendor catalogs, and patents), of the phrase "Southern blot/Southern blotting"—a method used to detect DNA sequences using gel electrophoresis (Southern, 1975). She reported that the eponym developed quickly (1.5 years) and generated its own noneponymic secondary terms for similar methods (northern and western blotting). Between 1978 and 1990, there was a steady trend of increasing percentages of implicit citations to Southern's paper (use of the eponym without citation), ending in 1990 with over 55% implicit citations. Thomas (1992) notes the existence of indirect citations to the Southern blot method but did not deal with them separately in her analyses, although she did check reference lists of the actual cited works to confirm that the Southern paper was in the bibliographies. Száva-Kováts (1994) studied the occurrence (but not trends over time) for "non-indexed eponymal citedness (NIEC)"—the use of eponymous technical terms without formal bibliographic references (the equivalent of implicit citations)—in selected physics journals. Száva-Kováts (1994) described NIEC as the "first stage" in OBI and asserted that, in order to be achieved, OBI must result in total incorporation through elimination of the creator's name from the concept.

Three other studies have based their analyses primarily on database searches of authors' names or eponyms. Marx and Cardona (2009) searched Web of Science, Chemical Abstracts, INSPEC, and Google Scholar for sources mentioning names of pioneers in chemistry and physics in titles, abstracts, or keywords. They distinguished between "informal citations" (names mentioned only = eponymic phrases) and "formal citations" (full reference to published work as part of a database record) and reported a very high proportion of informal citations to the named scientists. Gorraiz, Gumpenberger, and Wieland (2011) searched Web of Science, Scopus, and Google Scholar for citations to works by Francis Galton and, separately, sought out sources that mentioned Galton but did not cite him. They reported that about one-third of Galton's works were subject to OBI through use of eponyms and suggest that OBI may be more common in areas where named formulas, effects, etc., are prevalent. Finally, McCain (2011) searched the Web of Science and other relevant bibliographic databases (BIOSIS, INSPEC, EconLit, PsychINFO, Compendex) for the occurrence of the eponymic phrase "Nash Equilibrium/Nash Equilibria" and/or citations to one or both of Nash's two key papers in any part of the database record. Overall, the percent of implicit citations ranged between 60% and 70% annually from 1999 to 2008; across specific subject areas, the degree of observed OBI varied widely, while still fluctuating within a narrow range within subject.

Studies focusing on catchphrases rather than eponyms are much rarer. As noted earlier, Messeri (1978) studied the literature of "seafloor spreading," relying on subject indexing rather than database record content to identify relevant articles but using the phrase as an indicator of (potential) citation of the core literature. Bottom, Kong, and Zhang (2007) searched key journals in management and psychology for the terms *stereotype* and *schema* to test the degree of obliteration of Lippman's concept, reporting that by the 1950s, citations to Lippman had essentially disappeared, although the counts rebounded marginally in the 1970s and 1980s. Marx and Cardona (2009) included one catchphrase ("density functional theory") in their studies of formal versus informal citations. McCain (2012a) identified 1,040 articles that included a variant of the phrase "evolutionarily stable strategies" (ESS) in the database record and conducted both database-record-level and citation-in-context analyses of print articles to assess the degree of OBI—the lack of a citation to one of a small set of publications by John Maynard Smith. McCain reported record-level OBI generally increasing over time (early 1970s through 2008) with a peak of 62% in 2002 but that citation-in-context OBI percentages were lower with no clear pattern. This was due primarily to the addition of indirect citations (which cannot be detected at the record level) to the total formal citation count at the text level.

Studies reporting palimpsestic replacement are very scarce—perhaps because it is necessary to have a very clear notion of what *should* be cited along with the eponym or catchphrase (this is not the same thing as the MacRoberts' assertion [see MacRoberts & MacRoberts, 2010] that authors should cite all influences). In addition to his discussion of his use of the phrase "on the shoulders of giants," mentioned earlier, Merton (1972, p. 31) notes that he had "inadvertently contributed to a palimpsestic syndrome" in his attribution of an aphorism to Weber rather than Simmel. In a 1987 paper, Merton discussed the shift in terminology from "focussed group-interview" to "focus group" and the obliteration through palimpsestic replacement (my terminology, not his) by a relatively recent work on "focus groups" of his out-of-print handbook on "focussed interviews" (Merton, Fiske, & Kendall, 1956), which was itself an extension of a journal article published 10 years earlier (Merton & Kendall, 1946). In an extended discussion of the lack of attention to and awareness of intellectual heritage in the field of knowledge management (KM), Lambe (2011) stated that, while Leonard-Barton's book *Wellsprings of Knowledge* (Leonard-Barton, 1995) dealt well with prior theoretical work, "references to her work quickly substituted in the canonical literature of knowledge management for the sources on which she so closely depended" (Lambe, 2011, p. 176). Bottom et al. (2007) characterized the (temporary) obliteration of Lippman's connection to the concept of "stereotype" as an example of the "palimpsestic syndrome," but their discussion focused more on the changing patterns of citation counts to Lippman and context analysis demonstrating the mischaracterization of his original concept. I found no systematic discussion of works that were palimpsestic replacements.

Methodological Issues in Studying Obliteration by Incorporation

As the brief review of prior research shows, several variables should be considered when designing an OBI study. These include:

- Choice of concept symbol—catchphrase or eponym—and its text representation
- Whether the concept symbol is represented by citations to a single work or whether the citation can be to one of a small canonical set
- Choice of search venue—online database with bibliographic records only versus online database with searchable full text
- Level of analysis—counts based on the presence of concept text and references in the database record or within the full text
- How indirect citations are handled and OBI measured

Some of these choices may depend on the researcher's subject interest or ease in identifying the original source. Eponyms, because they are name-based, may be more easily connected to the author and his or her original contribution. The eponym may also be a key feature of the subject area (e.g., Southern blot in molecular biology, Nash equilibrium in game theory). It may be more difficult to identify a catchphrase that can easily be linked with one or a small, canonical set of publications. As examples of the latter, "evolutionarily stable strategies" can be reasonably connected to one of several publications of John Maynard Smith (McCain, 2012a), and Herbert Simon is cited both as sole and coauthor for publications discussing "bounded rationality" and "satisficing" (see McCain, 2012b, and the discussion in Kahneman, 2003). Examples of single-work attributions might include the phrases "community of practice" (Lave & Wenger, 1991) and "Brooks' law" (Brooks, 1975; an enhanced "silver anniversary" edition was published in 1995). On the other hand, a concept may have acquired several different text representations that must be traced by the researcher. For instance, the process underlying the empirical observation that highly cited works attract proportionally more additional citations than might be expected has been discussed as the "Matthew effect" (Merton, 1988), "cumulative advantage" (Price, 1976), and "preferential attachment" (Barabási & Albert, 1999) depending on the disciplinary venue (see the discussion in McCain, 2012c, and Newman, 2003).

When assessing OBI, the most basic measure calculates the percentage of implicit, informal citations in the dataset and reports the degree to which this percentage changes over time. But, as noted earlier, a later work may be cited in place of the original. Messeri (1978) speculated that these citations might exist, but did not report any empirical findings affecting his study of "seafloor spreading." Rousseau (1987) makes a strong case that indirect citations should be considered when studying a paper's total influence (but does not discuss palimpsestic replacement or OBI), and Dervos and Klimis (2008) discuss the importance of indirect citations in information retrieval. Thomas (1992) and McCain (2012a) combined indirect and explicit citations when determining the degree of OBI as a percentage of all papers having only implicit citations to, respectively, "Southern blot" and "evolutionarily stable strategies."

The remaining variables represent interrelated aspects of the use of bibliographic databases (with or without a subsequent text-level analysis) versus primary full-text searching—how to search and at what level to collect data. Bibliographic database searching allows a wide net to be cast. Searches may cross many disciplines in the case of a multidisciplinary database such as the Web of Science or Scopus, or keep a more disciplinary focus in areas such as the life sciences and medicine (BIOSIS, Medline), engineering (INSPEC, Compendex), economics (EconLit), computer science (ACM Digital Library), and psychology (PsychINFO). A very large number of database records may be retrieved with a combination of text field and citation searching (in the databases that permit the latter). But counts of database records using the eponym or catchphrase while lacking an appropriate reference may, as McCain (2012a) noted, miss the indirect citations that are still indicative of influence and overcount items where the apparently explicit citation is not textually connected to the eponym or catchphrase. One way to handle this challenge is to examine the text of the articles identified in the database record retrieval (Garfield, 1998). Needless to say, the task can quickly become onerous, both in terms of reading and in terms of storage if the articles are acquired, read, and maintained in hard copy. The researcher may need to resort to samples of the complete retrieval if the total source item count is high.

There may also be problems with variation in bibliographic database record structure and content. The record structure may change over time (see the discussion in Marx, 2011). For instance, abstracts, author keywords, and terms from titles of references (Keywords Plus) were not reliably included in the Web of Science until about 1993–1994. Also, when abstracts are available, they may not be written by the author. Many databases edit the author abstracts (e.g., INSPEC, EconLit) or write new ones (ABI Inform), while MathSci provides reviews rather than author abstracts. In all of the studies reviewed earlier, the authors found it necessary or useful to enhance a database retrieval set with searches of other sources for additional data to include in the analysis.

One potentially useful alternative to the "first search the database, then examine the retrieved articles" approach would be to turn to databases that allow initial full-text searching of the indexed articles. For instance, sources such as HighWire Press and JSTOR cover a variety of disciplines, although there may be subscription and time-window limitations on what is searchable and accessible. In August 2012, one could search over 6 million HighWire Press full-text articles and export the abstract and bibliographic information, but there was frequently a charge for downloading the PDF (HighWire Press, 2012). In the case of JSTOR, university libraries may subscribe to only some of JSTOR's title bundles and both sources may restrict access to current issues of journals. At the disciplinary level, full-text-searchable databases such as ABI Inform and Library Literature & Information Science Full-Text cover a range of professional and trade publications in their respective fields.

An experiment in using JSTOR via Drexel's Hagerty Library to collect information for an OBI study (McCain, 2012b) identified an additional and unexpected result of full-text searching. A full-text search of the phrase "bounded rationality" yielded approximately 3,700 articles published between 1962 and 2011 that were directly accessible as PDFs or accessible via a link to the Hagerty Library digital journal collection. A systematic sample of 10% of the results revealed that about one-fifth of all articles were retrieved because the catchphrase was only in a title of an item in the retrieved article's bibliography, and the article lacked the catchphrase "bounded rationality." This last set of articles may contribute to the overall study insofar as they are also indexed by JSTOR (part of the library subscription package) and would be in the full retrieved dataset, but should not be tallied on their own as contributing to an understanding of the catchphrase in the sample of the literature in hand.

Thus far we have seen that Merton's twin notions of obliteration by incorporation and the palimpsestic syndrome are accepted phenomena in studies of scholarly discourse and are, for the most part, unquestioned—possibly influencing but not totally negating assessments of scholarly influence via citation counts and histories. We have also seen that there is some limited empirical evidence that OBI and the use of indirect citations can be systematically observed and measured (whether palimpsestic replacement has been demonstrated empirically is less clear). However, what we still lack in this discussion is any notion of what underlies authors' choices to make implicit, rather than explicit, reference to a known observation, method, theory, etc., or their decision to cite a newer work rather than an older work for the same concept. In other words, why do authors invoke, but not cite, or cite by substitution, key older works in their current writing? To explore this, we must go beyond studies of citing texts and turn to research that focuses on authors—their citation behavior and motivations.

What Do We Know about Why Authors Cite or Don't Cite Older Literature?

Citation behavior studies have taken two approaches—a focus on the text through citation content and context analyses, and a focus on citing authors' choices and behaviors through surveys or interview studies (see reviews in Bornmann & Daniel, 2006; White, 2004). The first approach overlaps with the text-based research discussed previously—building and applying taxonomies of citation roles, functions, and content as they can be deduced from the analysis of the text associated with the embedded citation/footnote. The second elicits information directly from citing authors—seeking ratings or comments on papers cited, motives for citing, etc. The two approaches may be combined, as in Cano's 1989 study in which she asked 21 "elite" scientists to classify their cited references based on Moravcsik and Murugesan's (1975) citation content classification scheme.

In general, author surveys and interviews are designed to tap into the authors' memories of why cited works were chosen, their quality or prominence, and/or their rhetorical function in the citing article. In Cronin's words, we need to "step inside that individual's head" (Cronin, 1984, p. 50; see also the discussion in Bornmann & Daniel, 2006; Case & Higgins, 2000), with results that may or may not be reliable. As Cronin and others have noted, the act of citing is private and authors' memories fallible. Another problem, from the point of view of studying OBI (though perhaps not indirect citing or palimpsestic replacement), is that, in these studies, data gathering and discussion tend to focus on the explicit citations associated with the work under discussion, and not on what works were not cited, or why a citation was implicit rather than explicit. For instance, T. A. Brooks (1985, 1986) interviewed a convenience sample of 20 authors of scholarly journal articles, asking about their motives for citing the sources they did. He identified seven citer motivations but reported no discussion about author choices or motives not to cite. Shadish, Tolliver, Gray, and Sen Gupta (1995) surveyed citing authors in psychology, eliciting ratings of single, selected cited works using a 28–32 item list of article characteristics that may potentially have influenced their decision to cite the work. Case and Higgins (2000) used the 32-item list of reasons for citing in a survey of authors citing papers by two well-known communication science researchers, adding questions on the relationship(s) between the surveyed author and the cited author and document. Both Shadish et al. (1995) and Case and Higgins (2000) reported that some cited works were assessed as "classic references," "well-known," and "concept markers," but neither study dealt with the choice not to cite, to only cite implicitly, or to substitute one work for another in the reference list. Case and Higgins reported that 24% of the respondents cited a work as having a review function; Shadish et al. reported a lower number.

Author-based studies *can* be designed to address issues of OBI and indirect citation/palimpsestic substitution. MacRoberts and MacRoberts (1988) self-report on their choices of references to include with a brief ecological note—choosing three for reasons of "persuasion, currency, and social consensus" (p. 433) and rejecting others they still viewed as "influential." Cronin (1981) distributed a "de-referenced" article to a set of scholars with expertise in the area (but with embedded referenced author names and direct quotations retained). Respondents were asked to indicate where, in the text, a reference would be warranted. He reported limited agreement between the article's author and the survey respondents, and among the respondents themselves, suggesting "a (limited) shared understanding of the function of citations in particular contexts" (Cronin, 1981, p. 32).

Vinkler (1987) explicitly considered OBI in his discussion of the responses of 20 chemists who were surveyed and interviewed about the reference lists in recently published papers. One of the potential reasons for not citing offered to the chemists was "the author of the potentially citeable paper or the work, theorem or theory is publicly known (e.g. Planck constant, Friedel-Crafts reaction, etc.)" (Vinkler, 1987, p. 59). Of 201 potentially citable works that were not cited, 52 (26%) were coded for this reason (and frequently for other reasons at the same time). Vinkler (1987, p. 59) observed that eponyms are "labeled information," indicators that the original information has been "incorporated into science to an extent that their knowledge has become everyday routine" but that, for more modern methods, one would cite the work of the originator. He also mentioned the role of reviews, handbooks, and textbooks in "hiding" information, making these sources useful citations that can lead readers to the original papers but also resulting in the failure to cite those original papers (representing 80% of reasons not to cite).

Wang and White (White & Wang, 1997; Wang & White, 1999) report the results of a longitudinal interview study of writing and citing by agricultural economists—specifically focusing on whether documents that were selected and read early in the research were cited in the final papers. Coding of subjects' responses produced a list of document selection/citation criteria. Two criteria would seem to fit the notion of OBI in terms of an author choosing to cite or not cite a document:

- "Classic/founder" (Wang & White, 1999, p. 104), defined as the recognition of the document as "the first substantial work on a topic or methodology, or the author as the founder of a theory or method." If documents were judged classics, they were cited;

similarly, if the author was known as the originator of a concept, theory, or analytic approach, a rhetorically useful work would be cited

- "Standard reference" (Wang & White, 1999, p. 105), generally a text or handbook that is "the best accepted treatment of a particular topic." This was listed as a "meta-level documentation concern" by White and Wang (1997, p. 145).

These two criteria are reported as "applied" in citation decisions by several of the 25 subjects, with some suggestion that the ultimate decision was "cite" rather than "don't cite." In neither report is there a discussion of OBI in the sense of implicit citations that were deliberately decoupled from useful references. The citation of texts and standard references in lieu of the original reports fits the description of indirect citations.

Harwood's interview studies of citation function and citing behavior in two groups—computer scientists and sociologists—provide some additional insights into authors' citing decisions related to OBI (Harwood, 2008, 2009). The choice of publication outlet may affect the decision to include a reference (Harwood, 2008). One computer scientist stated that, in the case of a book for an audience that is seen as less specialized and knowledgeable, a citation to a broad reference might be included, where it would be omitted for a journal for specialists. Specialists would not need to be reminded of well-known names or to be provided with references for well-known algorithms (Merton and the notion of OBI are specifically discussed in Harwood's discussion). A sociologist interviewed suggested that he would use an American source as citation for an American readership but a British source for a British journal (depriving one or the other of citations). Harwood (2009) reports 11 citation functions extracted from these unstructured interviews of computer scientists and sociologists. Two relate to our goal of understanding citation practices that may result in OBI through the choice of using implicit or explicit citations. As noted earlier, the author may support less knowledgeable readers (as opposed to scholars in the field) by "signposting" (providing citations that point to background sources). "Tying citations" may be used in cases where the audience might expect references to "things that are well-known in the community"—the respondent felt that not citing would leave him open to criticism (but others may not share that concern).

On Indirect Citations and the Palimpsestic Syndrome

Text-level analysis can easily distinguish between explicit citations (eponym or catchphrase linked directly to the originator's work) and implicit citations (eponym or catchphrase not accompanied by any citation to a formal source). But how should we think about indirect citations? Are these examples of palimpsestic replacement, or an indication that the citing author recognized the need to give some support to the reader and chose a source that would be useful to that reader or would be a shorthand pointer to a more extended discussion than the space (and restrictions on number of references) allowed?

Merton's concern with later authors being given (or taking) credit for an earlier originator's contribution does not appear to be the same thing as reviews being cited rather than the papers they review. It may also not be the same thing as the use of a textbook reference to point to a developed explanation of a theory or development of an algorithm. While it is almost certainly true that later writers may deprive an author of citations to his or her original work by citing the review, monograph, or text, Merton's expressed concern is with the (permanent?) misattribution of the original "significant idea or formulation" to a later author, resulting in an erroneous awarding of citations to that later author and a loss of connection back to the originator.

But can palimpsestic replacement be identified without insights from the citing author, self-assessment of reference choices, a very careful analysis of the cited work's text, or recognition that the chain of references has been broken over time (possibly created by the "lifting" of citations without having read the cited works—see the review in Hoerman & Nowicke, 1995)? For large-scale studies, I suspect that we will have to take indirect citations "at their word"—perhaps after verifying whether or not the earlier author's work is cited in the indirectly cited publication. The presence of a citation is an indicator that the citing author felt that some pointer to the literature was needed and that may have to suffice for bibliometric purposes.

Taking Stock

Obliteration by incorporation differs from the more general obsolescence or aging of scholarly literature—the anticipated swift or more gradual decline in citation counts within some reasonable span of years after publication that has been described by Van Raan (2004) and others. It involves only a small percentage of all publishing authors and of all published works—those that are much more highly visible and highly cited than the "standard" scholarly publication. Zuckerman (1987) notes that "OBI occurs more often for 'elite' scientists and initially much cited papers than for the rest" (p. 331). Indeed, she suggests that it may be the fact that their work was "obliterated," becoming part of currently accepted knowledge, that confers elite status (the "advantage of being obliterated"; Garfield, 1975). Similarly, Marx and Cardona (2009) describe OBI as affecting "the greater achievements and, accordingly, a minority of scientists"

(p. 20). OBI may work to counteract the overcitation that also apparently comes with high visibility—a "counter-Matthew effect."[1] It is certainly worth detailed study if we want to gain a fuller understanding of this aspect of formal scholarly communication.

The studies reviewed in this chapter suggest the range and degree of OBI that may be encountered in the scholarly literature. No key contribution that has been explored has achieved complete obliteration—there are always authors who will attach a reference, explicit or indirect, to the eponym or catchphrase and, in some cases, OBI appears to have attained a steady state with between 20% and 30% of papers invoking the concept also citing a source. The few studies of authors' choices whether or not to cite point to some of the countervailing processes that can affect the extent and trend of OBI. Some authors will not reference a citation because the concept is well known, while others will reference a classic work because it is a classic (and thus expected?); reviews are cited in the natural course of covering prior contributions, reducing the citation counts to those works that are reviewed; and an author may cite a source for its expository value for a nonspecialist audience or one outside his or her home turf but refrain from citing when writing for scholarly peers—or may cite a book rather than an article if a more detailed or mature presentation is needed. Not discussed in the research to date, but possibly worthy of investigation, are questions like the following: To what extent does the general orientation of the source item (e.g., the type of research or scholarly approach taken or the degree of focus on the key concept) correlate with the presence or absence of an attached reference? Are some authors prone to cite whenever possible or practical, while others tend not to provide references for citable concepts (and are these traits that may be passed from mentor to student or differ from field to field)? Are more citations in the "steady-state" datasets coming from authors who are relatively new to the field and the relevant scholarship? Informative empirical studies will need to focus on text-level analyses if more than a rough impression of OBI trends is to be achieved—and these studies would benefit by being enhanced with interviews that address authors' perceptions of when it is appropriate and useful to link a key concept and a source and when it is not necessary. Eliciting these insights will be a challenge, but it will be necessary if we want to go beyond examination of trends and publication-based variables affecting OBI.

The interlocking issues surrounding indirect citations and palimpsestic replacement are more problematic. Merton's definition of the palimpsestic syndrome was specific on the citing author's actions (if not motives)—the citing author refers to the work through which the idea was encountered rather than the original source and, over time, the citations to ever-newer works erase the connections to the older ones. But if the newer works refer back to the old in the text, as appears to be the case with most indirect citations, how is one to know, when an indirect citation is encountered, what the citing author had in mind? Merton's examples are personal—recognition of his own errors of attribution or (in the case of the focused interview/focus group discussion) a solid knowledge of his contributions to the scholarly development of the field. At the citation-in-context level, we can say that the reference attached to the key concept explicitly cites an appropriate source work, indirectly cites the work, or cites a work that lacks a link to the original source. Interviews with authors (with very carefully worded questions) may be the only way to understand authors' perceptions of the rhetorical role of indirect citations in their publications, and careful text-level analysis of the treatment of the concept in the indirectly cited work may point to the possibility of ongoing palimpsestic replacement.

Note

1. I thank Blaise Cronin for this tagline.

References

Aversa, E. S. (1985). Citation patterns of highly cited papers and their relationship to literature aging: A study of the working literature. *Scientometrics, 7*, 383–389.

Barabási, A.-L., & Albert, R. (1999). Emergence of scaling in random networks. *Science, 286*, 509–512.

Björk, B.-C. (2007). A model of scientific communication as a global distributed information system. *Information Research, 12*(2). Retrieved from http://informationr.net/ir/12-2/paper307.html.

Borgman, C. L., & Furner, J. (2002). Scholarly communication and bibliometrics. *Annual Review of Information Science & Technology, 36*, 2–72.

Bornmann, L., & Daniel, H.-D. (2006). What do citation counts measure? A review of studies on citing behavior. *Journal of Documentation, 64*, 45–80.

Bottom, W. P., Kong, D., & Zhang, Z. (2007, January 2). *The palimpsestic syndrome in management research: Stereotypes and the obliteration process.* Retrieved from http://apps.olin.wustl.edu/workingpapers/pdf/2007-01-002.pdf.

Braun, T., Glänzel, W., & Schubert, A. (2010). On sleeping beauties, princes and other tails of citation distributions. *Research Evaluation, 19*, 195–202.

Brooks, F. P., Jr. (1975). *The mythical man-month.* Reading, MA: Addison, Wesley.

Brooks, T. A. (1985). Private acts and public objects: An investigation of citer motivations. *Journal of the American Society for Information Science, 36*, 223–229.

Brooks, T. A. (1986). Evidence of complex citer motivations. *Journal of the American Society for Information Science, 37*, 34–36.

Burrell, Q. L. (2012). Alternative thoughts on uncitedness. *Journal of the American Society for Information Science and Technology, 63*, 1466–1470.

Cano, V. (1989). Citation behavior: Classification, utility, and location. *Journal of the American Society for Information Science, 40*, 284–290.

Case, D. O., & Higgins, G. M. (2000). How can we investigate citation behavior? A study of reasons for citing literature in communication. *Journal of the American Society for Information Science, 51*, 635–645.

Cole, J. R. (2000). A short history of the use of citations as a measure of the impact of scientific and scholarly work. In B. Cronin & H. B. Atkins (Eds.), *The web of knowledge: A Festschrift in honor of Eugene Garfield* (pp. 281–300). Medford, NJ: Information Today.

Costas, R., Van Leeuwen, T. N., & Van Raan, A. F. J. (2010). Is scientific literature subject to a "sell-by-date"? A general methodology to analyze the "durability" of scientific documents. *Journal of the American Society for Information Science and Technology, 61*, 329–339.

Cozzens, S. E. (1989). What do citations count? The rhetoric-first model. *Scientometrics, 15*, 437–447.

Cronin, B. (1981). Agreement and divergence on referencing practice. *Journal of Information Science, 3*, 27–33.

Cronin, B. (1984). *The citation process: The role and significance of citations in scientific communication*. London: Taylor Graham.

De Bellis, N. (2009). *Bibliometrics and citation analysis: From the science citation index to cybermetrics*. Lanham, MD: Scarecrow Press.

Dervos, D. A., & Klimis, L. (2008). Exploiting cascading citations for retrieval. *Proceedings of the American Society for Information Science and Technology, 45*, 1–12.

Edge, D. (1979). Quantitative measures of communication in science: A critical review. *History of Science, 17*, 102–134.

Frank, G. (1999). Scientific communication—a vanity fair? *Science, 286*, 53–55.

Furner, J. (2003). Little book, big book: Before and after *Little Science, Big Science*. A review article, Part II. *Journal of Librarianship and Information Science, 35*, 189–201.

Garfield, E. (1975). The obliteration phenomenon. *Current Contents, 51/52*, 5–7.

Garfield, E. (1977). Introducing *Citation Classics:* The human side of scientific reports. *Essays of an Information Scientist, 3*, 1–2.

Garfield, E. (1979). *Citation indexing: Its theory and application in science, technology, and humanities*. New York: Wiley.

Garfield, E. (1980). Premature discovery or delayed recognition—why? *Current Contents, 21*, 5–10.

Garfield, E. (1985). In tribute to Derek John de Solla Price: A citation analysis of *Little Science, Big Science*. *Scientometrics, 7*, 487–503.

Garfield, E. (1998). Random thoughts on citationology—its theory and practice. *Scientometrics, 43*, 69–76.

Garvey, W. D., & Griffith, B. C. (1967). Scientific communication as a social system. *Science, 157*, 1011–1016.

Gorraiz, J., Gumpenberger, C., & Wieland, M. (2011). Galton 2011 revisited: A bibliometric journey in the footprints of a universal genius. *Scientometrics, 88*, 627–652.

Hargens, L., & Felmlee, D. (1984). Structural determinants of stratification in science. *American Sociological Review, 49*, 685–697.

Harwood, N. (2008). Publication outlets and their effect on academic writers' citations. *Scientometrics, 77*, 253–263.

Harwood, N. (2009). An interview-based study of the functions of citations in academic writing across two disciplines. *Journal of Pragmatics, 41*, 497–518.

HighWire Press. (2012). Free online full-text articles. Retrieved from http://highwire.stanford.edu/lists/freeart.dtl.

Hoerman, H. L., & Nowicke, C. E. (1995). Secondary and tertiary citing: A study of referencing behavior in the literature of citation analysis deriving from the Ortega hypothesis of Cole and Cole. *Library Quarterly, 65*, 415–434.

Hyland, K. (1999). Academic attribution: Citation and the construction of disciplinary knowledge. *Applied Linguistics, 20*, 541–567.

Kahneman, D. (2003). Maps of bounded rationality: Psychology for behavioral economics. *American Economic Review, 93*, 1449–1475.

Lambe, P. (2011). The unacknowledged parentage of knowledge management. *Journal of Knowledge Management, 15*, 175–197.

Lave, J., & Wenger, E. (1991). *Situated learning: Legitimate peripheral participation*. Cambridge: Cambridge University Press.

Leonard-Barton, D. (1995). *Wellsprings of knowledge: Building and sustaining the sources of innovation*. Cambridge, MA: Harvard Business Review Press.

MacRoberts, M. H., & MacRoberts, B. R. (1987). Testing the Ortega hypothesis: Facts and artifacts. *Scientometrics, 12*, 293–295.

MacRoberts, M. H., & MacRoberts, B. R. (1988). Author motivation for not citing influences: A methodological note. *Journal of the American Society for Information Science, 39*, 432–433.

8 A Network Approach to Scholarly Evaluation

Jevin D. West and Daril A. Vilhena

Introduction

As Derek de Solla Price famously noted in 1965, the scientific literature forms a vast network (Price, 1965). The nodes of this network are the millions of published articles, and the edges are the citations between them. There is a wealth of information—not only within the content of these nodes (the text), but also within the structure connecting these nodes (the network topology). In fact, the network topology by itself provides clues about the quality of the content. This is similar to how Google's PageRank algorithm harnesses the hyperlink structure of the web to evaluate web pages (Page, Brin, Motwani, & Winograd, 1998).

Surprisingly, scholarly evaluation over the last century has largely ignored this network property. The well-known Impact Factor simply counts the number of incoming links; it does not take into account the source of a citation and therefore ignores the extra information in the network (Garfield, 1955). So, why has it taken decades for network methods to become standard in the field of bibliometrics? Though researchers have long recognized the potential of a network approach, the field has suffered from a lack of computational resources and data. However, in this data-driven age, citation networks are now a staple of bibliometrics and are used as model systems in other disciplines.

In this chapter, we describe what the network approach offers over nonnetwork approaches, and how the network can be incorporated to assess scholarly journals, authors, and institutions. We explain what a network is, the difference between basic networks in bibliometrics, and the difference between network measures that incorporate the structural information of a network and measures that focus solely on the pairwise relationships between entities—be they papers, authors, journals, or institutions. We explain some of the technical differences and trade-offs between degree centrality (nonnetwork) measures and eigenvector centrality (network) measures. We also provide a brief history of network science and how scholarly networks have become a cornerstone for this burgeoning field.

This chapter is aimed at researchers interested in the use of citation networks for assessing scholarly output. Our hope is that the reader will gain some understanding of network measures and how they differ from nonnetwork measures. As a case study, we focus on two industry standard measures for journal quality, the Eigenfactor score, a network measure, and the classic Impact Factor, which ignores network information. We illustrate the difference between these two by example, and calculate both of these measures for small journal-level citation networks.

Big Data Facilitate Network Analysis

The amount of digital data created each year now exceeds the world's ability to store it (EMC, 2010). This increase in recorded human activity and technology has ushered in an age of "data science," where the sheer magnitude of available data is enough to reject or confirm scientific hypotheses (McKinsey, 2011). Nearly every discipline of science has been deluged with data, and some attention in each discipline has been redirected to the development of algorithms and methods that facilitate the storage, processing, and analysis of large data repositories.

The 21st-century data deluge has enabled bibliometricians to study entire systems instead of parts of systems. Bibliometric research was transformed by the foundational and groundbreaking publication by Price (1965), which first recognized that papers and their citations form a vast network. This methodological leap forced researchers to recognize and acknowledge systemwide, higher-level properties, which had previously been ignored by studies that focused exclusively on the pairwise relationships between citer and citee. Over the last several decades researchers have applied the network approach to citation networks (Pinski & Narin, 1976; Liebowitz & Palmer, 1984; Kalaitzidakis, Mamuneas, & Stengos, 2003; Palacios-Huerta & Volij, 2004; Kodrzycki & Yu, 2006; Bollen, Rodriguez, & Van de Sompel, 2006). However, only recently has the widespread availability of electronic data not only made network analysis of large datasets a possibility, but also made network analysis a cornerstone of bibliometric research.

What Is a Network?

A network is an abstract model of the relationships in a system. *Nodes*, which represent entities, such as papers, authors, journals, and institutions, are connected to one another by *links*, such as citations and collaborations. Accurately modeling a real-world system requires that the nodes and links be chosen carefully, such that they capture meaningful relationships in the system under study. Therefore, the first step of any network analysis is to identify what data will be nodes and links in the network. For example, in a paper-level citation network, nodes are papers, and links are the citations between papers, while in a coauthorship network, nodes are authors, and links denote the number of times a pair of authors has coauthored.

The reader should note that the relationships between nodes in a paper-level citation network and a coauthorship network are fundamentally different. In a paper-level citation network, every link is weighted equally, and the links are directional—a citation is rarely reciprocal. However, in the coauthorship network, all links are reciprocal, and the links have different *weights*, such that some links have higher values—more joint authorships—than other links. We call the paper-level citation network *unweighted* because all links are equal, and *directed* because all links point in only one direction. The coauthorship network is *weighted* because some links have higher values than others, and *undirected* because every link must be reciprocal. Network models can have any combination thereof (figure 8.1): journal-level citation networks, where nodes are journals and links are the number of citations between journals, are directed and weighted;

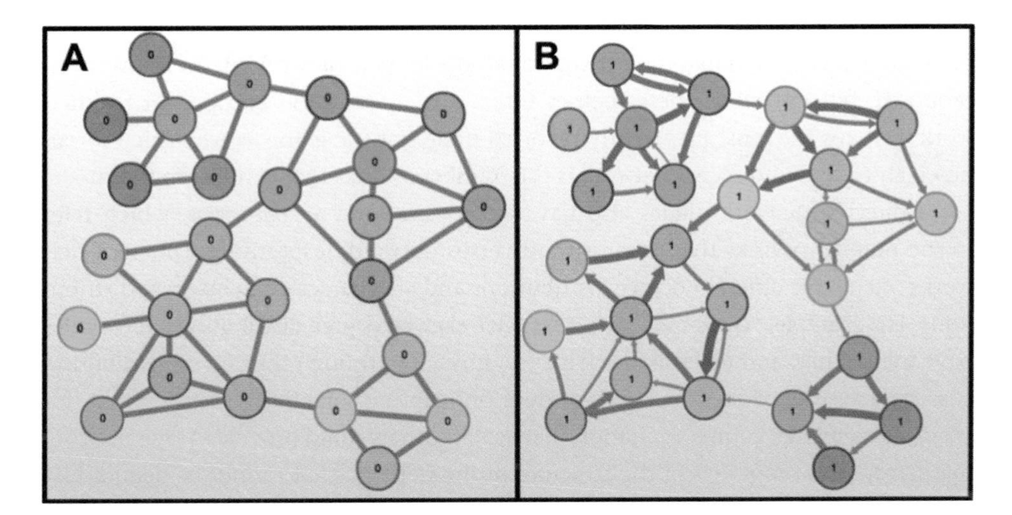

Figure 8.1
Different types of networks. Panel (A) shows an unweighted, undirected network (nodes labeled as zeros). In this network, the links have no direction or weights. (B) In bibliometrics, we generally deal with networks that have direction and weights like the example shown in the panel (nodes labeled as ones). Citations between journals have direction and weight (e.g., Journal A cites Journal B 10 times). The same applies to author-level citation networks. However, article citations have direction but rarely have weight. Articles generally cite another article once in the reference list.

article-level citation networks have direction but rarely have weight, because articles generally only list an article once in the reference list. Directionality and type of node can have a profound effect on the network topology. If one were to follow citations in an article-level citation network, they would move backward in time, which makes article-level networks more difficult to handle mathematically than journal-level citation networks, which do not have this property (Walker, Xie, Yan, & Maslov, 2007).

The second step of a network analysis is to deduce the temporal scale of study. A network can be an *aggregate*, such that all nodes and links irrespective of timestamp are gathered into a single network, or it can be *temporal*, such that nodes and/or links have time durations and the network changes as time progresses. The way a network changes over time can be more interesting than a snapshot of the network. Recently, in a study of a journal-level citation network, Rosvall and Bergstrom (2010) measured the amount of citation between journals for 2001, 2003, 2005, and 2007, and restricted the links in each of the by-year journal networks to citations made in those years. They found that the patterns of citation between journals had shifted over time, and discovered that one major shift of citation in these temporal journal networks was the birth of the new, stand-alone field of neuroscience, whose academic parents include neurology and molecular and cellular biology.

Once the nodes and links have been defined, the data can be converted into a network, which is suitable for network analysis. Myriad network metrics have been proposed, but not all of these metrics utilize the information encoded in the network. Perhaps the most prevalently reported node statistic is the *degree*, which for an unweighted and undirected network is the number of links that connect to that node. For a directed network, nodes also have an *in-degree* and an *out-degree*, which refer to the number of links that point inward versus outward, respectively. Different networks often have different degree distributions and are characterized by these distributions. For example, "scale-free" networks have skewed degree distributions, few nodes have many links, and many nodes have few links. Yet, though the degree distribution is a characteristic of every network, it does not capture something beyond what, for example, tabulated counts of citations between journals could provide. A true network metric captures properties of the structure of the network that cannot be deduced by the pairwise relationships between nodes alone. Most scientists have never cited Albert Einstein or Charles Darwin, but both of these scientists have had an undeniable influence on many scientists. A network measure should capture these *higher-order* effects.

The measurement of scholarly impact is a primary focus of bibliometric research. In a network, nodes that have high impact tend to have high *centrality*. As the word implies, centrality measures how central an author, paper, journal, or institution is in the topology of the network. There are many forms of centrality (e.g., betweenness centrality, closeness centrality) (Newman, 2001). Degree centrality, which is the number of links to a node, is perhaps best known in the sciences as the Impact Factor, or the average number of citations given to papers in a journal that were published in the previous two years. While degree centrality can be measured in a network context, it does not measure the higher-order properties of a given network topology, such as variation in the quality of citation sources. We argue that the difference between these approaches—measures that incorporate higher-order properties versus those that do not—is essentially the difference between a network and a nonnetwork approach.

One common method of measuring network properties is called eigenvector centrality. Bonacich (1972) first introduced eigenvector centrality as a way to quantify the status of a person in a friendship network. Bonacich's aim was to use social network structure to identify important people. How do we tell who the important people are? He suggested that important people have important friends. While this answer may sound circular, it is well defined mathematically and easy to compute. The most famous commercial application of eigenvector centrality is Google's PageRank algorithm, which ranks the importance of websites with the hyperlink structure of the World Wide Web (Page et al., 1998).

In the following analysis, we show that centrality measures that incorporate the properties of high-order structural effects provide a more accurate assessment of scholarly impact. We begin by example, with a detailed calculation of both the Impact Factor (degree centrality) and the Eigenfactor score (eigenvector centrality) (West, Bergstrom, & Bergstrom, 2010b). Eigenvector centrality can be extended to authors and institutions (West, Jensen, Dandrea, Gordon, & Bergstrom, 2013), but we focus on journal-level citations for this chapter.

How to Calculate a Network Measure[1]

The Adjacency Matrix

The first step in calculating network-based measures is to build an adjacency matrix from the network. In this case, the adjacency matrix, $\mathbf{Z}_{ij}$ indicates the number of times that articles published in journal, j, cite articles published in journal, i. The dimension of this square matrix is $n \times n$, where n is the number of journals. For example, suppose there are journals A, B, C, D, E, and F.

In the adjacency matrix in table 8.1, journal A cites itself once, journal B three times, journal C twice, and so on. Journal F receives no citations. Many methods insert zero in the diagonal so that journals do not receive credit for self-citations.

Table 8.1

	A	B	C	D	E	F
A	1	0	2	0	4	3
B	3	0	1	1	0	0
C	2	0	4	0	1	0
D	0	0	1	0	0	1
E	8	0	3	0	5	2
F	0	0	0	0	0	0

Table 8.2

	A	B	C	D	E	F
A	0	0	2/7	0	4/5	3/6
B	3/13	0	1/7	1	0	0
C	2/13	0	0	0	1/5	0
D	0	0	1/7	0	0	1/6
E	8/13	0	3/7	0	0	2/6
F	0	0	0	0	0	0

The second step is to normalize the outgoing citation of each journal by its total outgoing citations (render it a "column stochastic matrix"). Specifically, we divide each entry in a column by the sum of that column to get the Ç matrix: $\boldsymbol{H}_{ij} = \boldsymbol{Z}_{ij} / \boldsymbol{Z}_j$. After the steps above, the raw adjacency matrix in table 8.1 is transformed as in table 8.2.

There may be columns that sum to zero (i.e., journals that cite no other journals). These are "dangling nodes," because they may receive but do not give citations. These so-called dangling nodes pose an issue mathematically, so we need to take special steps to handle them.

Dangling Nodes

As mentioned in the previous section, there will be journals that do not cite any other journals. These journals are called dangling nodes and can be identified by looking for columns that contain all zeros. We need special notation to handle these nodes for our ranking calculation, so we introduce a special binary vector d of length n. The ones indicate that a journal is a dangling node; the zeros indicate a nondangling node. For the example above, d would be the row vector shown in table 8.3.

Table 8.3

	A	B	C	D	E	F
d_i	0	1	0	0	0	0

Table 8.4

	a
A	3/14
B	2/14
C	5/14
D	1/14
E	2/14
F	1/14

The Influence Vector and Teleportation

The next step is to construct a transition matrix **P** and compute its leading eigenvector. This step includes a "teleportation probability" between all nodes, which is a small weight added to each element in **H** that renders the entire matrix nonzero, and therefore eliminates dangling nodes. This step is important mathematically, because we are only able to calculate eigenvector centrality scores for a matrix without dangling nodes.

The eigenvector vector, normalized so that its components sum to 1, will be called the influence vector $ð^*$. This vector gives us the journal weights that we will use in assigning Eigenfactor scores. To calculate the influence vector $ð^*$, we need six inputs: the matrix **H** that we just created, an initial start vector $ð^{(0)}$, the constants α and ε, the dangling node vector d, and an article vector a.

Several teleportation methods have been proposed, most commonly uniform teleportation (teleport with equal probability to any journal), but this has been shown to overweight smaller journals (West, Bergstrom, & Bergstrom, 2010a). For the Eigenfactor score, teleportation is proportional to journal size. To represent this, let a_i be the proportion of total articles that journal i has in the journal dataset. Table 8.4 shows an example article vector, a, with 14 total articles.

The start vector, $ð^{(0)}$, is used in iterating the influence vector. Set each entry of this column vector to $1/n$ and it would look like the vector in table 8.5.

Calculating the Influence Vector

We can now calculate the influence vector, $ð^*$, which is the leading eigenvector (normalized so that its terms sum to one) of the matrix **P**, defined as[2]

Table 8.5

	$\pi^{(0)}$
A	1/6
B	1/6
C	1/6
D	1/6
E	1/6
F	1/6

Table 8.6

	A	B	C	D	E	F
A	0	3/14	2/7	0	4/5	3/6
B	3/13	2/14	1/7	1	0	0
C	2/13	5/14	0	0	1/5	0
D	0	1/14	1/7	0	0	1/6
E	8/13	2/14	3/7	0	0	2/6
F	0	1/14	0	0	0	0

$$\boldsymbol{P} = á\boldsymbol{H}' + (1-á)a \cdot e^T, \tag{1}$$

where $e^T$is a transposed all-one row vector, used as a linear algebra trick to make the product $a \cdot e^T$ a matrix. The matrix $\boldsymbol{H}'$ is a modification of the matrix **H** with each dangling node column replaced by the article vector a, which causes dangling node journals to teleport out proportional to journal size. It would look as in table 8.6.

In the jargon of stochastic matrix theory, the teleportation matrix **P** will be an irreducible aperiodic Markov chain and has a guaranteed unique leading eigenvector by the Perron-Frobenius theorem (MacCluer, 2000). We could compute the normalized leading eigenvector of the matrix **P** directly using the power method (multiplying the **P** matrix many times), but this involves repeated matrix multiplication operations on the dense matrix **P** and is thus computationally intensive. Instead, we can use an alternative approach that involves only operations on the sparse matrix **H** and is thus faster (Langville & Meyer, 2006). To compute the influence vector rapidly, we can iterate the following equation

$$ð^{k+1} = á \cdot H \cdot ð^k + \left[á \cdot d \cdot ð^k + (1-á)\right] \cdot A, \tag{2}$$

where $ð^k$ can be set to $ð^0$ on the first iteration. To find the influence vector, we iterate repeatedly. This iteration will converge uniquely to the leading eigenvector of **P**, normalized so that its terms sum to 1. After each iteration, check to see if the residual

$(\hat{o} = ð^{k+1} - ð^{k})$ is less than $\mathring{a}$, which is an arbitrary stop criterion. If it is, then $ð^{*} \sim ð^{k+1}$ is the influence vector. Typically, this process does not take more than 100 iterations with $\mathring{a} = 0.00001$. With the raw adjacency matrix example above and the corresponding article vector a, the stationary vector converges after 16 iterations to the influence vector with $\acute{a} = 0.85$ and $\mathring{a} = 0.00001$ (Table 8.7).

From Influence Vector to Eigenfactor Score

The vector of Eigenfactor values for each journal is given by the dot product of the **H** matrix and the influence vector $ð^{*}$, normalized to sum to 1 and then multiplied by 100 to convert the values from fractions to percentages:

$$EF = 100\frac{H \cdot ð^{*}}{\sum_i [\boldsymbol{H} \cdot ð^{*}]_i}. \quad (3)$$

If everything is calculated correctly, the Eigenfactor values for our example are

Table 8.7

	π^{*}
A	0.3040
B	0.1636
C	0.1898
D	0.0466
E	0.2753
F	0.0206

Table 8.8

	EF
A	34.0510
B	17.2037
C	12.1755
D	3.6532
E	32.9166
F	0.000

Interpreting the Rankings

The Eigenfactor score is a network measure, which has a different interpretation from a first-order measure like the Impact Factor. If a researcher randomly follows citations from journal to journal infinitely, the Eigenfactor score can be interpreted as the fraction of total time the researcher spends at each journal. An interactive applet is available online to explore this process, created by Martin Rosvall and Daniel Edler.[3]

Network Metrics: How Are They Different from Traditional Metrics?

This section focuses on two key differences (of many) between the Impact Factor and the Eigenfactor. Our goal is to explain by example the difference between a network (Eigenfactor) and a nonnetwork (Impact Factor) measure. A network approach incorporates the fact that not all citation sources are equal (figure 8.2). The blue nodes in figure 8.2A each have one link. These blue nodes collectively share 40% of the total citations. Note that the degree is the same for each of these blue nodes. However, under eigenvector centrality, each node no longer has the same score, because each blue node has a different position in the network. The source of each citation matters under eigenvector centrality.

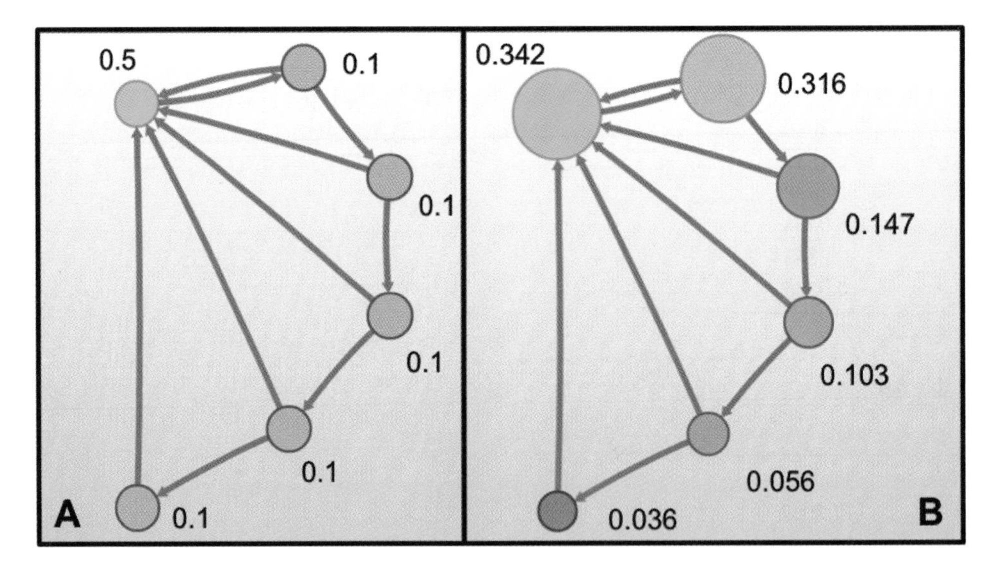

Figure 8.2
Citation source matters—difference between degree centrality and eigenvector centrality. The numbers indicate the ranking of the nodes according to both approaches. In (A) the blue nodes all have one incoming link. Under degree centrality, all blue nodes are ranked the same and are indistinguishable. In (B) the size of the nodes corresponds to the ranking by eigenvector centrality measures. Because the source of the citation matters under a network measure, the ranking of nodes in (B) progressively lowers with citation distance from the high-impact orange node. This example is available at http://www.mapequation.org/apps/mapdemo.html.

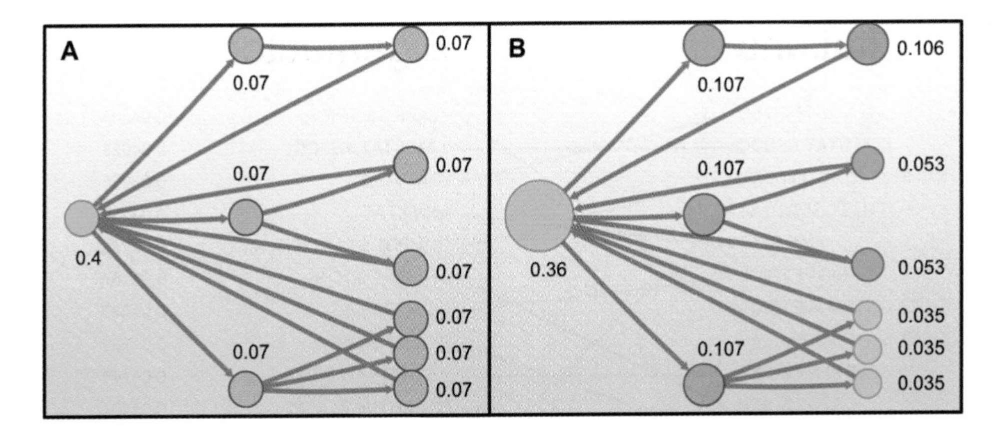

Figure 8.3
Difference between degree centrality (Panel A) and eigenvector centrality (Panel B)—differences in the number of outgoing citations from the citation source, shown numerically adjacent to each node. In (A) all the blue nodes are exactly the same under a degree centrality ranking because each receives one citation. Unlike the blue nodes in figure 8.1, however, the source nodes (the nodes marked 0.07) all have one citation from the same source, the orange node. But the number of outgoing citations from the red nodes is different (red in Panel B), which affects the nodes downstream. You can see this effect in (B). The end nodes (the far-right nodes) have different scores depending on the source of their citations.

For eigenvector centrality, the number of outgoing citations from a journal also matters. The more citations a journal gives, the less each citation is worth individually (figure 8.3). The blue nodes in figure 8.3A all have one citation. By degree centrality, all the blue nodes have the same centrality. However, when a network approach is used the purple, green, and light-blue nodes are ranked differently—not because the source nodes are different as in figure 8.2, but because the red source nodes all give a different number of citations (figure 8.3B).

Counting Citations vs. Eigenfactor

Substantial ranking differences arise when Eigenfactor rankings are compared with the ranking of citations received (West et al., 2010a). This is best illustrated with data. Figure 8.4 compares the rankings in 2010 of the top 40 statistics journals, with citation data extracted from Thomson Reuters' Journal Citation Reports (JCR). The journals in the left column are ranked by raw citations received (degree centrality). The column on the right ranks journals by Eigenfactor. We find a number of alterations in ranking

Citations	Eigenfactor	
STAT MED	ECONOMETRICA	0.04564
J AM STAT ASSOC	J AM STAT ASSOC	0.04028
COMPUT STAT DATA AN	STAT MED	0.03808
FUZZY SET SYST	ANN STAT	0.03459
ANN STAT	COMPUT STAT DATA AN	0.02275
BIOMETRICS	J R STAT SOC B	0.02067
ECONOMETRICA	BIOMETRICS	0.02032
CHEMOMETR INTELL LAB	BIOMETRIKA	0.01782
J STAT PLAN INFER	J STAT PLAN INFER	0.01741
J R STAT SOC B	STOCH PROC APPL	0.01497
BIOMETRIKA	ANN PROBAB	0.01491
J COMPUT BIOL	FUZZY SET SYST	0.01244
BIOSTATISTICS	ANN APPL PROBAB	0.01219
STAT PROBABIL LETT	PROBAB THEORY REL	0.01202
STOCH PROC APPL	BIOSTATISTICS	0.01171
J STAT SOFTW	J MULTIVARIATE ANAL	0.01146
J MULTIVARIATE ANAL	STAT PROBABIL LETT	0.01137
INSUR MATH ECON	J BUS ECON STAT	0.009791
ANN PROBAB	J COMPUT BIOL	0.008414
J CHEMOMETR	ELECTRON J PROBAB	0.008337
STOCH ENV RES RISK A	BERNOULLI	0.008293
ANN APPL PROBAB	STAT SCI	0.008073
STATA J	COMB PROBAB COMPUT	0.007532
J R STAT SOC A STAT	J COMPUT GRAPH STAT	0.007461
PROBAB THEORY REL	STAT SINICA	0.00739
COMMUN STAT-THEOR M	J R STAT SOC A STAT	0.007317
BIOMETRICAL J	CHEMOMETR INTELL LAB	0.007171
IEEE ACM T COMPUT BI	J STAT SOFTW	0.007103
J BIOPHARM STAT	INSUR MATH ECON	0.006978
MULTIVAR BEHAV RES	ANN APPL STAT	0.006713
J BUS ECON STAT	J APPL PROBAB	0.00655
J COMPUT GRAPH STAT	STATA J	0.006173
STAT APPL GENET MOL	STAT COMPUT	0.005882
STAT SCI	SCAND J STAT	0.005806
STAT METHODS MED RES	BIOMETRICAL J	0.005615
STAT SINICA	TECHNOMETRICS	0.005582
ANN APPL STAT	COMMUN STAT-THEOR M	0.00555
TECHNOMETRICS	BAYESIAN ANAL	0.005512
J APPL PROBAB	MULTIVAR BEHAV RES	0.00528
BERNOULLI	FINANC STOCH	0.005123

order. One journal that increases its rank substantially is *Econometrica*, which reports on novel statistical approaches in economics. However, a journal that ranks worse under a network approach is *Chemometrics and Intelligent Laboratory Systems*.

The reader should note that we have not compared the Eigenfactor score directly with the Impact Factor. We note that the two cannot be directly compared without additional steps, so we refer the reader to a full analysis of the subject (West et al., 2010a). This incomparability is due to normalization. The Impact Factor is degree centrality for the previous two years, divided by the number of articles published in the previous two years. Thus, the total number of received cites is directly comparable with the Eigenfactor because both are not normalized by journal size. To amend this, the Eigenfactor can be divided by the number of articles in a journal (referred to as the *Article Influence* score), which renders the Impact Factor comparable to Article Influence.

The Limitations of Citations as a Measure of Impact

As scientists, we sometimes become so enamored of conventional procedure in our disciplines that we forget what we study is often an oversimplified model of the real world. For example, we might be tempted to think that papers with more citations have more positive impact—yet this simple number ignores the positive impact that a paper could have on the participants in a journal club, the fact that citations are often negative, the variance in discipline size, and the media coverage of that paper (Priem, Taraborelli, Groth, & Neylon, 2010). In this example, citation is a *proxy* for impact. Undoubtedly citation correlates positively with impact, but citation is still only an approximation of impact, and is limited by the resolution that citation provides. If we study impact, we do not study citation, we study *impact.* If a better framework to estimate impact is available, then we should discard the previous framework in favor of the new one. Networks are now a cornerstone of bibliometric research and we have discarded tabulations of citations, but we should not forget that networks are models

Figure 8.4
Ranking differences between degree centrality and eigenvector centrality for statistics journals. The statistics journals are pulled from Thomson Reuters 2010 Journal Citation Reports (JCR). The rankings on the left are based on the number of incoming citations, while the rankings on the right are based on the Eigenfactor score, an eigenvector centrality measure. The green lines connect journals that increase in rank when the network is taken into context. The red lines connect journals that drop in the rankings, while the black lines connect journals that do not change.

with limitations as well. As the amount of data continues to grow and our academic knowledge improves, the networks we use may change and improve—eventually all links in a paper-level citation network may include the context of the citation. We are in the midst of a revolution inspired by network thinking and fueled by data, which means the future of the science of science is bright.

Conclusion

In this chapter, we explained what a citation network is; we demonstrated how to calculate a network measure using a small example; and we illustrated how network measures differ from nonnetwork measures, both with examples and data. Network-based measures are more complicated than nonnetwork measures, but the richness gained with such a measure is worth the extra effort.

Notes

1. Code and the example network data in this section can be found at Eigenfactor.org.

2. This matrix describes a stochastic process in which a random walker moves through the scientific literature; it is analogous to the "Google matrix" that Google uses to compute the PageRank scores of websites. The stochastic process can be interpreted as follows: a fraction *á* of the time the random walker follows citations and a fraction 1-*á* of the time the random walker "teleports" to a random journal chosen at a frequency proportional to the number of articles published.

3. http://www.mapequation.org/apps/MapDemo.html

References

Bollen, J., Rodriguez, M. A., & Van de Sompel, H. (2006). Journal status. *Scientometrics, 69*(3), 669–687.

Bonacich, P. (1972). Factoring and weighting approaches to clique identification. *Journal of Mathematical Sociology, 2*, 113–120.

EMC. (2010). IDC white papers: The digital universe decade—Are you ready? Retrieved from www.emc.com/leadership/programs/digital-universe.htm.

Garfield, E. (1955). Citation indexes for science. *Science, 122*, 108–111.

Kalaitzidakis, P., Mamuneas, T. P., & Stengos, T. (2003). Rankings of academic journals and institutions in economics. *Journal of the European Economic Association, 1*(6), 1346–1366.

Kodrzycki, Y. K., & Yu, P. (2006). New approaches to ranking economics journals. *Contributions to Economic Analysis & Policy, 5*(1), 1–40.

Langville, A. N., & Meyer, C. D. (2006). *Google's pagerank and beyond: The science of search engine rankings*. Princeton, NJ: Princeton University Press.

Liebowitz, S. J., & Palmer, J. P. (1984). Assessing the relative impacts of economics journals. *American Economic Association, 22*(1), 77–88.

MacCluer, C. (2000). The many proofs and applications of Perron's theorem. *SIAM Review, 42*(3), 487–498.

McKinsey & Company. (2011). Big data: The next frontier for innovation, competition, and productivity. Retrieved from http://www.mckinsey.com/insights/business_technology/big_data_the_next_frontier_for_innovation.

Newman, M. E. J. (2001). Scientific collaboration networks. II. Shortest paths, weighted networks, and centrality. *Physical Review E: Statistical Physics, Plasmas, Fluids, and Related Interdisciplinary Topics, 64*(1), 016132–1–016132–7.

Page, L., Brin, S., Motwani, R., & Winograd, T. (1998). *The pagerank citation ranking: Bringing order to the web*. Technical report, Stanford Digital Library Technologies Project. Retrieved from http://ilpubs.stanford.edu:8090/422.

Palacios-Huerta, I., & Volij, O. (2004). The measurement of intellectual influence. *Econometrica, 73*(3), 963–977.

Pinski, G., & Narin, F. (1976). Citation influence for journal aggregates of scientific publications: Theory, with application to the literature of physics. *Information Processing & Management, 12*, 297–326.

Price, D. J. de S. (1965). Networks of scientific papers. *Science, 149*, 510–515.

Priem, J., Taraborelli, D., Groth, P., & Neylon, C. (2010). Alt-metrics: A manifesto. Retrieved from http://altmetrics.org/manifesto.

Rosvall, M., & Bergstrom, C. T. (2010). Mapping change in large networks. *PLoS ONE, 5*(1), e8694.

Walker, D., Xie, H., Yan, K., & Maslov, S. (2007). Ranking scientific publications using a model of network traffic. *Journal of Statistical Mechanics*. Retrieved from http://iopscience.iop.org/1742-5468/2007/06/P06010.

West, J. D., Bergstrom, T. C., & Bergstrom, C. T. (2010a). Big macs and Eigenfactor scores: Don't let correlation coefficients fool you. *Journal of the American Society for Information Science and Technology, 61*(9), 1800–1807.

West, J. D., Bergstrom, T. C., & Bergstrom, C. T. (2010b). The Eigenfactor™ metrics: A network approach to assessing scholarly journals. *College & Research Libraries, 71*(3), 236–244.

West, J. D., Jensen, M. C., Dandrea, R. J., Gordon, G. L., & Bergstrom, C. T. (2013). Author-level Eigenfactor metrics: Evaluating the influence of authors, institutions and countries within the SSRN community. *Journal of the American Society for Information Science and Technology, 64*(4), 787–801.

9 Science Visualization and Discursive Knowledge

Loet Leydesdorff

Introduction

There is a rich tradition in the visualization of scientific developments and historical events, accelerated recently by the growth and availability of large-scale datasets, software, and computational approaches (see Börner's [2010] *Atlas of Science* for a visual chronicling of this history and Börner, Chen, & Boyack, 2003, for a review). Science visualizations are often predicated on a map metaphor, so much so that the term *science visualization* has become interchangeable with *mapping science*. However, unlike the geographic map, science has no natural baselines (see Day, chapter 4, this volume). Scientific domains are not bounded like nations or states, particularly in interdisciplinary areas (Small & Garfield, 1985). Given the complexity of knowledge organization and interaction, there is always some degree of reductionism that must occur in order to project the knowledge space onto a two- or three-dimensional landscape. Furthermore, if the variable of time is included (e.g., if a scholar wishes to animate evolving dynamics), additional care must be taken to stabilize the representation so that the results can be captured as a mental map (Liu & Stasko, 2010; Misue, Eades, Lai, & Sugiyama, 1995).

The intellectual space of science can be mapped in terms of words (e.g., title words) and authors, and co-occurrences of these variables (Callon, Courtial, Turner, & Bauin, 1983; White & Griffith, 1982; White & McCain, 1998). At a higher level of aggregation, journal-journal citation relations—available from the *Science Citation Index*—have been used since the mid-1980s for mapping developments in and among disciplines (Doreian & Fararo, 1985; Leydesdorff, 1986; Tijssen, De Leeuw, & Van Raan, 1987). Small and others further developed the mapping of cocitations (e.g., Garfield, 1978; Small & Sweeney, 1985).

In this chapter, I argue that observable network relations organize the sciences under study into historical instantiations that can be statically visualized. The development of scholarly discourse, however, can be considered self-organizing in terms of fluxes of communication along the various dimensions that operate within different (e.g., disciplinary) codes. Over time, this adds evolutionary differentiation to the historical integration; a richer structure can process more complexity. Latent Semantic Analysis (LSA) focuses on these latent dimensions in textual data, and social network analysis (SNA) on the networks of observable relations. However, the two coupled topographies of information processing in the network space and meaning processing in the vector space operate with different (nonlinear) dynamics.

Multidimensional Scaling

Computer-aided visualization of multivariate data predated the advent of the personal computer and the Internet. Based on Kruskal (1964), scholars in psychometrics developed spatial representations of sets of variables by multidimensional scaling (MDS) (e.g., Kruskal & Wish, 1978; Schiffman, Reynolds, & Young, 1981). Among other forms of output, MDS can generate a two-dimensional map. The first large-scale MDS program ALSCAL ("alternating least square analysis") is still available in current versions of statistical packages such as SPSS.

Table 9.1 provides distances in terms of flying mileages among 10 American cities (SPSS, 1993; Leydesdorff & Vaughan, 2006). MDS enables us to regenerate the map from which these distances were obtained by minimizing the stress (S) in the projection (figure 9.1). Feeding these data into ALSCAL, for example, leads not surprisingly to an almost perfect fit ($S = 0.003$).

These data measure *dissimilarity*, because the larger the numbers, the further apart the cities are—that is, the more "dissimilar" they are in location. One can also use similarity measures for mapping, such as correlation coefficients. Options that might be added to a next generation of such maps include:

1. The ability to visualize the network of connections among the cities
2. Measures of distance other than Euclidean ones—for example, correlations in a multidimensional (vector) space provide a different topology
3. Groupings of nodes using different colors based on attribute values
4. The ability to scale nodes and links with the values of attributes; etc.

A large number of current network visualization and analysis programs provide these features and can be downloaded from the Internet.

Table 9.1
Flying mileages among 10 American cities

	Atlanta	Chicago	Denver	Houston	Los Angeles	Miami	New York	San Francisco	Seattle	Washington, D.C.
Atlanta	0	.	.	.	.	.	.	.	.	.
Chicago	587	0	.	.	.	.	.	.	.	.
Denver	1212	920	0	.	.	.	.	.	.	.
Houston	701	940	879	0	.	.	.	.	.	.
Los Angeles	1936	1745	831	1374	0	.	.	.	.	.
Miami	604	1188	1726	968	2339	0	.	.	.	.
New York	748	713	1631	1420	2451	1092	0	.	.	.
San Francisco	2139	1858	949	1645	347	2594	2571	0	.	.
Seattle	2182	1737	1021	1891	959	2734	2408	678	0	.
Washington, D.C.	543	597	1494	1220	2300	923	205	2442	2329	0

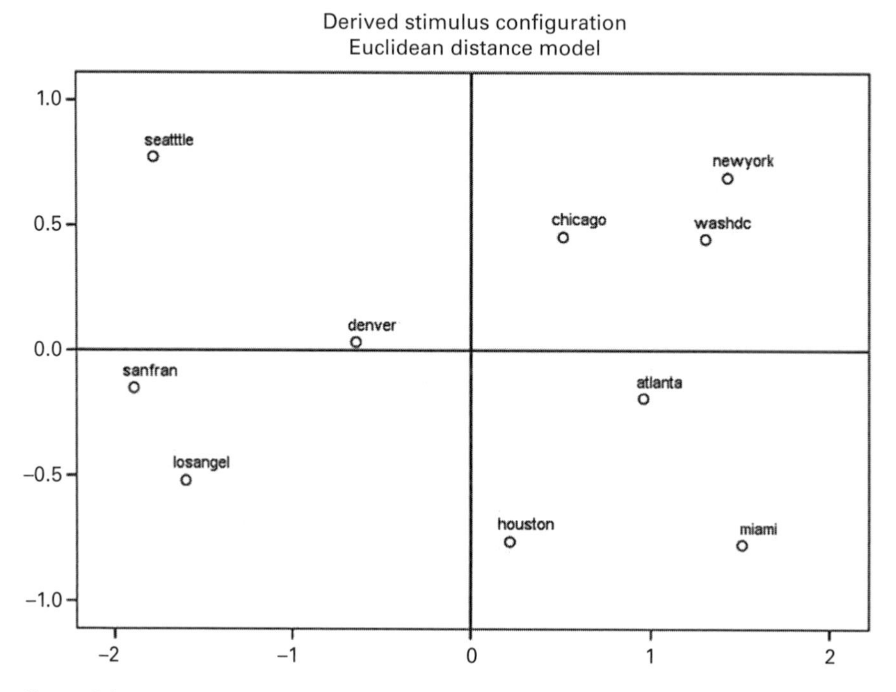

Figure 9.1
MDS mapping (ALSCAL) of 10 American cities using the distance matrix in table 9.1 (normalized raw stress = 0.003).

Graph Theory and Network Analysis

During the 1980s, graph theory emerged as a theoretical basis for network analysis. In the original programs (such as GRADAP) the links had to be drawn by hand. UCINet 2.0 (1984) provided the first network analysis program that integrated a version of MDS (MINISSA),[1] but the number of variables was at the time limited to 52: 26 uppercase and 26 lowercase characters could be indicated (Freeman, 2004). These programs allowed for the use of similarity measures other than Euclidean distances. For example, Leydesdorff (1986) used Pearson correlations to visualize factor structures in aggregated journal-journal citation matrices using UCINet 2.0.

Graphical interfaces became available during the 1990s with the further development of Windows (Windows 95) and the Apple computer. Pajek followed as a visualization and analysis tool for large networks in 1996 (De Nooy, Mrvar, & Batagelj, 2005). Pajek also allows for non-Western characters such as Chinese and Arabic (Leydesdorff & Jin, 2005).[2]

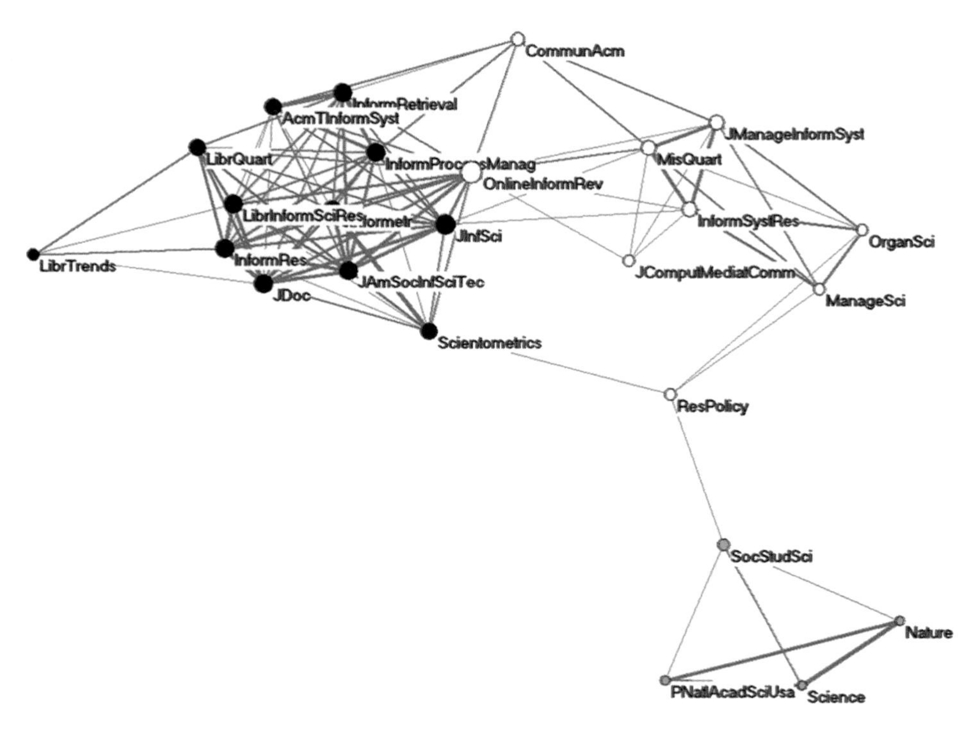

Figure 9.2
Twenty-five journals most cited by authors in *JASIST* during 2010; Kamada and Kawai (1989) used for the layout; node sizes proportional to degree centrality; node colors according to modularity (Q = 0.328); edge width proportional to *cosine* values (*cosine* > 0.2).

Figure 9.2 provides an example of the current state of the art: the aggregated citation network of the *Journal of the American Society for Information Science and Technology* (*JASIST*) as mapped in 2010. (These 25 journals are cited in *JASIST* to the extent of more than 1% of its total citations.) The matrix is analyzed using both Pajek and Gephi;[3] links are indicators of *cosine* similarities between the citing patterns of these journals; the vertices are colored according to the modularity algorithm (Q = 0.328; Blondel, Guillaume, Lambiotte, & Lefebvre, 2008), and sized according to their degree centrality (De Nooy et al., 2005).

Research Policy, positioned between the three components in this map, has accordingly the highest betweenness centrality (0.305). Although different in some details, both the factor analysis[4] and the modular decomposition classify *Research Policy* as belonging to the information systems group of journals (within this context). The visualization adds a network of relations among the nodes. As noted, one is able to use attributes of nodes and links in order to further enrich the visual.

Relational and Positional Maps of Science

Using MDS, one visualizes the variables as a system (e.g., a word-document matrix). In spatial terms, the words attributed to documents are considered as vectors that are vector-summed into a vector space (Salton & McGill, 1983). Given parameter choices (such as the similarity measure), the projection of the variables in MDS is deterministic. For example, the Euclidean distance between San Francisco and New York does not change depending on the intensity of the network relations (e.g., flights) between these two cities.

In network analysis, one is often as interested in a representation that uses the intensity of the relations as the distance on the map. For instance, two authors who frequently coauthor should be positioned next to each other in a coauthorship map. In this case, it is not the *correlations* among the distributions, but the *relations* among the nodes that are used for the mapping. Graph-analytic algorithms (e.g., Kamada & Kawai, 1989) optimize the network in terms of relations. The choice of starting point can be random, and each run may lead to a somewhat different outcome.

Let us compare the two approaches to optimizing the vector space versus the network topology. In figures 9.3 and 9.4, 43 title words are included that occurred more than 10 times among the 455 titles in the 2010 and 2011 volumes of *JASIST*. A five-factor solution in the underlying data matrix is used for coloring the nodes in the vector space (figure 9.3) and the network space (figure 9.4), respectively.

Factor 1, for example, is composed of the words *impact, factor, journal, citation,* and *source*. These (green-colored) words are grouped in both figures: they not only entertain strong relations to one another (figure 9.4), but also co-occur in similar *patterns* among the other title words in the sample (figure 9.3). Factor 4, however, with primary factor loadings for the words *effect, image, study, online,* and *behavior,* can more easily be distinguished in figure 9.3 than in figure 9.4. These words co-occur with other words in the set more diffusely, yet they form a latent dimension of the data.

In other words, there is no necessary relationship between co-occurrences in the observable network of relations, and correlations among co-occurrence patterns. The co-occurrence patterns can be mapped using the correlation coefficients among the distributions, whereas the values of co-occurrence relations provide us with a symmetrical (affiliations) matrix that can be visualized directly. In the latter case, one visualizes the network of observable *relations,* whereas in the former, one visualizes the latent structure in these data. For example, two synonyms may have (statistically) similar *positions* in a semantic map, but they will rarely co-occur in a single title.

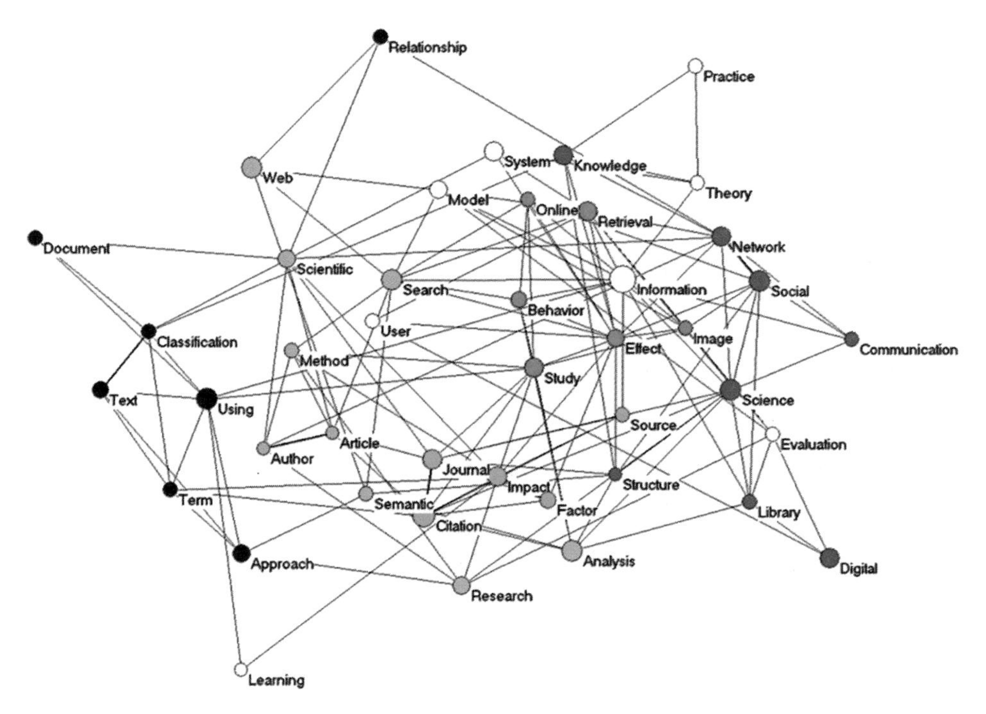

Figure 9.3
Cosine-normalized map of 43 words occurring more than 10 times during 2010 and 2011 in titles of *JASIST*. (*Cosine* ≥ 0.1; Kamada & Kawai, 1989.) The nodes are colored according to the five-factor solution of this network (Varimax rotated; SPSS), and scaled in accordance with their degree centrality.

These two perspectives on the data have led to two different research traditions in textual analysis and social network analysis, respectively. As noted, LSA focuses on the latent dimensions in the data, while SNA focuses on the observable relations in networks. In SNA, for example, eigenvector centrality—that is, factor loading on the first factor—can be used as an attribute of the nodes, whereas in LSA the factors (eigenvectors) in different directions organize the semantic maps (Landauer, Foltz, & Laham, 1998). The factor-analytic approach has been further developed using Singular Value Decomposition (SVD), whereas graph theory has provided an alternative paradigm for developing algorithms in SNA.

A star in a graph can be in the center of the multidimensional space, and therefore not load strongly on any of the dimensions. In figure 9.4, for example, the word *information*, which occurs 94 times in this set (followed by *citation*, 44 times), did not load positively on any of the five factors distinguished; this variable is factor-neutral and

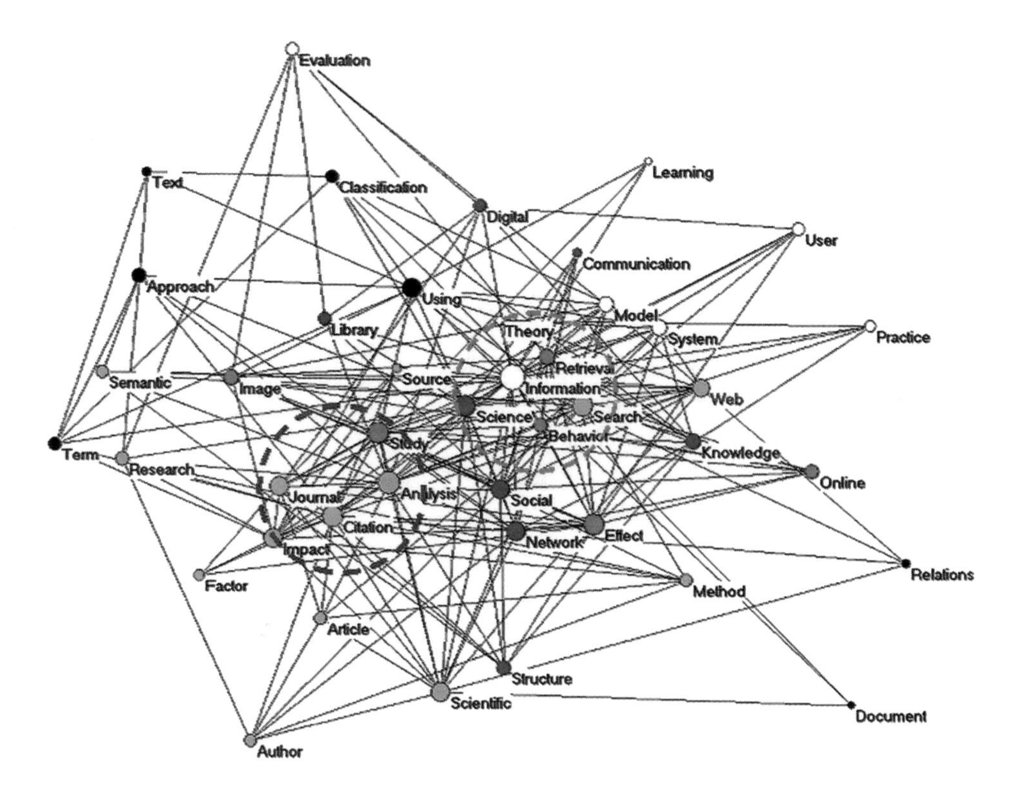

Figure 9.4
Co-occurrence map of 43 words occurring more than 10 times during 2010 and 2011 in titles of *JASIST*. (Co-occurrence values ≥ 2; Kamada & Kawai, 1989.) The nodes are colored according to the five-factor solution of this network (Varimax rotated; SPSS), and scaled in accordance with their degree centrality.)

therefore colored white. However, using the degree distribution for sizing the nodes in figure 9.4, *information* has the highest degree, co-occurring with 37 of the 43 title words, followed by *analysis* with a degree of 33. A core set of words surrounding *information* (circled red in figure 9.4) belongs to the center of the field of the information sciences. *Citation* (Factor 1) and *analysis* (Factor 3) are part of a secondary grouping of the relations (gray circled).

Interpreting Science Visualizations

When a network is spanned in terms of relations, this process shapes an architecture in which all components have a position. The analysis of this architecture (that is, the set of relations) enables us to specify what the relations mean in the network as a system. For example, the word *information* was most central in this network (figure 9.4), but it was not colored in terms of having meaning in any of the relevant dimensions indicated at the systems level. Yet the word as a variable carries Shannon-type information (uncertainty; Shannon, 1948).

The graph-analytic approach informs us, as analysts, about the network of relations, but not about what these relations mean in terms of the discourse(s) under study. However, graph-theoretic concepts such as centrality also have meaning in social network analysis. The analyst's (meta) discourse can be distinguished from the communication among the words under study. The latter communication can represent scholarly discourses, political discourses, and media information.

Within each of these discourses, codes of communication can span dimensions that provide the communicated words with meaning. Both the developments in the observable networks (vectors) and the hypothesized dimensions (eigenvectors) can be theorized. The relations among nodes can be considered attributes of the nodes, but the dimensions of the communication are attributes of the links. SNA focuses on the positions of nodes in terms of vectors, whereas LSA focuses on the positions of links in terms of these next-order structures.

This scheme can be generalized: the relations among authors can also be considered as a system of links and therefore another semantic domain. Any system that can position its components as a system provides itself and its elements with meaning (Maturana, 1978). A discourse, for example, provides meaning to the words that are communicated.

The two perspectives of meaning processing and information processing can be considered feedback mechanisms operating on each other. The shaping of the networks of relations causes structures that can provide feedback evolutionarily as a next-order system on the networks of relations from which they emerge. Meaning is provided from the perspective of hindsight, but with reference to other possibilities ("horizons of meaning;" cf. Husserl, 1929/1973). The next-order meaning processing cannot continue without information processing; otherwise, the systems would no longer be historical. The historical instantiation can from this perspective be considered a retention mechanism of the semantic systems that evolve over time (Leydesdorff, 2011a).

The Network and the Vector Space

The multidimensional (vector) space can be regarded as a *system* of relations including interaction terms, and the network space as an *aggregate* of observable relations among nodes. One can also call the network relations first-order (being observable) and the vector space second-order because the latent dimensions of the system are not given but hypothesized—for example, in a factor-analytic model. Whereas observable variation is stochastic, latent structure is deterministic. The deterministic selection mechanism(s), however, can be expected to be further developed over time in parallel to the networks of relations because of the feedback mechanisms involved.

Accordingly, the systems view of MDS is deterministic, whereas the graph-analytic approach can also begin with a random or arbitrary choice of a starting point. Using MDS, the network is first conceptualized as a multidimensional space that is then reduced stepwise to lower dimensionality. At each step, the stress increases; Kruskall's stress function is formulated as follows:

$$S = \sqrt{\frac{\sum_{i \neq j} (\|x_i - x_j\| - d_{ij})^2}{\sum_{i \neq j} d_{ij}^2}} \tag{1}$$

In this formula $||x_i - x_j||$ is equal to the distance on the map, while the distance measure d_{ij} can be, for example, the Euclidean distance in the data under study. As noted, one can use MDS to illustrate factor-analytic results (in tables), and in this case the Pearson correlation obviously provides the best match.

Spring-embedded or force-based algorithms can be considered a generalization of MDS but were inspired by the above-mentioned developments in graph theory during the 1980s. Kamada and Kawai (1989) were the first to reformulate the problem of achieving target distances in a network in terms of energy optimization. They formulated the ensuing stress in the graphical representation as follows:

$$S = \sum_{i \neq j} s_{ij} \text{ with } s_{ij} = \frac{1}{d_{ij}^2} (\|x_i - x_j\| - d_{ij})^2 \tag{2}$$

Equation 2 differs from equation 1 by taking the square root in equation 1, and because of the weighting of *each* term in the numerator with $1/d_{ij}^2$ in equation 2. This weight is crucial for the quality of the layout but defies normalization with $\sum d_{ij}^2$ in the denominator of equation 1; hence the incomparability between the two stress values.

The ensuing difference at the conceptual level is that spring embedding is a graph-theoretic concept developed for the topology of a network. The weighting is achieved for each individual link. MDS operates on the multivariate space as a system, and hence refers to a different topology. In the multivariate space, two points can be close to each other without entertaining a relationship (Granovetter, 1973). For example, they can be close or distant in terms of the correlation between their *patterns* of relationships (cf. Burt, 1992).

In the network topology, Euclidean distances and geodesics (shortest distances) are conceptually more meaningful than correlation-based measures. In the vector space, correlation analysis (factor analysis, etc.) is appropriate for analyzing the main dimensions of a system. The *cosines* of the angles among the vectors, for example, build on the notion of a multidimensional space. In bibliometrics, Ahlgren, Jarneving, and Rousseau (2003) have argued convincingly in favor of the *cosine* as a nonparametric similarity measure because of the skewedness of the citation distributions and the abundant zeros in citation matrices. Technically, one can also input a *cosine*-normalized matrix into a spring-embedded algorithm. The value of (1 – *cosine*) is then considered a distance in the vector space (Leydesdorff & Rafols, 2011). In sum, there is a wealth of possible combinations in a parameter space of clustering algorithms and similarity criteria.

The Visualization of Heterogeneous Networks

The two coupled topographies of information processing in the network space and meaning processing in the vector space operate with different (nonlinear) systems dynamics (Luhmann, 1995). The historical dynamics of information processing in instantiations organizes the system, and thus interfaces with and tends to integrate, the (analytically orthogonal) dynamics along each eigenvector. The systems dynamics, however, can be considered self-organizing in terms of fluxes along the various dimensions—used as codifiers of the communication—and with potentially different speeds. This development over time adds evolutionary differentiation to the historical integration; a richer structure can process more complexity.

Integrating retention can be organized in dimensions other than differentiating expansion. For example, archives and reflexive authors historicize and thus stabilize the volatile networks of new ideas, metaphors, and concepts. Relations among words can be regarded as providing us with access to the variation, whereas cited references anchor new knowledge claims in older layers of texts (Lucio-Arias & Leydesdorff, 2009). Authors and institutions may provide historical stability because differences are reflected and locally integrated in communicative actions.

The textual domain provides us with options to combine these different layers in visualizations and animations. The sciences evolve as heterogeneous networks of words, references, authors, and at different levels of aggregation. The composing subdynamics, for example, of specialties and disciplines are not organized neatly in terms of specific variables, but in terms of configurations of variables, such as specific resonances among cognitive horizons (paradigms), social identities, and corpora of literature. The human beings involved (and their organizations) cannot be reduced to literature, and

	au_1	au_2	...	au_m
doc_1	a_{11}	a_{21}	...	a_{m1}
doc_2	a_{12}	...	...	a_{m2}
doc_3	...	...	...	...
...	...	...	...	...
...	...	...	...	...
doc_n	a_{1n}	...	...	a_{mn}

+

	w_1	w_2	...	w_k
doc_1	b_{11}	b_{21}	...	b_{k1}
doc_2	b_{12}	...	...	b_{k2}
doc_3	...	...	...	...
...	...	...	...	...
...	...	...	...	...
doc_n	b_{1n}	...	...	b_{kn}

=

	v_1	v_2	...	$v_{(m+k)}$
doc_1	c_{11}	c_{21}	...	$c_{(m+k)1}$
doc_2	c_{12}	...	...	$c_{(m+k)2}$
doc_3	...	...	...	...
...	...	...	...	...
...	...	...	...	...
doc_n	c_{1n}	...	...	$c_{(m+k)n}$

Figure 9.5
Two matrices for n documents with m authors and k words can be combined to a third matrix of n documents vs. $(m + k)$ variables.

cognitive development can be considered a latent dimension emerging in networks of texts and people (Leydesdorff, 1998). This thesis of the heterogeneity of the technosciences was first proposed by authors in the semiotic tradition (Callon et al., 1983).

Because the different dynamics at interfaces within and between knowledge-based systems (such as science, technology, and innovation) are documented in texts, the texts can provide us with access to the different dimensions. In SNA, for example, these various dimensions of the data can be mapped as modalities. Another option for mapping hybrid networks was suggested by Leydesdorff (2010). All relevant variables can be attributed to (sets of) documents as units of analysis. The various asymmetrical matrices of n documents versus, for example, k words and m authors can be aggregated as visualized in figure 9.5.

The resulting matrix can be factor-analyzed or—using matrix algebra—transformed into a symmetrical affiliations matrix. In figure 9.6, 33 of the 36 coauthors of these same documents are positioned in a semantic map (as in figure 9.3). (Three other authors were not connected at *cosine* > 0.1.) I added a dashed circle around the coauthorship network of Mike Thelwall as an example. Other variables (e.g., cited references, institutional addresses, country names) can be made equally visible, and colored or sized accordingly.

Animation of the Visualizations

Can the maps for different years (or other time intervals) also be animated? Several network visualization programs are available that enable the user to smooth the transitions based on interpolations among the solutions at different moments in time. The dynamic problem is then reduced to a comparatively static one: the differences among maps for different years are assumed to provide us with a representation of the

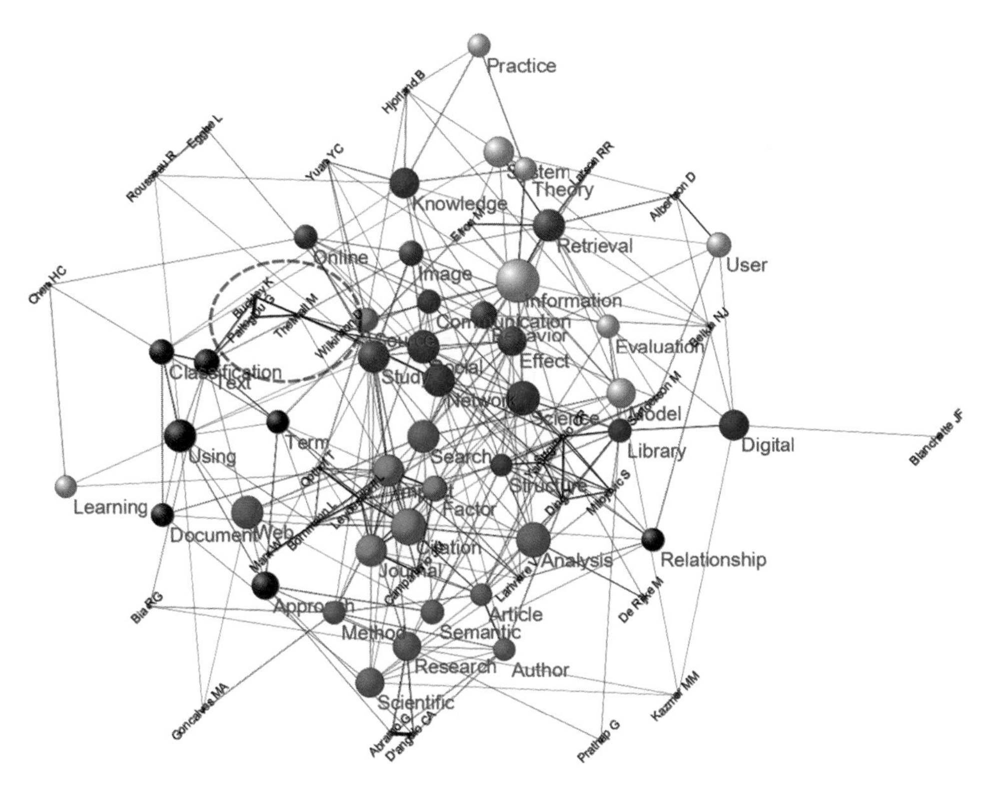

Figure 9.6
43 words (from figure 9.3) and 33 authors related at *cosine* > 0.1.

evolution of the system. However, the solution for each year is already an optimization of a higher-dimensional configuration into the two-dimensional plane. It can thus be difficult to distinguish between the development of the system and error.

An analytic solution of the system of partial differential equations provided by all the changing vectors and eigenvectors is impossible, and a numerical one computationally too intensive. Using MDS, however, Gansner, Koren, and North (2005) proposed minimizing not the stress, but the *majorant* of the stress, as a computationally more effective and methodologically more promising optimization. Baur and Schank (2008) extended this algorithm to layout dynamic networks (cf. Erten, Harding, Kobourov, Wampler, & Yee, 2004). The corresponding dynamic stress function is provided by the following equation:

$$S = \left[\sum_t \sum_{i \neq j} \frac{1}{d_{ij,t}^2} (\|x_{i,t} - x_{j,t}\| - d_{ij,t})^2 \right] + \left[\sum_{1 \leq t < |T|} \sum_i \omega \|x_{i,t} - x_{i,t+1}\|^2 \right] \quad (3)$$

In equation 3, the term on the left is equal to the static stress (in Equation 2), while the term on the right adds the dynamic component, namely the stress over subsequent years. This dynamic extension penalizes drastic movements of the position of node *i* at time *t* ($\vec{x}_{i,t}$) toward its next position ($\vec{x}_{i,t+1}$) by increasing the stress value. Thus, stability is provided in order to preserve the mental map between consecutive layouts (Liu & Stasko, 2010).

In other words, the configuration for each year can be optimized in terms of the stress in relation to the solutions for previous years and in anticipation of the solutions for following years. In principle, the algorithm allows us (and the dynamic version of *Visone*—available at http://www.leydesdorff.net/visone—enables us) to extend this method to more than a single time step. Using a single year in both directions, Leydesdorff and Schank (2008) animated, for example, the aggregated journal-journal citations in "nanotechnology" during the transition of this field at the end of the 1990s.[5]

Note that this approach is different from taking the solution for the previous moment in time as a starting position for a relative optimization. The nodes are not repositioned given a previous configuration, but the previous and the next configurations are included in the algorithmic analysis for each year. More recently, Leydesdorff (2011b) further elaborated this approach by projecting the eigenvectors as constructs among the variables into the animations.[6] Thus, one can make visible not only the evolution of observable variables, but also the evolution of latent structures. In principle, it would be possible to decompose the resulting stress into dynamic and static components.

Conclusion and Future Directions

The relations between semantic maps and social networks have been central to my argument because when visualizing the sciences as bodies of knowledge, the multimodal network of words, authors, etc., has to be specified. Discursive knowledge is communicated, and thus a network visualization is possible in different dimensions. However, knowledge can be considered a latent dimension of meaning processing in a network: discursive knowledge emerges in configurations of words, authors, references, etc., and can then be codified and institutionalized, for example, in journals, specialties, departments, and disciplines. The self-organization of the sciences in latent dimensions conditions and enables the observable relations in networks of authors, words, and citation relations.

The sciences are first shaped by the communicating agents, but textual communications can then develop a dynamic of their own as the communications are further codified by theorizing. The sciences develop as systems of rationalized expectations in this codified dimension. However, the development of ideas leaves footprints in the texts (Fujigaki, 1998). The dynamics of texts and authors are different, and the dynamics of communication are (co)determined by the feedback from emerging knowledge dimensions. In figure 9.2, for example, the knowledge dimension was operationalized as three groups of journals belonging to different specialties.

The visualization of the sciences as a research program thus requires distinguishing among semantic maps, social networks, and the latent sociocognitive structures that can emerge on the basis of the interactions among people and texts. Three layers (people, texts, cognitions) coevolve in terms of observable variables and latent eigenvectors. Because of the next-order organization, the variables can be expected to interact among themselves and to shape and reproduce structures that can both recur on previous states and anticipate further developments of the system(s) (Luhmann, 1995; Maturana, 1978).

Visualization and animation of the sciences constitute an active research front in the development of the information sciences and bibliometrics. In the future, animations using multiple perspectives can be expected to replace models of multivariate analysis in which independent factors explain the data. Configurations of variables generate different synergies (Leydesdorff, Rotolo, & De Nooy, 2013). These implications follow from considering not only the communication of information, but also its meaning (Krippendorff, 2009; Leydesdorff, 2010). Horizons of meaning can be expected to generate redundancy—that is, new and more possibilities that change the value of existing ones.[7]

Animations enable us to capture different perspectives analogously as visualizations capture different arrangements of variables. The development of animations in the coupled layers of information and meaning processing can be expected to raise new questions for the further development of bibliometrics, network analysis, statistics, and relevant neighboring specialties.

Acknowledgments

I am grateful to Katy Börner for comments on a previous version, and to Thomson Reuters for access to relevant data.

Notes

1. MINISSA is an acronym for "Michigan-Israel-Nijmegen Integrated Smallest Space Analysis"; it became available around 1980 (Schiffman, Reynolds, & Young, 1981).

2. Pajek is a freeware program for network visualization and analysis available at http://vlado.fmf.uni-lj.si/pub/networks/pajek.

3. Gephi is an open-source program for network analysis and visualization, available at https://gephi.org.

4. Three factors explain 49.2% of the variance in this matrix.

5. Available at http://www.leydesdorff.net/journals/nanotech.

6. See http://www.leydesdorff.net/eigenvectors/commstudies.

7. The mutual information in three dimensions (μ^*; cf. Yeung, 2008, pp. 59–60) among the three main factors structuring the coword network (figure 9.3) is –122.2 mbits, whereas this redundancy virtually disappears when the 33 coauthors are added to the network: $\mu^* = -7.0$ mbit (figure 9.6). For the social network among the 36 coauthors, this value of μ is positive. In other words, the coauthor network itself does not communicate meaning in this case (Leydesdorff, 2010, 2011b; Leydesdorff & Ivanova, in press).

References

Ahlgren, P., Jarneving, B., & Rousseau, R. (2003). Requirement for a cocitation similarity measure, with special reference to Pearson's correlation coefficient. *Journal of the American Society for Information Science and Technology, 54*(6), 550–560.

Baur, M., & Schank, T. (2008). *Dynamic graph drawing in Visone*. Technical University Karlsruhe, Karlsruhe. Retrieved from http://i11www.iti.uni-karlsruhe.de/extra/publications/bs-dgdv-08.pdf.

Blondel, V. D., Guillaume, J. L., Lambiotte, R., & Lefebvre, E. (2008). Fast unfolding of communities in large networks. *Journal of Statistical Mechanics, 8*(10), 10008.

Börner, K. (2010). *Atlas of science: Visualizing what we know*. Cambridge, MA: MIT Press.

Börner, K., Chen, C., & Boyack, K. W. (2003). Visualizing knowledge domains. *Annual Review of Information Science & Technology, 37*(1), 179–255.

Burt, R. S. (1992). *Structural holes: The social structure of competition*. Cambridge, MA: Harvard University Press.

Callon, M., Courtial, J.-P., Turner, W. A., & Bauin, S. (1983). From translations to problematic networks: An introduction to co-word analysis. *Social Sciences Information. Information Sur les Sciences Sociales, 22*(2), 191–235.

De Nooy, W., Mrvar, A., & Batagelj, V. (2005). *Exploratory social network analysis with Pajek*. New York: Cambridge University Press.

Doreian, P., & Fararo, T. J. (1985). Structural equivalence in a journal network. *Journal of the American Society for Information Science, 36*, 28–37.

Erten, C., Harding, P. J., Kobourov, S. G., Wampler, K., & Yee, G. V. (2004). GraphAEL: Graph animations with evolving layouts. In G. Liotta (Ed.), *Graph drawing. Lecture Notes in Computer Science 2912*, 96–110. Berlin: Springer Verlag.

Freeman, L. C. (2004). *The development of social network analysis: A study in the sociology of science.* North Charleston, SC: BookSurge.

Fujigaki, Y. (1998). Filling the gap between discussions on science and scientists' everyday activities: Applying the autopoiesis system theory to scientific knowledge. *Social Sciences Information. Information Sur les Sciences Sociales, 37*(1), 5–22.

Gansner, E. R., Koren, Y., & North, S. (2005). Graph drawing by stress majorization. In J. Pach (Ed.), *Graph drawing Lecture notes in computer science* (pp. 239–250). Berlin: Springer.

Garfield, E. (1987). Launching the *ISI Atlas of Science*: For the new year, a new generation of reviews. *Current Contents,* (1), 3–8.

Granovetter, M. S. (1973). The strength of weak ties. *American Journal of Sociology, 78*(6), 1360–1380.

Husserl, E. (1973). *Cartesianische Meditationen und Pariser Vorträge* [Cartesian meditations and the Paris lectures]. The Hague: Martinus Nijhoff. (Original work published 1929)

Kamada, T., & Kawai, S. (1989). An algorithm for drawing general undirected graphs. *Information Processing Letters, 31*(1), 7–15.

Krippendorff, K. (2009). Information of interactions in complex systems. *International Journal of General Systems, 38*(6), 669–680.

Kruskal, J. B. (1964). Multidimensional scaling by optimizing goodness of fit to a nonmetric hypothesis. *Psychometrika, 29*(1), 1–27.

Kruskal, J. B., & Wish, M. (1978). *Multidimensional scaling.* Beverly Hills, CA: Sage.

Landauer, T. K., Foltz, P. W., & Laham, D. (1998). An introduction to latent semantic analysis. *Discourse Processes, 25*(2), 259–284.

Leydesdorff, L. (1986). The development of frames of references. *Scientometrics, 9*(3–4), 103–125.

Leydesdorff, L. (1987). Various methods for the mapping of science. *Scientometrics, 11*, 291–320.

Leydesdorff, L. (1998). Theories of citation? *Scientometrics, 43*(1), 5–25.

Leydesdorff, L. (2010). Redundancy in systems which entertain a model of themselves: Interaction information and the self-organization of anticipation. *Entropy, 12*(1), 63–79.

Leydesdorff, L. (2011a). "Meaning" as a sociological concept: A review of the modeling, mapping, and simulation of the communication of knowledge and meaning. *Social Sciences Information. Information sur les sciences sociales, 50*(3–4), 1–23.

Leydesdorff, L. (2011b). "Structuration" by intellectual organization: The configuration of knowledge in relations among scientific texts. *Scientometrics, 88*(2), 499–520.

Leydesdorff, L., & Ivanova, I. A. (in press). Mutual redundancies in inter-human communication systems: Steps towards a calculus of processing meaning. *Journal of the American Society for Information Science and Technology*.

Leydesdorff, L., & Jin, B. (2005). Mapping the Chinese Science Citation Database in terms of aggregated journal-journal citation relations. *Journal of the American Society for Information Science and Technology, 56*(14), 1469–1479.

Leydesdorff, L., & Rafols, I. (2011). Indicators of the interdisciplinarity of journals: Diversity, centrality, and citations. *Journal of Informetrics, 5*(1), 87–100.

Leydesdorff, L., Rotolo, D., & De Nooy, W. (2013). Innovation as a nonlinear process, the scientometric perspective, and the specification of an "innovation opportunities explorer." *Technology Analysis and Strategic Management, 25*(6) 641–653.

Leydesdorff, L., & Schank, T. (2008). Dynamic animations of journal maps: Indicators of structural change and interdisciplinary developments. *Journal of the American Society for Information Science and Technology, 59*(11), 1810–1818.

Leydesdorff, L., & Vaughan, L. (2006). Co-occurrence matrices and their applications in information science: Extending ACA to the web environment. *Journal of the American Society for Information Science and Technology, 57*(12), 1616–1628.

Liu, Z., & Stasko, J. T. (2010). Mental models, visual reasoning and interaction in information visualization: A top-down perspective. *IEEE Transactions on Visualization and Computer Graphics, 16*(6), 999–1008.

Lucio-Arias, D., & Leydesdorff, L. (2009). The dynamics of exchanges and references among scientific texts, and the autopoiesis of discursive knowledge. *Journal of Informetrics, 3*(2), 261–271.

Luhmann, N. (1995). *Social systems*. Stanford, CA: Stanford University Press.

Maturana, H. R. (1978). Biology of language: The epistemology of reality. In G. A. Miller & E. Lenneberg (Eds.), *Psychology and biology of language and thought: Essays in honor of Eric Lenneberg* (pp. 27–63). New York: Academic Press.

McCain, K. W. (1990). Mapping authors in intellectual space: A technical overview. *Journal of the American Society for Information Science, 41*(6), 433–443.

Misue, K., Eades, P., Lai, W., & Sugiyama, K. (1995). Layout adjustment and the mental map. *Journal of Visual Languages and Computing, 6*(2), 183–210.

Salton, G., & McGill, M. J. (1983). *Introduction to modern information retrieval*. New York: McGraw-Hill.

Schiffman, S. S., Reynolds, M. L., & Young, F. W. (1981). *Introduction to multidimensional scaling: Theory, methods, and applications*. New York: Academic Press.

Shannon, C. E. (1948). A mathematical theory of communication. *Bell System Technical Journal, 27*, 379–423, 623–656.

Small, H., & Garfield, E. (1985). The geography of science: Disciplinary and national mappings. *Journal of Information Science, 11*(4), 147–159.

Small, H., & Greenlee, E. (1986). Collagen research in the 1970s. *Scientometrics, 19*(1–2), 95–117.

Small, H., & Sweeney, E. (1985). Clustering the Science Citation Index using co-citations I. A comparison of methods. *Scientometrics, 7*(3–6), 391–409.

SPSS. (1993). *SPSS professional statistics 6.1*. Chicago: SPSS.

Tijssen, R., De Leeuw, J., & Van Raan, A. F. J. (1987). Quasi-correspondence analysis on square scientometric transaction matrices. *Scientometrics, 11*(5–6), 347–361.

White, H. D., & Griffith, B. C. (1982). Authors as markers of intellectual space: Co-citation in studies of science, technology and society. *Journal of Documentation, 38*(4), 255–272.

White, H. D., & McCain, K. (1998). Visualizing a discipline: An author cocitation analysis of information science, 1972–1995. *Journal of the American Society for Information Science, 49*(4), 327–355.

Yeung, R. W. (2008). *Information theory and network coding*. New York: Springer.

10 Measuring Interdisciplinarity

Vincent Larivière and Yves Gingras

Introduction

As other chapters in this book show, bibliometric indicators have been used to measure many aspects of the dynamics of science. Every time a new question has emerged about the changing practices of scientific research, indicators have been designed to try to answer it with empirical data. When the question of the extent of the internationalization of science became a topic of interest to governments and university managers in the 1990s, one could provide an indicator of international collaboration by looking at the presence of different countries in the address field of scientific publications and analyze the evolution of its proportion over time. An even more recent issue for decision and opinion makers in higher education and research is the extent of interdisciplinary research and its supposed necessity in a world in which problems are complex and multifaceted. To go beyond buzzwords and performative discourses that say interdisciplinarity is inevitable and so should be widely adopted (Gibbons et al., 1994; Nowotny, Scott, & Gibbons, 2001), bibliometricians have proposed different ways it could be operationalized, thus contributing to the recent wave of interest in the nature and growth of interdisciplinary research (Weingart & Stehr, 2000; Frodeman, Thompson Klein, & Mitcham, 2010). Despite the fact that this idea has been repeated and promoted by many university managers and higher education "gurus," solid data confirming the supposed trend toward increased interdisciplinarity are hard to find. In an effort to go beyond vague references to "inter," "multi," or even "trans" disciplinarity, bibliometric indicators have focused on measures of interactions between disciplines and specialties. A bibliometric approach to the contested issue of interdisciplinarity creates a unique opportunity to analyze trends over an extended period, and to test assertions about its increased role in different scientific disciplines (Gibbons et al., 1994; Hessels & Van Lente, 2008).

The most complete bibliometric study of interdisciplinarity is that of Porter and Rafols (2009), who analyzed its evolution in six research areas over a 30-year period. Although they found an increase in interdisciplinarity, it was quite small (about 5%). Another recent study by Levitt, Thelwall, and Oppenheim (2011) analyzed the evolution of interdisciplinarity in the social sciences using 14 Social Sciences Citation Index (SSCI) categories for three specific years: 1980, 1990, and 2000. They showed that the median level of interdisciplinarity of these fields had *decreased* between 1980 and 1990, but then climbed back in 2000 to its 1980 level. Van Leeuwen and Tijssen (2000) analyzed changes in specialties' level of interdisciplinarity between 1985 and 1995 and found that very few disciplines displayed significant changes during that time.

Other studies on the evolution of interdisciplinarity either focused on one discipline over a few decades (Tomov & Mutafov, 1996; Rinia, Van Leeuwen, and Van Raan, 2002b; Rafols & Meyer, 2007) or used a few years of data for many disciplines (Adams, Jackson, & Marshall, 2007). Recent work in the visualization of science has provided a global view of the relationships between scientific disciplines (Börner, Chen, & Boyack, 2003; Boyack, Klavans, & Börner, 2005). Although they shed some interesting light on the changing relations between disciplines and specialties, none of these studies provides a complete overview of changes over a long historical period. Using macrolevel data for the period 1900–2010, this chapter provides the first historical overview of the relationships between all scientific disciplines in the natural and medical sciences as well as in the social sciences and humanities.

After a brief discussion of the necessary distinctions between the discourse *on* and the practice *of* interdisciplinarity, we review the relevant literature on the operationalization of the concept of interdisciplinarity and describe the specific methods used in this chapter. The third section presents the results obtained for each of the disciplines, while the final section discusses the results.

Distinguishing Discourses from Practices

Before analyzing the *practice* of interdisciplinarity, as defined through bibliometrics, it is worth looking at the evolution of interest in this topic over the course of the 20th century. Andrew Abbott (2001) has suggested that interest in interdisciplinarity has in fact been almost stable over time. To support this surprising assertion, he calculated the ratio of the number of items from the SSCI having the word *interdisciplinary* in their title to the number of papers using the word *national* in their title. Because that ratio has been quite stable (around 0.07 or 0.08), he concluded that, contrary to what most people think, no real upsurge of interest in interdisciplinarity has occurred. Given our intuition that discourses on interdisciplinarity were much in vogue in the 1960s and again in the 1990s, there may be reasons to doubt the validity of the rough indicator used by Abbott to support his claim. Indeed, why use a ratio with *national* and not with *race, gender,* or any other term? An increase in interdisciplinarity could be hidden behind a parallel increase in "national" research. It seems obvious that the best way to measure the relative interest in interdisciplinarity is to compare the place of interdisciplinarity in titles across all papers in the SSCI and AHCI (as well as SCI for comparison)—which is a stable base—instead of comparing it with the proportion of papers that use other terms and that are likely to fluctuate from one year to another.

As figure 10.1 shows, discourse on interdisciplinarity in the social sciences and humanities (SSH) emerged after the Second World War but only became fashionable from the mid-1960s to the end of the 1970s. A second wave of interest began in the 1990s, stabilized at the end of the decade, and then increased again at a very fast pace after 2006. As might be expected, there is much less in the natural sciences and engineering (NSE) where titles refer to content and object studied and rarely to methods.[1] Fluctuating interest in the topic of interdisciplinarity underscores the importance of distinguishing between the discourses *on* and rhetoric *of* interdisciplinarity from its actual practice, which rarely commands the use of the word itself. The subject is

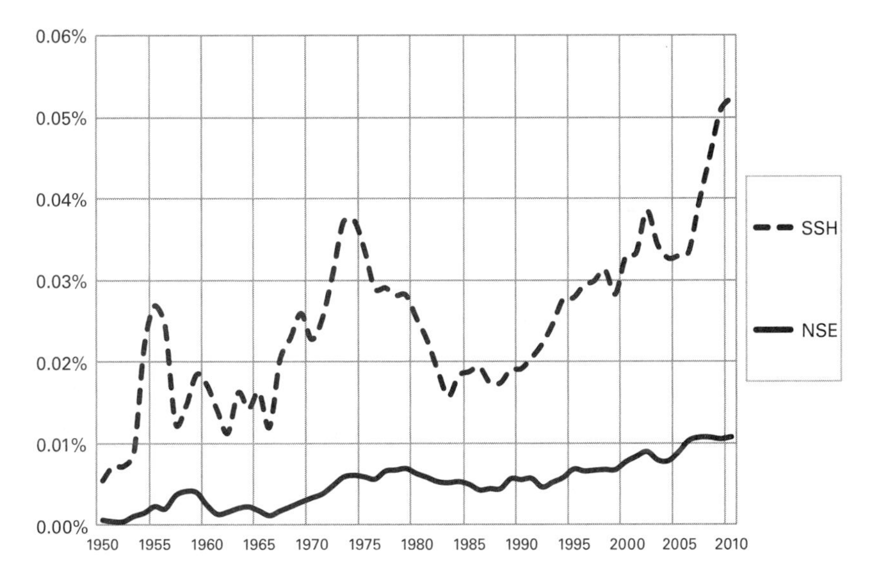

Figure 10.1
Percentage of Web of Science papers with "Interdisciplinar*" in their titles, by domain, 1950–2010. Three-year moving averages.

typically addressed in editorials in general-interest science periodicals or in papers published in social science journals that consider bringing disciplines together for the purpose of producing new knowledge on a given object of study, be it social or natural. In the latter case, one does not expect to find explicit mention of *interdisciplinarity*, a term more suited for use in papers discussing epistemology or methodology. For example, a paper in *Science* in 1944 discussed "General Aspects of Interdisciplinary Research in Experimental Human Biology" (Brozek & Keys, 1944). In 1948, the Harvard economist Wassily Leontief published a paper in the *Journal of Philosophy* titled "Note on the Pluralistic Interpretation of History and the Problem of Interdisciplinary Cooperation" (Leontief, 1948). In 1952, the presidential address of Dorothy Swaine Thomas at the annual meeting of the American Sociological Society concerned "Experiences in Interdisciplinary Research" (Thomas, 1952). Many others in the 1950s suggested a "framework" or "programs" for interdisciplinary research, while some raised "problems," "obstacles," and "challenges" associated with interdisciplinarity in the 2000s. The rhetoric thus seems more recurrent than constant, coming back every 20 to 25 years, but it is not our intention to trace in detail the historical shifts in the rhetoric of interdisciplinarity. We focus instead on measuring the evolution of interdisciplinarity on the basis of various indicators and see if the results converge to show a tendency toward greater interaction between disciplines (as suggested by many), a stable situation, or a cyclic pattern typical of fads.

Background and Methods

From a bibliometric point of view, the concept of interdisciplinarity is generally operationalized on the basis of authors' disciplinary affiliations (departments), using the references papers contain or the citations they receive (Wagner et al., 2011). Most studies, such as Adams et al. (2007), Tomov and Mutafov (1996), and Morillo, Bordons, and Gómez (2001), follow Porter and Chubin (1985) and measure, for a set of papers (or journals), the percentage of citations made by the papers (or journals) outside their discipline or specialty (which they label as the *Citations Outside Category*). On the other hand, Rinia, Van Leeuwen, Bruins, Van Vuren, and Van Raan (2001, 2002a) and Rinia, Van Leeuwen, and Van Raan (2002b) define interdisciplinarity as the percentage of papers from a group of researchers published outside their "main" discipline. Others, like Levitt and Thelwall (2008), operationalize the concept using articles published in journals that are classified in more than one field by Thomson Reuters' Web of Science or by Elsevier's Scopus. However simple this approach may sound, it is doubtful that it captures interdisciplinarity since the fact that a journal is attributed to more than one discipline does not necessarily imply that the papers published in that journal are actually "interdisciplinary." Such a journal could, in fact, be publishing papers from different disciplines, with very little interaction between them, as is the case with multidisciplinary journals such as *Science* and *Nature*. Finally, using the researcher as a unit of analysis, Le Pair (1980) constructed a different indicator, based on the migration of scientists from one discipline to another throughout the course of their careers. Though interesting, this indicator is quite difficult to compile because of the lack of systematic data.

The analysis presented here uses the references contained in papers as a basis for constructing indicators of interdisciplinarity and interspecialty (within a given discipline). It is based on Thomson Scientific's databases, which are the only ones covering over a century of both papers and references. For the 1900–1944 period, data are drawn from the Century of Science, which indexes 266 distinct journal titles covering most natural sciences and medical fields. For the social sciences, data between 1900 and 1956 come from the Century of Social Sciences Index, which indexes 308 journals from these disciplines. From 1945 to 2010, data are from the Web of Science (WoS), which includes the Science Citations Index Expanded, the Social Sciences Citation Index, and the Arts and Humanities Citation Index. The disciplinary classification of journals used in this paper is that of the U.S. National Science Foundation (NSF). This classification categorizes each journal into one discipline and one specialty. For the social sciences and humanities, the NSF categorization was completed with our own classification. The classification defines 14 disciplines divided into 143 specialties. For the sake of graphical representation, these 14 disciplines have been regrouped into 4 larger domains: medical fields (MED), natural sciences and engineering (NSE), social sciences (SS), and arts and humanities (A&H).

For each document indexed in Thomson's databases (source items), a list of references is included. Following Porter and Chubin (1985), we measure the degree of interdisciplinarity of a given paper using the relationship between the discipline of that paper and those of its cited documents. Two dimensions of interdisciplinarity were measured: the interdisciplinarity of references made and the interdisciplinarity of citations received. However, given that the tendencies observed were almost identical for the two measures at this level of aggregation—they are two sides of the same coin—the interdisciplinarity measures presented here are only those based on the interdisciplinarity of references made.

Following Rinia's (2007) typology, we calculated two types of interdisciplinarity: (a) references made to journals classified in a discipline different from that of the paper, and (b) references made to journals classified in the same discipline but in a specialty different from that of the paper. We call the first measure "interdisciplinarity" since it measures the link with other disciplines (e.g., links between biology and physics) and the second "interspecialty" since it measures the relationships between different specialties within a given discipline (e.g., links between optics and nuclear physics). More specifically, the measure of interdisciplinarity presented in the figures is the percentage of references made to papers published in journals categorized into a specialty of another discipline and the measure of interspecialty is the percent of references made to another specialty of the same discipline. The other measure presented is the percentage of references made to journals of the same specialty. For example, an article published in a particle physics journal that includes 12 references to papers published in journals from the same specialty, 8 to journals in other specialties (optics, nuclear physics, etc.) of the same discipline (physics), and 10 to journals of specialties of other disciplines—for a total of 30 references—will obtain an interdisciplinarity score of 33.3% (10/30), an interspecialty index of 26.7% (8/30), and a same-specialty index of 40.0% (12/30).

Our method of measuring interdisciplinarity has its limitations. First, despite important changes in the structure of scientific disciplines during the last century, our list of disciplines remains the same throughout the period studied. However, given that very few fields have ceased to exist and that many new fields have emerged, using today's disciplinary and specialty categories should not cause significant anachronisms. For example, the *American Journal of Surgery* is categorized in the specialty of surgery throughout the period, and has indeed always published papers related to surgery. On the other hand, there are no papers in the field of cancer or computer science before, respectively, the 1940s and the 1950s. As a consequence, the references they cited during the first couple of years of their existence can only come from outside their specialty. Similarly, journals that change scope generally change their name and, hence, are reclassified accordingly.

Another limitation is that we have discipline and specialty information only for references made to articles published in journals that are also indexed in the Web of Science (source items). Hence, a more or less significant percentage—depending on the discipline and the publication year—of the cited literature is excluded from the analysis and this proportion changes over time. In all disciplines combined (figure 10.2C), we see that, in 2010, about 70% of the references were made to source items. These results are similar to those obtained by Larivière et al. (2006), which showed that serials' share of cited literature globally increased steadily since the early 1980s. However, at the beginning of the period, a lower share of the references was made to WoS-covered material. This is normal, given the fact that older papers cite a larger proportion of pre-1900 papers that are not indexed in the WoS. Also, the fact that the WoS indexes a smaller number of journals and papers for the period 1900–1945 lessens the chance that references made by articles indexed are indeed made to other source items. One might argue that these references to non–source items would reveal a different trend than the one observed here. However, we also analyzed the trends using citations received, which by definition has 100% coverage, and the overall trends are almost identical, a finding that supports our argument that the results obtained for the references made to source items are a representative sample of the whole population of references. The different panels of figure 10.2 also show that the number of papers (A) and number of cited references (B) have been increasing steadily in all domains except in A&H. Using the CD-ROM version of the SCI, Larivière, Archambault, and Gingras (2008) argued that the rate of exponential growth of publications declined in the 1970s; figure 10.2A shows that this trend is also valid for the expanded version of

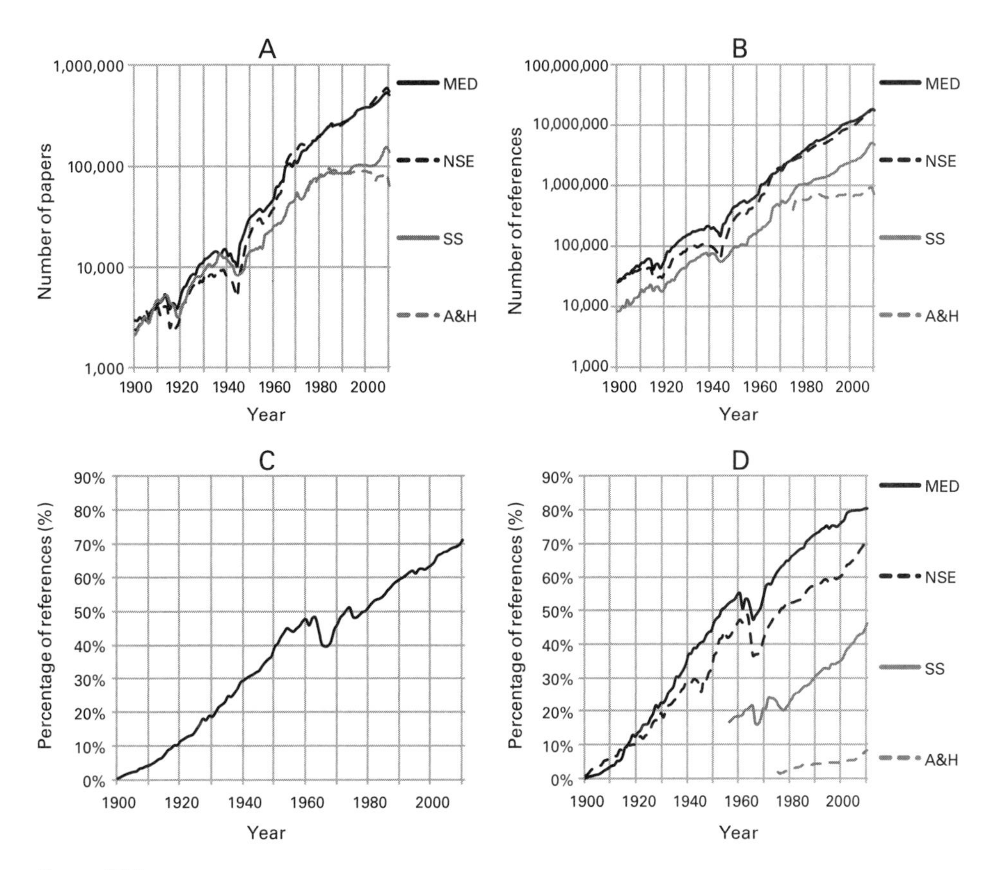

Figure 10.2
Yearly number of papers (A), references (B), and percentage of references made to source items, all domains combined (C), and by domain (D), 1900–2010.

the SCI, which confirms that these global trends do not heavily depend on the sample used. The results presented here are based on 768 million references made by about 35 million papers. Out of these 768 million references, about 470 million were made to these 35 million source items, meaning that 61% of references were covered.

Rinia et al. (2001) showed that there is a delay in interdisciplinary knowledge exchange such that one has to wait a couple of years for interdisciplinary citations to accumulate. That result is interesting because it suggests that it takes more time for discoveries to permeate disciplinary boundaries and be cited in other disciplines and specialties than is the case for knowledge circulating within the discipline or specialty. The data presented in figure 10.3 show that the same phenomenon occurs in all domains, albeit at a lower level in MED—which could be expected given the shorter half-life of papers in these disciplines (Larivière et al., 2008)—and that the percentage of references made to papers published in journals outside the discipline or specialty of the citing document rises steadily as one increases the citation window to include

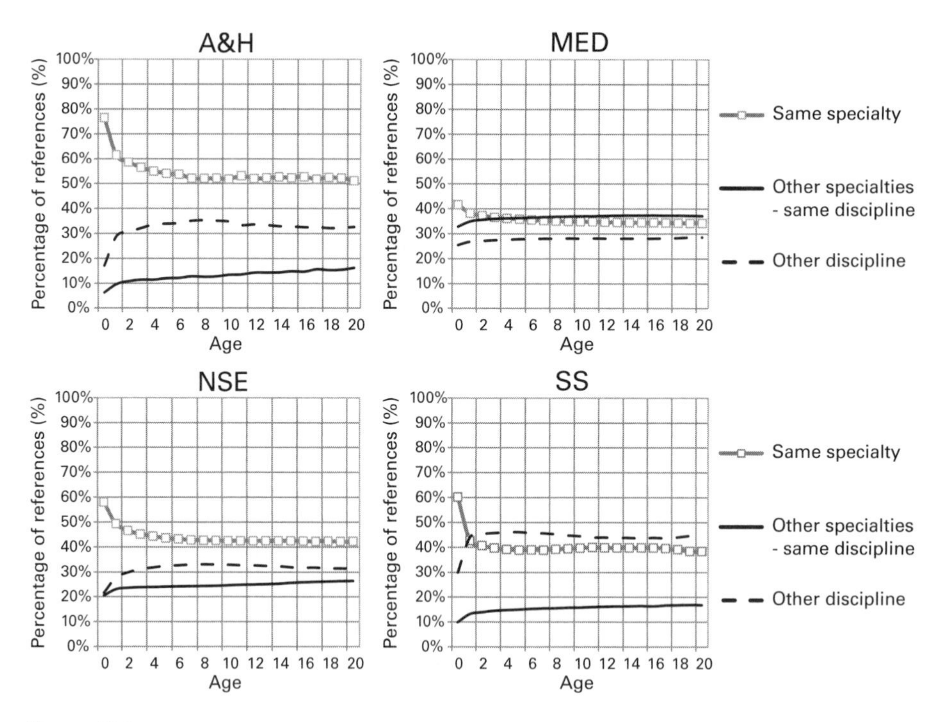

Figure 10.3
Percentage of references to a different discipline, to the same specialty, and to other specialties of the same discipline, by age of cited paper and domain, 1998–2007.

older documents. On the other hand, references to more recent material are more often being made to papers published in the same specialty. On the whole, this figure shows that, in order to have a good measure of interdisciplinarity, one cannot limit the analysis to references made to papers published during the two previous years. For that reason, our measures of interdisciplinarity use citing years for which at least five years of reference data exist. In other words, because cited papers published before 1900 are not source items and therefore do not have a field associated with them and because we used a five-year citation window, data for MED, NSE, and SS start in 1905, and data for A&H start in 1980.

Varying Relations between Disciplines and Specialties

Figure 10.4 presents, for each of the four broad fields of science, three measures of the relations between disciplines and specialties, defined at the level of the citing paper:

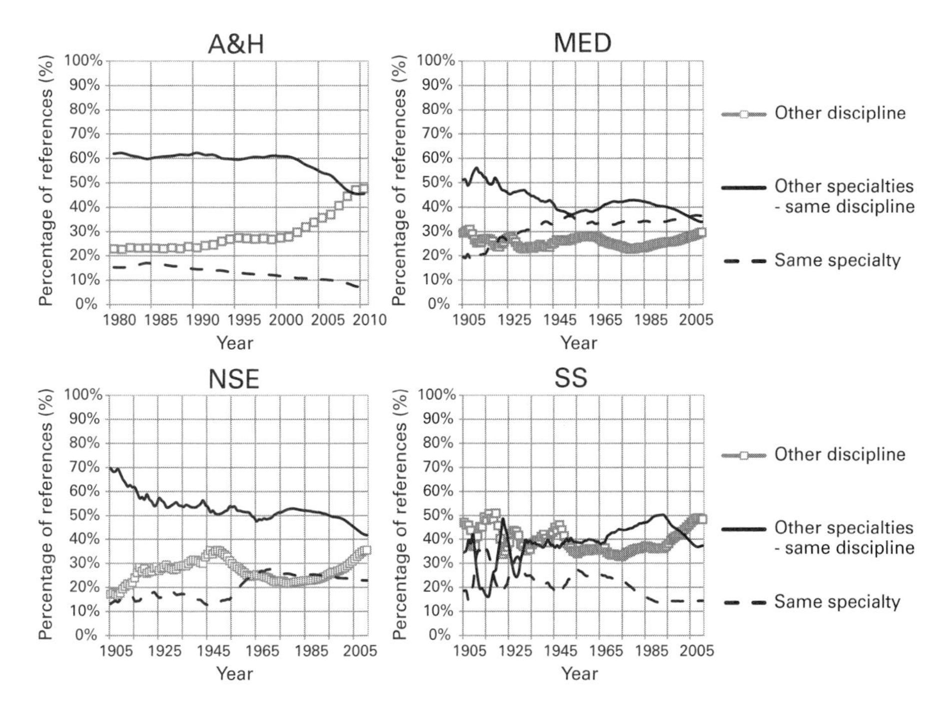

Figure 10.4
Percentage of references made to papers from other disciplines, to other specialties of the same discipline, and to the same specialty, by domains, 1900–2010. Three-year moving averages.

1. The percentage of references made to papers outside the discipline of the citing paper, which provides an indicator of interdisciplinarity

2. The percentage of references made to papers from other specialties but in the same discipline, which provides a measure of the relations between specialties within a given discipline (interspecialty)

3. The percentage of references made to papers from the same specialties, which measures the internal focus of the specialty

For NSE and MED, three broad periods can be distinguished: (a) 1900–1945, when a decline in the degree of specialization accompanies a rise in interdisciplinarity for NSE and an increase in interspecialty for MED; (b) 1945 to about 1980, characterized in MED by a decline in interdisciplinarity and a growing emphasis on specialties, which tend to refer more to themselves, while the interactions between specialties are stable; in NSE, the same period also sees a decline in interdisciplinarity accompanied by a rise in references to other discipline specialties between 1945 and 1965, followed by a period of stability up to the mid-1980s; and (c) mid-1980s to 2010, where the links between disciplines increase again at the expense of the internal focus of the specialties, while the proportion of references made to other specialties of the same discipline remains stable.

As could be expected, the social sciences and humanities follow a quite different pattern. In the social sciences, the level of specialization, measured by the proportion of intraspecialty references, remains stable between 1935 and 1965 (discounting the strong fluctuations in the data before the 1930s) and then increases until the mid-1990s to about 50%, to drop again to just below 40% in the face of growing interdisciplinarity. By 2010, about 50% of the references were to disciplines others than that of the paper, while interspecialty was at its lowest point (35%). This means that after the mid-1990s, a paper in a given specialty is more open to other disciplines than to other specialties of its own discipline. Though not shown, the evolution of the interdisciplinarity of *citations received* follows the same pattern.

The trends in the humanities are much simpler: we observe a surge in interdisciplinary references around 2000, at the expense of intraspecialty references. Before that period, there is a quite stable practice and about 60% of the references are to papers from the same specialty, with a slow but continuous decline of references to other specialties in the same discipline. It is worth mentioning that, although SSH researchers discuss the notion of interdisciplinarity five times more often than their NSE colleagues (figure 10.4), both disciplines have similar levels of interdisciplinarity (around 25% to 30% before the mid-1990s) when measured in terms of their referencing practices.

Discussion and Conclusion

Over the course of a century, we observe that for the NSE as well as for the MED disciplines, the percentage of references made to the literature published in journals categorized in the same specialty has been decreasing, from 70% to 40% in NSE and from 50% to about 35% in MED. This decline in the internal focus of specialties has taken two paths: in the NSE it first corresponded to a rise in interdisciplinarity until 1945, while for the same period the MED disciplines saw a rise in interspecialty relations. This latter trend makes sense as an effect of the creation of the various medical specialties. In the 30-year period following the end of the Second World War, we observe a decline in interdisciplinarity in NSE accompanied by a greater focus on the specialties of the discipline, which again is consistent with the multiplication of specialties in most disciplines during that period. In MED, we also observe a slight decline in interdisciplinarity until the 1980s and a small rise in interspecialty. From about the mid-1980s, both NSE and MED raised their level of interdisciplinarity at the expense of a focus on other specialties of the same discipline.

Taken globally, these results suggest that, while specialties within disciplines multiplied during the first two-thirds of the 20th century and then maintained a certain level of stability, exchanges between different disciplines started to increase again thereafter, particularly after the mid-1980s. By the end of the century the referencing practices in MED disciplines were roughly equally distributed among the specialty of the paper, other specialties of the same discipline, and other disciplines, with, in general and for most of the century, intraspecialty references tending to dominate. As figure 10.4 shows, the social sciences as well as the arts and humanities are most open to other disciplines, while the MED disciplines tend to stay within themselves while being receptive to other specialties of the same discipline. For their part, NSE papers are mostly focused on their own discipline, something that can be related to the use of highly specialized instrumentation in narrow research domains.

It is to be noted that, though interdisciplinarity has risen significantly since the mid-1980s, the level attained by the end of the century was not much higher than what it was in the 1930s and never declined below 20% even in SNE and MED. Though a more detailed historical analysis would be needed to confirm our hypothesis, everything suggests that the growth in the availability of research money during the "golden" years from 1945 to 1975 brought a stop to interdisciplinarity and a turn toward the growth of disciplines through the multiplication of specialties. Additionally, the new wave of interdisciplinarity visible in the data from the mid-1980s could be the effect of specific government programs and discourses promoting interdisciplinarity as a good thing in itself. Scientists have been urged to collaborate with colleagues from different disciplines and be attentive to the kind of knowledge created outside their own specialty in order to solve "complex" problems. This new influx of money, with *interdisciplinary* and other collaborative strings attached, also seems to be instrumental in making researchers more open to neighboring disciplines.

The changing relations between disciplines and specialties are obviously complex and can be affected as much by the internal development of new concepts or instruments as by monetary pressures. Whereas specialties tend to emerge from the internal dynamics of disciplines (Mullins, 1972), the recent drive toward greater integration through interdisciplinarity seems to depend more on discourse and policies than on internal forces. Whatever the case, and even taking into account the inherent limitation of bibliometric indicators, our analysis shows that the process of disciplinarization and specialization is complex and that interdisciplinarity is itself influenced by the complex internal dynamics of knowledge growth.

As discussed in the chapters by Bornmann and colleagues (chapter 11) and Gingras (chapter 6), there is a need for standard bibliometric indicators in research evaluation, and our analysis of interdisciplinarity sheds light on another technical aspect of their use in such contexts. The observed reconfiguration of the relationships between disciplines and specialties has methodological consequences for the measurement of the scientific impact of papers. As papers increasingly refer to different specialties having different citation practices and are cited by papers also coming from different disciplines, the usual normalization of citations, based only on the discipline of the journal in which the paper appears, is increasingly biased and should be replaced by new kinds of normalization that take into account the mix of disciplines and specialties present in the references as well as in the citations (Zitt & Small, 2008; Moed, 2010; Zitt, 2010).

Note

1. A similar trend is obtained using the JSTOR database.

References

Abbott, A. (2001). *Chaos of disciplines*. Chicago: University of Chicago Press.

Adams, J., Jackson, L., & Marshall, S. (2007). *Bibliometric analysis of interdisciplinary research*. Report to the Higher Education Funding Council for England. Leeds, UK: Evidence.

Börner, K., Chen, C., & Boyack, K. W. (2003). Visualizing knowledge domains. *Annual Review of Information Science & Technology*, *37*, 179–255.

Boyack, K. W., Klavans, R., & Börner, K. (2005). Mapping the backbone of science. *Scientometrics, 64*(3), 351–374.

Brozek, J., & Keys, A. (1944). General aspects of interdisciplinary research in experimental human biology. *Science, 100*(2606), 507–512.

Frodeman, R., Thompson Klein, J., & Mitcham, C. (Eds.). (2010). *The Oxford handbook of interdisciplinarity*. Oxford: Oxford University Press.

Gibbons, M., Limoges, C., Nowotny, H., Schwartzman, S., Scott, P., & Trow, M. (1994). *The new production of knowledge: The dynamics of science and research in contemporary societies*. London: Sage.

Hessels, L. K., & Van Lente, H. (2008). Re-thinking new knowledge production: a literature review and a research agenda. *Research Policy, 37*(4), 740–760.

Larivière, V., Archambault, É., & Gingras, Y. (2008). Long-term variations in the aging of scientific literature: From exponential growth to steady-state science (1900–2004). *Journal of the American Society for Information Science and Technology, 59*(2), 288–296.

Larivière, V., Archambault, É., Gingras, Y., & Vignola-Gagné, É. (2006). The place of serials in referencing practices: Comparing natural sciences and engineering with social sciences and humanities. *Journal of the American Society for Information Science and Technology, 57*(8), 997–1004.

Leontief, W. (1948). Note on the pluralistic interpretation of history and the problem of interdisciplinary cooperation. *Journal of Philosophy, 45*(23), 617–624.

Le Pair, C. (1980). Switching between academic disciplines in universities in the Netherlands. *Scientometrics, 2*(3), 177–191.

Levitt, J. M., & Thelwall, M. (2008). Is multidisciplinary research more highly cited? A macrolevel study. *Journal of the American Society for Information Science and Technology, 59*(12), 1973–1984.

Levitt, J. M., Thelwall, M., & Oppenheim, C. (2011). Variations between subjects in the extent to which the social sciences have become more interdisciplinary. *Journal of the American Society for Information Science and Technology, 62*(6), 1118–1129.

Moed, H. F. (2010). Measuring contextual citation impact of scientific journals. *Journal of Informetrics, 4*(3), 265–277.

Morillo, F., Bordons, M., & Gómez, I. (2001). An approach to interdisciplinarity through bibliometric indicators. *Scientometrics, 51*(1), 203–222.

Mullins, N. (1972). The development of a scientific specialty: The Phage Group and the origins of molecular biology. *Minerva, 10*, 51–82.

Nowotny, H., Scott, P., & Gibbons, M. (2001). *Re-thinking science: Knowledge and the public in an age of uncertainty*. London: Polity Press.

Porter, A. L., & Chubin, D. E. (1985). An indicator of cross-disciplinary research. *Scientometrics, 8*(3–4), 161–176.

Porter, A. L., & Rafols, I. (2009). Is science becoming more interdisciplinary? Measuring and mapping six research fields over time. *Scientometrics, 81*(3), 719–745.

Rafols, I., & Meyer, M. (2007). How cross-disciplinary is bionanotechnology? Explorations in the specialty of molecular motors. *Scientometrics, 70*(3), 633–650.

Rinia, E. J. (2007). *Measurement and evaluation of interdisciplinary research and knowledge transfer.* Unpublished doctoral dissertation, Universiteit Leiden.

Rinia, E. J., Van Leeuwen, T. N., Bruins, E. E. W., Van Vuren, H. G., & Van Raan, A. F. J. (2001). Citation delay in interdisciplinary knowledge exchange. *Scientometrics, 51*(1), 293–309.

Rinia, E. J., Van Leeuwen, T. N., Bruins, E. E. W., Van Vuren, H. G., & Van Raan, A. F. J. (2002a). Measuring knowledge transfer between fields of science. *Scientometrics, 54*(3), 347–362.

Rinia, E. J., Van Leeuwen, T. N., & Van Raan, A. F. J. (2002b). Impact measures of interdisciplinary research in physics. *Scientometrics, 53*(2), 241–248.

Thomas, D. S. (1952). Experiences in interdisciplinary research. *American Sociological Review, 17*(6), 663–669.

Tomov, D. T., & Mutafov, H. G. (1996). Comparative indicators of interdisciplinarity in modern science. *Scientometrics, 37*(2), 267–278.

Van Leeuwen, T. N., & Tijssen, R. (2000). Interdisciplinary dynamics of modern science: Analysis of cross-disciplinary citation flows. *Research Evaluation, 9*(3), 183–187.

Wagner, C. S., Roessner, J. D., Bobb, K., Klein, J. T., Boyack, K. W., Keyton, J., et al. (2011). Approaches to understanding and measuring interdisciplinary scientific research (IDR): A review of the literature. *Journal of Informetrics, 5*(1), 14–26.

Weingart, P., & Stehr, N. (2000). *Practicing interdisciplinarity.* Toronto: University of Toronto Press.

Zitt, M. (2010). Citing-side normalization of journal impact: A robust variant of the audience factor. *Journal of Informetrics, 4*(3), 392–406.

Zitt, M., & Small, H. (2008). Modifying the journal impact factor by fractional citation weighting: The audience factor. *Journal of the American Society for Information Science and Technology, 59*(11), 1856–1860.

11 Bibliometric Standards for Evaluating Research Institutes in the Natural Sciences

Lutz Bornmann, Benjamin F. Bowman, Johann Bauer, Werner Marx, Hermann Schier, and Margit Palzenberger

Introduction

Bibliometrics is an established field within scientometrics (Andres, 2011; De Bellis, 2009; Moed, 2005; Vinkler, 2010), though uniformity in the conduct of bibliometric analyses has yet to be achieved. Experts share many tacit assumptions regarding appropriate practice, but the field lacks an established set of standards to guide newcomers and outsiders. Even within the field, consensus on several critical issues has not been reached. The aim of this chapter, therefore, is to describe standards for applying bibliometric methods in the evaluation of research institutes in the natural sciences. These standards inform the selection of the underlying data from citation databases/indexes, statistical analysis of the data, and presentation of results.

Our formulation of the standards for metrics-based evaluation of institutes is based on many years of experience in conducting analyses at the Max Planck Society (MPG) and on our bibliometric research more generally (e.g., Bornmann, Mutz, Marx, Schier, & Daniel, 2011; Bornmann, Mutz, Neuhaus, & Daniel, 2008). In developing these standards, we have been careful to focus on the essential elements—that is, on indicators necessary and meaningful in an institute-level evaluation. This chapter covers only a subset of the large number of indicators used in bibliometrics. An overview of common indicators is provided by Rehn, Kronman, and Wadskog (2007). We have also tried to keep the standards as simple as possible, so that users with limited knowledge of bibliometrics and empirical research can apply them.

In the sections that follow, we describe indicators and procedures that can be used for measuring the productivity of a research institute and the impact of its research via publications and citations, including methods for analyzing productivity and citation impact diachronically. To show how our standards can be applied, we make use of data from six institutes that conduct research in similar areas. We would emphasize that the data are being utilized here solely to illustrate our standards. For this reason, the names of the research institutes (RI) have been anonymized (RI1, RI2, RI3, RI4, RI5, RI6).

Establishing Study Parameters

To make reliable statements about the research performance of institutes based on bibliometric indicators, the following points should be taken into account:

1. Undertake a bibliometric analysis of an institute's research performance only if it has published at least 100 original research papers or reviews in scientific journals annually. This number is based on published guidelines for bibliometric research (e.g., NORIA-net, 2011). For bibliometric evaluation of individual researchers, Lehmann, Jackson, and Lautrup (2008) propose a minimum number of 50 papers: "It is possible to draw reliable conclusions regarding an author's citation record on the basis of approximately 50 papers" (p. 384). In an extensive study at the level of individual faculties, Larivière and Gingras (2011) included those with a minimum of 50 publications and, at the level of institutes, those with a minimum of 500.
2. When deciding which years of an institute's publication record should be included in the analysis, care should be taken that the majority of the papers (if not all) are at least two years old. In most fields, it can take three years before a paper's citations reach their peak; after that, citations usually decrease relatively quickly (Seglen, 1992). Thus, only after three years do we have a sense of a paper's likely long-term impact. Care should, however, be taken to avoid increasing the time period beyond this, because it will reduce the currency of the report (Glänzel, Thijs, Schubert, & Debackere, 2009). In all bibliometric studies, the publication years of the papers that you wish to study (publication window), and the timespan for which you wish to count citations to the publications (citation window), should be specified.
3. When evaluating the impact of publications there is the possibility of including or excluding authors' self-citations. Studies have reported different percentages of self-citations: in a study of researchers in Norway, Aksnes (2003) found 36% of citations to be self-citations; Garfield (1979) reported a figure of 10%; Snyder and Bonzi (1998) showed that the percentage of self-citations in the natural sciences is 15%, higher than in the social sciences (6%) and arts and humanities (6%). We recommend including self-citations for two reasons: (a) The percentage of self-citations will vary among the authors (and the publications), but in most cases institutes conducting research in similar areas are not likely to have very different self-citation percentages. In cases where there are reasons to think otherwise, self-citations will have to be taken into

account—for example, by conducting a separate bibliometric analysis at the research group level. (b) In addition, following Glänzel, Debackere, Thijs, and Schubert (2006), we believe self-citations are usually an important feature of the scientific communication and publication process: "A self-citation indicates the use of own [sic] results in a new publication. Authors do this quite frequently to build upon own [sic] results, to limit the length of an article by referring to already published methodology, or simply to make own [sic] background material published in 'grey' literature visible" (p. 265).

4. Many publications have more than one author, and the authors are often from different institutes. When quantifying an institute's productivity and the impact of its publications, it is possible to give each institute listed on a publication a proportional ("fractional") share of the paper and its citations (this is known as fractional as opposed to whole counting). The greater the number of institutions listed for the authors, the smaller the share given to each individual institute. These shares can be equally distributed across the institutes, or certain institutes (for example, the institute at which the first author works) can be given a larger share than the others. The formulas used are not uniform, because customs concerning the order of coauthors differ greatly across disciplines. Vinkler (2010, chap. 10) and Huang, Lin, and Chen (2011) have provided overviews of different ways to assign authors credit for a share of a paper and the advantages and disadvantages of each. Since there is no standard practice in bibliometrics concerning the use of whole or fractional counting and most studies do not use fractional counting, we recommend using whole counting. Each institute listed on a paper receives a whole count for that paper.

Using a Citation Database

Included in the bibliometric analysis are all publications of an institute where at least one author lists the institute as his or her affiliation. The optimal method for precise collection of the publications is to obtain and process the personal publication lists of all research staff. However, this is often impractical. We recommend searching for the institute's publications using citation databases (or "indexes"). To avoid errors and incompleteness, the publications found through the search should be checked against the institute's official lists of publications. In our experience, omissions and errors are commonly found both in the search results and in the institute's lists of publications. Unfortunately, a simple comparison of paper counts does not necessarily reveal these differences. We recommend checking both sources for nonmatching references. Missing references on either side give valuable hints to improve the compilation of the respective list.

Since a number of citation databases can be used for bibliometric analyses, the report should list the databases used and describe their basic features. In evaluative bibliometrics, the databases utilized are usually those provided by Thomson Reuters (Web of Science, InCites) and Elsevier (Scopus). In the natural sciences the number of publications appearing in core journals indexed by Web of Science (approximately 10,000 journals) has become the standard measure for quantification of scientific productivity. Thomson Reuters' Web of Science provides access to various citation databases (including Science Citation Index Expanded and Social Sciences Citation Index), which are available (generally and for specified time periods) by subscription. Since the databases included and the years covered depend on the specific license agreement, this information should be documented in any evaluation report. The data shown in this chapter are from searches in the Web of Science databases Science Citation Index Expanded, Social Sciences Citation Index, Arts & Humanities Citation Index, and Conference Proceedings Citation Index–Science and Conference Proceedings Citation Index–Social Science & Humanities (including all currently available back files for each database).

Outside of the core natural science disciplines, in particular in areas like computer science or engineering and technology, coverage based on the journals in Web of Science is inadequate. Publication activity in these areas can be better determined using discipline-specific citation databases (such as Chemical Abstracts or INSPEC for physics and related areas). Due to wider coverage in some areas, the multidisciplinary database Scopus might offer some advantages over Web of Science.

We do not recommend undertaking a bibliometric analysis based on Google Scholar (GS). Several studies (e.g., Bornmann et al., 2009; García-Pérez, 2010; Jacsó, 2009, 2010) have pointed out numerous shortcomings of GS, such as extensive deficiencies in the metadata (e.g., "ghost authors") and frequently inflated citation counts (e.g., 17% incorrect citations in GS as opposed to less than 1% in Web of Science).

Systematic Effects Caused by Choice of Database

When using bibliometric data from different databases/indexes for an evaluation study, it should be noted that this can cause systematic effects on the indicators being presented: publications that are in one database may not be in another (and vice versa). Table 11.1 shows a comparison of the paper counts found for the six research institutes that we examined in National Citation Report and Web of Science. The differences between the databases range from about 5% (RI2, RI3, RI4) to about 20% (RI1 and RI5). Two factors most likely account for these differences. First, in contrast to Web of Science, National Citation Report classifies publications of the document type "conference

Table 11.1
Difference in number of publications of the six research institutes captured in Web of Science and National Citation Report—Germany

Research institute (RI)	National Citation Report (NCR)	Web of Science (WoS)	Difference between the number (NCR—WoS)	Difference from NCR, in percent
RI1	472	389	83	18
RI2	684	683	1	0
RI3	695	730	–35	5
RI4	347	357	–10	3
RI5	216	252	–36	17
RI6	829	736	93	11
Total	3243	3147	96	3

proceedings" that have also been published in journals as "articles"—the document type "conference proceedings" does not exist in National Citation Report. Second, the selection of papers in the databases via their addresses takes place in different ways: in Web of Science the user must develop and apply an optimized search query, whereas in National Citation Report the user selects from variations of the address as provided by the database producer. Each search has its own possible sources of error when capturing publications for a research institute.

Selecting Document Types

To focus on publications of substance, we recommend limiting the bibliometric analysis to specific types of documents. However, it should be noted that database providers classify document types according to their own criteria (they categorize the publications differently). Moreover, their classifications often do not match the classifications used by journals (Meho & Spurgin, 2005).

Database providers usually call publications that report original research "articles," and extensive literature overviews are called "reviews." Short papers reporting original research are usually termed "brief communications" ("notes") (Moed, Van Leeuwen, & Reedijk, 1996). For this reason, the standard practice when considering document types in bibliometric analyses is as follows: "Use journal items that have been coded as regular discovery accounts [articles], brief communications [notes], and review articles—in other words, those types of papers that contain substantive scientific information. Traditionally left to the side are meeting abstracts (generally not much cited), letters to the editor (often expressions of opinion), and correction notices" (Pendlebury, 2008, p. 3).

Since "brief communications" ("notes") are now no longer categorized separately in Web of Science, we have included only "articles" and "reviews." The term *publication* in what follows always refers exclusively to these two document types.

Delineating Fields of Research

In the Thomson Reuters databases/indexes, a research field is defined by a set of journals that covers a specialized area. The journal set "Neuroscience" contains the journals *Nature Reviews Neuroscience* and *Behavioral and Brain Sciences*, for example. When we refer to "subject category," we mean the journal set. The size of the journal set can vary greatly. There are sets containing only a few journals (such as "Andrology," with 5 journals) and others that contain more than 250 journals (such as "Biochemistry & Molecular Biology," with 286 journals). The use of journal sets for classifying publications in specific fields is disputed (Bornmann et al., 2008; Boyack, 2004; Strotmann & Zhao, 2010). Many journals publish a broad range of papers that cannot be adequately categorized by the sets. Use of these journal sets comes up against limits, particularly in specialized areas within the natural sciences. The problem is especially serious with journals such as *Science* and *Nature*; these journals are grouped in a journal set named "Multidisciplinary," which means that the individual papers cannot be categorized in their specific fields at all. We used the NCR database for citation analysis, which calculates its indicators (such as relative citation rates and percentiles) based on journal sets; therefore, for bibliometric analyses with these databases we recommend using the corresponding journal sets (possibly with the exception of the journal set "Multidisciplinary") to assign publications to research fields.

Normalizing Citation Impact

Normalized citation impact is based on a comparison of observed and expected citations. The expected number of citations of a publication depends on the year of publication (its "age") and the scientific field (Abramo, Cicero, & D'Angelo, 2011; Radicchi, Fortunato, & Castellano, 2008). To the best of our knowledge, Thomson Reuters is the only provider of relative, age-normalized, and field-normalized citation counts, which can be used for bibliometrics-based research evaluation. The relative citation counts used here were derived from the subscription-based products National Citation Report (for Germany) and InCites. For overviews of current methods of normalizing citation counts, see Rehn et al. (2007) and Vinkler (2010).

Sampling Data Using Statistical Tests of Significance

The bibliometric data used in an evaluation study are usually a sample of the institute's entire data. The complete data would comprise all publications of an institute and their citations within a frame relevant to the rationale of a given evaluation setting. Using statistical tests of significance, it can be determined whether the results obtained from a sample are, with high probability, valid for all publications (Bornmann et al., 2008). Statistical tests of significance are used on a sample of data selected randomly from the entire dataset or on data generated using a sampling procedure that tries to minimize systematic errors and dependencies with regard to the question being investigated in the study (in the present context it is the differences among the research institutes in terms of bibliometric indicators). Here, we did not select the data randomly; we chose publication periods and citation windows for which we can assume that there are no systematic advantages or disadvantages for an institute. If the result of a test that examines the difference between two institutes in terms of the impact of their publications is significant, it can be assumed that the difference is not the result of chance and applies beyond the sample.

One benefit of using a sample as opposed to all publications is that it reduces the cost of conducting an evaluation study (Ruegg & Feller, 2003): the set of publications that must be put together for the study is smaller. Another benefit is that it is not necessary to use current publication years (that is, very recent years) for the study. When institutes are evaluated there are often requests for bibliometric indicators relating to the most recent years. As we show, current publication years can, however, lead to unreliable distortions of the entire data structure. If the result of a statistical test is significant, results based on less recent publication years can, with great probability, be applied to current publication years (provided no major changes, such as the appointment of a new director with a new research direction, occurred).

In general, presentation of results follows the guidelines of the American Psychological Association (2009)—the standard in empirical social sciences. If statistical tests are applied, statistical power considerations associated with the tests of hypotheses should be taken seriously (American Psychological Association, 2009).

Analyzing Productivity, Subject Categories, and Citation Impact

Extensive information on the statistical tests used in our evaluations can be found in Sheskin (2007). For the analysis of citation rates, nonparametric procedures are frequently better than parametric tests. Parametric tests assume a normal distribution (Calver & Bradley, 2009), which is often not the case with these data. We used Stata statistical software (StataCorp., 2011). The analyses can be conducted using other statistical packages (such as SPSS or R) (see Gagolewski, 2011).

Here, we included papers published between 2003 and 2008. For the trend analysis of the impact of these publications, the citation windows were defined as indicated. For further analyses of the impact of the publications, the citation window was defined as the time of publication up to the end of 2008 (or 2007).

Productivity

Table 11.2 shows the results of the productivity analyses for the six research institutes (RI) that we examined (RI1 to RI6) for the years 2003 to 2008. As mentioned above, when interpreting the results, the number of research staff should always be taken into consideration: the greater the number of active researchers at an institute, the greater the number of publications that the institute can be expected to produce. For this reason, the lower section of table 11.2 shows the average number of research staff for the time period analyzed. The 3,147 papers published by the six institutes appeared in a total of 821 journals (on average approximately four publications per journal). With more than 50 published papers each, *journal A* ($n = 96$) and *journal B* ($n = 54$) published most (table 11.2 does not show the distribution of papers across journals).

The upper section of table 11.2 shows the number of papers published annually by the institutes. These numbers are shown as a graph in the middle section of the table to illustrate the trend. Looking at the development of the individual institutes over time, we see that RI1 and RI2 statistically significantly increased their publications year after year. For these two institutes there is a statistically significant association between number of publications and publication years (Pearson's correlation, $p < 0.05$).

In addition to showing productivity over time, the lower section of table 11.2 includes several productivity indicators for the institutes: the sum total of publications over the entire time period, average productivity (median and arithmetic mean), and range. Range is a measure of statistical dispersion and is the difference between the highest and lowest number of publications in the years analyzed. Number of publications is shown as paper counts for the document types "article" and "review." As the results in the lower section of table 11.2 show, RI3 ($n = 730$) and RI6 ($n = 736$) published the most papers over the years in total and also on average (RI3: arithmetic mean = 122, median = 123; RI6: arithmetic mean = 123, median = 123.5). Whereas RI6 had the largest number of research staff of the six institutes, RI3 had a relatively low number. In relation to number of research staff, RI3 thus had high productivity over the entire time period. RI2 had the highest productivity in a single year (2008), with 176 publications; however, it had a comparatively high number of research staff. RI2 had the

Table 11.2
Productivity indicators of six research institutes for the years 2003–2008

Publication year	Research institute (RI): Number of publications					
	RI1*	RI2*	RI3	RI4	RI5	RI6
2003	34	64	102	49	55	128
2004	45	96	121	37	42	145
2005	67	104	135	43	45	119
2006	68	105	116	77	29	110
2007	71	138	125	65	32	106
2008	104	176	131	86	49	128

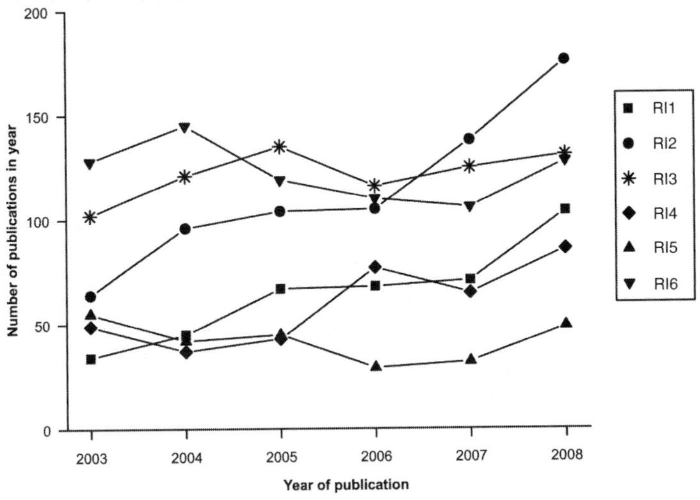

Number of publications in year
200
150
100
50
0
2003 2004 2005 2006 2007 2008
Year of publication
RI1
RI2
RI3
RI4
RI5
RI6

Indicator	RI1	RI2	RI3	RI4	RI5	RI6	Total
Sum total	389	683	730	357	252	736	3147
Articles	367	643	693	328	213	666	2910
Reviews	22	40	37	29	39	70	237
Arithmetic mean	65	114	122	60	42	123	
Median	67.5	104.5	123	57	43.5	123.5	
Maximum	104	176	135	86	55	145	176
Minimum	34	64	102	37	29	106	29
Range	70	112	33	49	26	39	
Average number of research staff across all years	240	470	200	350	180	670	2110

Note: *The association between numbers of publications and publication years is statistically significant (Pearson's correlation, $p < 0.05$).

biggest difference in productivity between individual publication years (range = 112); RI5 had the smallest difference (range = 26).

Finally, we used analysis of covariance to test whether the mean paper counts of the six research institutes differed statistically significantly across publication years. With analysis of covariance the number of research staff at an institute can be controlled for and the number of publications can be compared "directly." This gave us a test result on the mean numbers of publications, "neutralized" with regard to the number of research staff. As the results show, there was a statistically significant difference between the institutes ($p < 0.05$) in the average number of publications for the years 2003 to 2008. According to the test results, at least one research institute differed statistically significantly from the rest of the institutes in the average number of publications.

As an alternative to analysis of covariance, it is also possible to test statistically for differences in researcher productivity (for this the mean number of publications of an institute is divided by the number of research staff) using one-factor analysis of variance (ANOVA) and a post hoc test (Tukey HSD procedure, $p < 0.05$). This analysis revealed that an effect was recognizable (ANOVA, $R^2 = 0.83$, $p < 0.001$), but only RI3 differed statistically significantly from the other institutes. For RI3 the mean paper count per year and researcher was 0.6 and thus statistically significantly higher than for the other institutes, with mean values of 0.2 and 0.3. There were no statistically significant differences among the other institutes—that is, the observed differences could be due to random fluctuations.

Subject Categories

In all, the publications of the six institutes (n = 3,147) fell into 164 different subject categories (two publications were not assigned to any journal set by Thomson Reuters). The distribution of the publications across the subject categories provides an indication of how similar their areas of research are. Since many publications from the six institutes belong to more than one journal set, each was assigned on average to two categories. For example, the journal *Behavioral and Brain Sciences* is assigned to the category "Neuroscience" as well as to the category "Behavioral Sciences." Table 11.3 shows the eight (anonymized) subject categories (SC1 to SC8), to which at least 5% of all publications of the six research institutes could be assigned. To create clear tables with subject categories in which institutes have published, we recommend showing no more than 10 categories. To do so it may be necessary to modify (increase) the 5% limit.

As table 11.3 (see Total column) shows, 32% of the publications of all six institutes belonged to SC1 and 13% to SC2. The Pearson χ^2 test can be used to determine the extent to which the number of publications in the subject categories differs statistically

Table 11.3
Publications of a research institute that were assigned to one of eight subject categories (in percentages)

Subject category (SC)	Research institute (RI)						
	RI1	RI2	RI3	RI4	RI5	RI6	Total
SC1*	39	2	56	9	37	41	32
SC2*	11	9	35	14	2	1	13
SC3*	1	0	3	3	10	32	9
SC4*	1	0	7	1	29	16	8
SC5*	6	0	15	1	21	2	6
SC6*	6	13	1	0	5	7	6
SC7*	3	13	5	5	2	4	6
SC8*	6	2	13	9	2	2	6

Note. Because a publication can usually be assigned to more than one subject category and the table does not show all subject categories to which all of the papers belong, sums of the rows and columns were not calculated.
* The difference between the research institutes regarding the assignment of their publications to the subject category is statistically significant (Pearson χ^2 test, Bonferroni-adjusted $p = 0.05$).

significantly between the institutes. When calculating the χ^2 test for these data, it should be noted that a publication can be assigned to several subject categories. In such cases, Jann (2005) suggests calculating an χ^2 test for each individual row in the table (here: for each subject category). Repeating the test for each row makes it necessary to correct the level of significance; we used the (conservative) Bonferroni correction. For this correction, the significance level of $\alpha = 0.05$ is divided by the number of repeated tests.

As the results of the Pearson χ^2 test in table 11.3 for each subject category show, the institutes differed statistically significantly in the assignment of their publications to subject categories. Their areas of research seem to be very different overall. For instance, 35% of the publications of RI3 belong to SC2, as opposed to only 1% of publications of RI6 (across all research institutes the figure is 14% of publications). And whereas 21% of the publications of RI5 were assigned to SC5, no publications of RI2 are in that category. The big differences between the research institutes in classification of the publications in the individual subject categories show, inter alia, the necessity of considering the subject area when analyzing the citations of the publications of the six research institutes. Many studies have shown that the expected values for citations vary greatly across research fields (e.g., Bornmann & Daniel, 2009; Bornmann et al., 2008; Radicchi et al., 2008).

Citation Impact

For each of the approximately 10,000 journals covered by Web of Science (with just a few exceptions), Thomson Reuters Journal Citation Reports provides the "Journal Impact Factor"—along with many other indicators—to capture the impact of a scientific journal (Garfield, 1972, 2006). In bibliometric studies, the Journal Impact Factor (JIF) has been used to measure the quality of individual papers. Although the JIF makes it possible to roughly evaluate journals as a whole within a narrow subject category, it does not say much about the actual impact of individual papers published in those journals. Distributions of citations are highly skewed across papers, with a few highly cited papers contributing heavily to the calculation of the Impact Factor. Therefore, the mean citation rate of a journal is not typical of most of the papers published in it. In short, a paper published in a high-impact journal is not necessarily a high-impact paper. We therefore recommend not using JIF for analysis of the impact of individual publications.

Analyzing Trends in Citation Impact

Table 11.4 shows the number of citations for the publications of the six research institutes. The citations were collected from Web of Science. The citation counts, as reported in table 11.4, are easily determined. For comparing research institutes, however, the numerical value of a citation count is not meaningful due to the different distribution of subject categories in the institutes. But if we look at changes in the citation count of an institute over a sufficiently long time period as a trend curve (see the graph in table 11.4), we can draw conclusions about its development and compare its development with that of other institutes (Marx, Schier, & Andersen, 2006). The advantage of this comparative presentation is that it is independent of the research fields represented in the institutes, provided the research profile of an institute has not changed significantly during the period in question.

For trend curves displaying the impact of publications, the time period is divided into overlapping windows using the moving-average method. We used time windows covering three publication years with each time window starting one year later. For each time window the average citation count for all publications is calculated. For the window 2003–2005 shown in table 11.4, for example, all papers published between 2003 and 2005 were included; for each paper in the publication window, the number of citations received between its publication date and the end of 2005 was determined. The three-year time window reduces the yearly fluctuations in the number of publications and their citations, allowing for accurate trend analysis.

Table 11.4

Citation impact indicators of six research institutes for the years 2003–2008

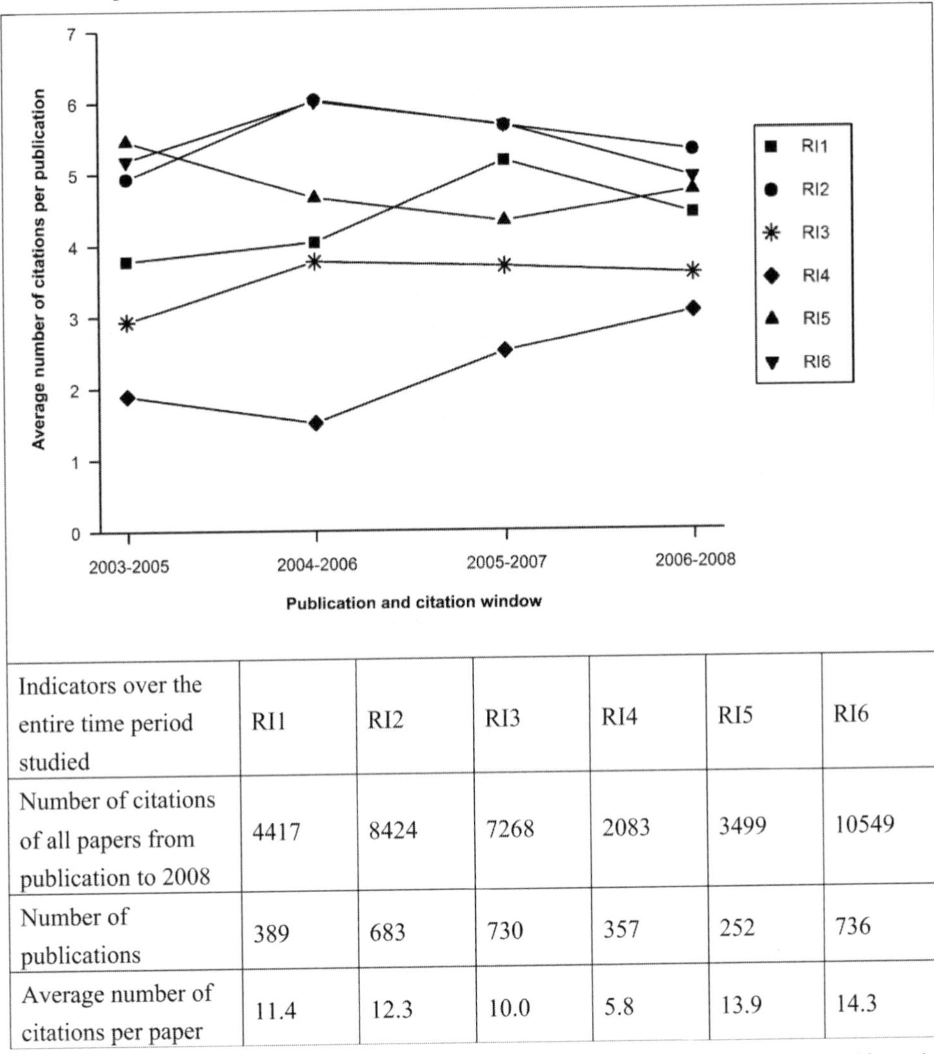

Indicators over the entire time period studied	RI1	RI2	RI3	RI4	RI5	RI6
Number of citations of all papers from publication to 2008	4417	8424	7268	2083	3499	10549
Number of publications	389	683	730	357	252	736
Average number of citations per paper	11.4	12.3	10.0	5.8	13.9	14.3

Note: For all of the research institutes, the association between development over time (the publication and citation windows) and the average number of citations is not statistically significant (Pearson's correlation, $p > 0.05$).

As the results in table 11.4 show, the association between the temporal development (the publication and citation windows) and the average citation counts for each institute are not statistically significant (Pearson's correlation, $p > 0.05$). Statistically, then, none of the institutes shows a trend with clearly increasing or decreasing citation counts over the years in question.

Using Percentiles to Measure Citation Impact

According to the current state of research in bibliometrics, percentiles are the best way to normalize the citation counts of individual papers with regard to research field and year of publication (Bornmann et al., 2011; Leydesdorff, Bornmann, Mutz, & Opthof, 2011): "Given the skewed nature of citation count distributions, it keeps a few highly cited papers from dominating citation statistics" (Boyack, 2004, p. 5194). According to findings by Albarrán, Crespo, Ortuño, and Ruiz-Castillo (2011), the distributions of citations in all subject categories are very skewed: "The mean is 20 points above the median, while 9-10% of all articles in the upper tail account for about 44% of all citations" (p. 385). By standardizing the citations based on percentiles, the impact of publications in different research fields and publication years can be compared directly.

The percentile indicates how a paper has performed relative to other papers (in the same subject category and publication year). Based on the citation frequency distribution (arranged in descending order), all papers in the same field and publication year as the reference paper are split into 100 percentile ranks. The maximum value is 100, indicating 0 cites received.[1] The smaller the percentile value, the higher the number of citations received by the paper compared to all papers in the same subject category and publication year. A percentile value of 10, for example, means that the paper belongs to the 10% most cited papers; the other 90% have had less impact. A percentile of 50 is the median and thus indicates average impact in comparison with the other papers.

Table 11.5 shows the result of the impact analysis based on percentiles for the publications of the six institutes. The upper section of the table shows the distribution of the percentiles for the individual institutes by publication year. The boxplots consist of boxes whose upper borders indicate the first quartile (25% of the values) and whose lower borders indicate the third quartile (75% of the values). The plus sign within the box shows the median (50% of the values lie above or below this value). The position of the median in the box conveys a sense of the skewness of the values. A comparison of the institutes shows that it is mainly the percentiles for the publications of RI2 that have shifted toward lower ranks.

Table 11.5

Percentile impact indicators of six research institutes for the years 2003–2008

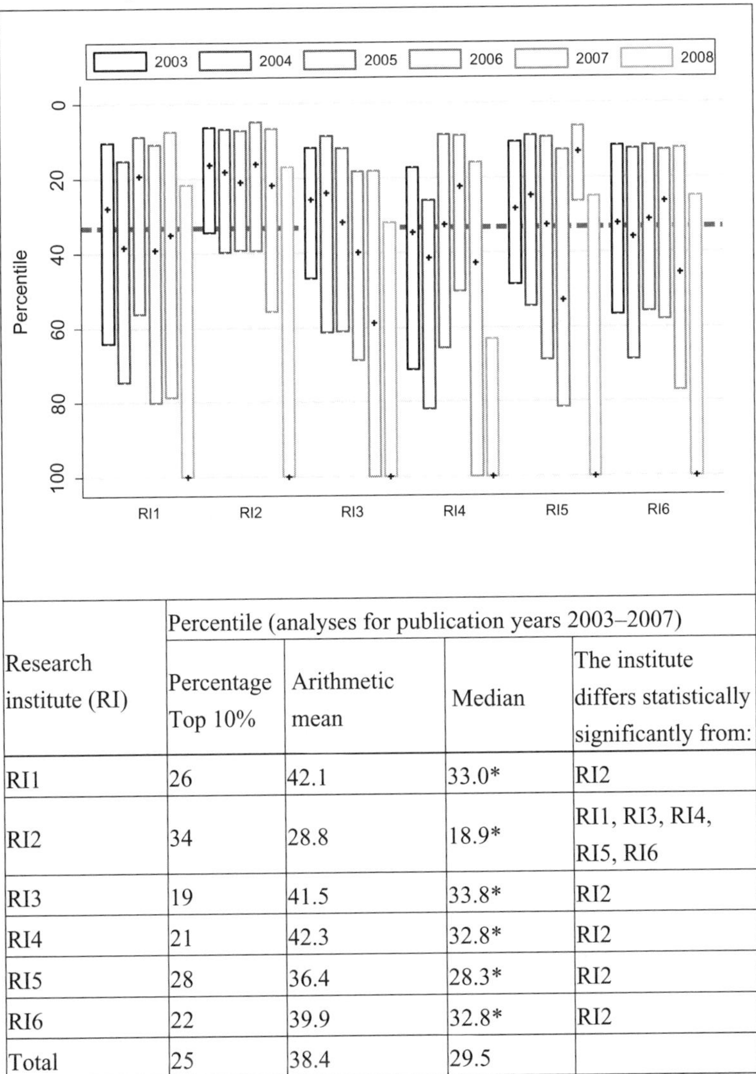

Research institute (RI)	Percentile (analyses for publication years 2003–2007)			
	Percentage Top 10%	Arithmetic mean	Median	The institute differs statistically significantly from:
RI1	26	42.1	33.0*	RI2
RI2	34	28.8	18.9*	RI1, RI3, RI4, RI5, RI6
RI3	19	41.5	33.8*	RI2
RI4	21	42.3	32.8*	RI2
RI5	28	36.4	28.3*	RI2
RI6	22	39.9	32.8*	RI2
Total	25	38.4	29.5	

Note: The horizontal line in the graph shows the median (33.27) across all institutes and years. The plus sign in the box shows the median for an institute in that year.

* The average percentile of a research institute differs statistically significantly ($p < 0.05$) from 50 (Wilcoxon rank-sum test)

For all six institutes it is noticeable that in 2008 there were predominantly high percentile numbers. Many publications from 2008 have not yet received any citations or only very few, indicating that the citation window for this last publication year before the time point for collecting citations was too short (from time of publication in 2008 to the end of 2008) to provide reliable information on the impact of the publications (see Radicchi & Castellano, 2011). Since the year 2008 led to an inadmissible distortion of the whole data structure, we did not include it in the analyses of the percentiles that follow; these are, therefore, based on the years 2002 to 2007.

The lower section of table 11.5 shows some indicators that were calculated based on percentiles for the individual papers. Since, according to Tijssen and Van Leeuwen (2006) and Tijssen, Visser, and Van Leeuwen (2002), publications that belong to the 10% most highly cited papers in their field should be called highly cited papers, the table shows for each institute the percentage of publications that belong to the top 10% in their field. As can be seen, the percentage for each institute is clearly higher than 10%. This result can be seen as a sign of quality, since, based on a random sample from the population of papers in the Thomson Reuters database, we would expect to find 10% of publications that belong to the 10% most cited publications in their field. This means that each institute published more papers in the top 10% than would be expected. In particular, the research staff at RI2, with a percentage of 34%, published three times as many highly cited papers as would be expected.

As measures of central tendency, table 11.5 provides the arithmetic mean and the median: the lower the arithmetic mean for an institute, the larger the average citation impact. Using the Wilcoxon rank-sum test (Sheskin, 2007, pp. 225–239), it can be determined to what extent the average percentile (the median) of a research institute differs statistically significantly from 50. For a research institute with publications having an average impact, we would expect a median of approximately 50. As the results show, all the institutes differed statistically significantly from 50. The average citation impact for each of the research institutes is thus statistically significantly higher than average.

Table 11.5 also shows the result of pairwise individual comparisons of the impact of papers published by the research institutes. These individual comparisons were conducted after the Kruskal-Wallis *H* test (Sheskin, 2007, pp. 981–1006), which indicated that there was a statistically significant difference between all the institutes regarding the median of the percentiles. The individual comparisons yield information concerning the differences between two institutes that led to this general finding (the statistically significant result across all institutes). Repeating the statistical test for each pairwise comparison makes it necessary to apply the Bonferroni correction of the level of significance mentioned above of $\alpha = 0.05$. As the results of the individual comparisons in the table show, RI2 differed statistically significantly from all other institutes. We can therefore consider it highly probable that the average impact of papers that RI2 publishes is higher than the average impact of papers published by the other research institutes.

Limitations of Research Evaluation

While the use of paper counts as an indicator of researcher productivity is widely recognized today, several reservations have been raised when it comes to the use of citation counts as an indicator of the quality of research papers (discussed in more detail in Day, chapter 4; Furner, chapter 5; and Gingras, chapter 6). It has been said that citation counts are not "objective" measures of scientific quality but rather a measurement construct that is open to criticism. Some of the criticisms are as follows:

1. Scientific quality is a complex phenomenon that cannot be measured on a one-dimensional scale (that is, based on citation counts) (e.g., Barbui, Cipriani, Malvini, & Tansella, 2006; Berghmans et al., 2003).
2. Authors cite publications that have had little intellectual influence on their paper.
3. Authors do not cite many of the publications that have had an intellectual influence on their paper.
4. Authors overlook relevant publications (Wright & Armstrong, 2007).
5. Researchers use publication strategies to distort the results of bibliometric analyses (Bornmann, 2010).
6. Citation databases/indexes used for citation analyses cover only a part of the publications that are published worldwide, and they also contain many errors (Marx, 2011).
7. In general, when evaluating the content of a publication, the publication should be read and assessed and not only viewed in terms of impact: "Formal measurements of scientists based on their past research may be precarious and need not reflect the true potential of a scientist. If one is to evaluate accomplishments of scientists one ought to understand what is in his/her publications" (Randić, 2009, p. 810).

We recommend that the results of a bibliometric analysis always be interpreted in light of these concerns. For instance, it should be taken into consideration that possibly not all publications of an institute were included in the analysis and that important items may have been omitted. Further, a bibliometric analysis should always be part of an informed peer review process (Daniel, Mittag, & Bornmann, 2007) that includes content evaluation by peers. A number of studies have demonstrated a high correlation not only between quantity (productivity) and quality (impact) (Abramo, D'Angelo, & Costa, 2010; Hemlin, 1996), but also between the results of bibliometric analyses and the judgments of peers (Bornmann, 2011); however, the high correlations are insufficient reason to carry out one (judgments of peers) without the other (bibliometrics). Peers can identify and appreciate dimensions of quality that cannot be captured through citations: "The 'impact' of a publication describes its actual influence on surrounding research activities at a given time" (Martin & Irvine, 1983, p. 70). According to Martin and Irvine (1983), a distinction should be made between impact and importance ("the influence on the advance of scientific knowledge") and the actual quality of research ("how well the research has been done"), which might be better accomplished via peer evaluation rather than bibliometrics.

In this chapter, we introduced a set of bibliometric methods that can be used to compare research institutes. The set can be modified depending on the particular application. The methods and indicators do not have to be used in every case. We did not, for example, deal with the *h*-index because we consider it unsuited to the evaluation of institutes (more on this in Gingras, chapter 6, this volume). The *h*-index was proposed as an indicator for evaluating individual researchers: "A scientist has index h if h of his or her N_p papers have at least h citations each and the other $(N_p - h)$ papers have $\leq h$ citations each" (Hirsch, 2005, p. 16569). It provides insufficient information to use with the typically extensive set of publications associated with individual institutes, because it does not adequately capture the distribution of the citations across the publications. If it is to be used, we recommend combining it with additional indicators that provide information on the distribution of the citations (see Bornmann, Mutz, & Daniel, 2010).

Conclusion

Reports containing bibliometric results on research institutes should end with a short summary of the most important findings. Regarding the productivity of the institutes examined here, we found that RI3 published more papers per research staff member than all the others. RI1 and RI2 statistically significantly increased their paper counts over the years. As far as publication impact was concerned, we found no significant trend in terms of increasing or decreasing citation counts. The average impact of the publications of all institutes is statistically significantly higher than the average in their respective research fields, and all research institutes (notably RI2) published more highly cited papers than could be expected. The publications of RI2 achieved a significantly higher impact than the publications of all other institutes.

The validity of standards for using bibliometrics in the evaluation of research institutes changes over time, as existing methods are refined and new methods introduced. For example, the use of percentiles in research evaluation is relatively new and further insights can be expected in the near future. In this regard, Bornmann (2013a) deals with the question of how percentile citation impact data can be meaningfully analyzed using statistical tests. Other research has focused on the calculation of percentiles. Various proposals for allocating publications to percentiles (percentile rank classes) have been published (Bornmann, 2013b; Leydesdorff et al., 2011; Rousseau, 2012; Schreiber, 2012a, 2012b). The use of samples in bibliometric evaluations of research institutes has also been proposed; Bornmann and Mutz (2013) suggest using a cluster sample instead of including all papers from the institutes under review.

The results of research on bibliometric methods (in particular on performance indicators) are important—new knowledge drives the development of new procedures in evaluation—but they should only be incorporated in evaluation standards if the methods in question have been tested successfully and are shown to be value-adding.

Note

1. Thomson Reuters' definition of percentile for subject areas used in InCites is available at http://incites.isiknowledge.com/common/help/h_glossary.html.

References

Abramo, G., Cicero, T., & D'Angelo, C. A. (2011). Assessing the varying level of impact measurement accuracy as a function of the citation window length. *Journal of Informetrics, 5*(4), 659–667. doi:10.1016/j.joi.2011.06.004.

Abramo, G., D'Angelo, C. A., & Costa, F. D. (2010). Testing the trade-off between productivity and quality in research activities. *Journal of the American Society for Information Science and Technology, 61*(1), 132–140.

Aksnes, D. W. (2003). A macro study of self-citation. *Scientometrics, 56*(2), 235–246.

Albarrán, P., Crespo, J., Ortuño, I., & Ruiz-Castillo, J. (2011). The skewness of science in 219 sub-fields and a number of aggregates. *Scientometrics, 88*(2), 385–397. doi:10.1007/s11192-011-0407-9

American Psychological Association. (2009). *Publication manual of the American Psychological Association* (6th ed.). Washington, DC: American Psychological Association.

Andres, A. (2011). *Measuring academic research: How to undertake a bibliometric study*. New York: Neal-Schuman.

Barbui, C., Cipriani, A., Malvini, L., & Tansella, M. (2006). Validity of the impact factor of journals as a measure of randomized controlled trial quality. *Journal of Clinical Psychiatry, 67*(1), 37–40.

Berghmans, T., Meert, A. P., Mascaux, C., Paesmans, M., Lafitte, J. J., & Sculier, J. P. (2003). Citation indexes do not reflect methodological quality in lung cancer randomised trials. *Annals of Oncology, 14*(5), 715–721.

Bornmann, L. (2010). Mimicry in science? *Scientometrics, 86*(1), 173–177.

Bornmann, L. (2011). Scientific peer review. *Annual Review of Information Science & Technology, 45*, 199–245.

Bornmann, L. (2013a). How to analyze percentile citation impact data meaningfully in bibliometrics: The statistical analysis of distributions, percentile rank classes and top-cited papers. *Journal of the American Society for Information Science and Technology, 64*(3), 587–597.

Bornmann, L. (2013b). The problem of percentile rank scores used with small reference sets. *Journal of the American Society for Information Science and Technology, 64*(3), 650.

Bornmann, L., & Daniel, H.-D. (2009). Universality of citation distributions. A validation of Radicchi et al.'s relative indicator $c_f = c/c_0$ at the micro level using data from chemistry. *Journal of the American Society for Information Science and Technology, 60*(8), 1664–1670.

Bornmann, L., Marx, W., Schier, H., Rahm, E., Thor, A., & Daniel, H. D. (2009). Convergent validity of bibliometric Google Scholar data in the field of chemistry: Citation counts for papers that were accepted by Angewandte Chemie International Edition or rejected but published elsewhere, using Google Scholar, Science Citation Index, Scopus, and Chemical Abstracts. *Journal of Informetrics, 3*(1), 27–35. doi:10.1016/j.joi.2008.11.001.

Bornmann, L., & Mutz, R. (2013). The advantage of the use of samples in evaluative bibliometric studies. *Journal of Informetrics, 7*(1), 89–35. doi:10.1016/j.joi.2012.08.002.

Bornmann, L., Mutz, R., & Daniel, H.-D. (2010). The *h* index research output measurement: Two approaches to enhance its accuracy. *Journal of Informetrics, 4*(3), 407–414. doi:10.1016/j.joi.2010.03.005

Bornmann, L., Mutz, R., Marx, W., Schier, H., & Daniel, H.-D. (2011). A multilevel modelling approach to investigating the predictive validity of editorial decisions: Do the editors of a high-profile journal select manuscripts that are highly cited after publication? *Journal of the Royal Statistical Society. Series A, (Statistics in Society), 174*(4), 857–879. doi:10.1111/j.1467-985X.2011.00689.x.

Bornmann, L., Mutz, R., Neuhaus, C., & Daniel, H.-D. (2008). Use of citation counts for research evaluation: Standards of good practice for analyzing bibliometric data and presenting and interpreting results. *Ethics in Science and Environmental Politics, 8*, 93–102. doi:10.3354/esep00084

Boyack, K. W. (2004). Mapping knowledge domains: characterizing PNAS. *Proceedings of the National Academy of Sciences of the United States of America, 101*, 5192–5199.

Calver, M., & Bradley, J. (2009). Should we use the mean citations per paper to summarise a journal's impact or to rank journals in the same field? *Scientometrics, 81*(3), 611–615.

Daniel, H.-D., Mittag, S., & Bornmann, L. (2007). The potential and problems of peer evaluation in higher education and research. In A. Cavalli (Ed.), *Quality Assessment for Higher Education in Europe* (pp. 71–82). London: Portland Press.

De Bellis, N. (2009). *Bibliometrics and citation analysis: From the Science Citation Index to cybermetrics*. Lanham, MD: Scarecrow Press.

Gagolewski, M. (2011). Bibliometric impact assessment with R and the CITAN package. *Journal of Informetrics, 5*(4), 678–692. doi:10.1016/j.joi.2011.06.006.

García-Pérez, M. A. (2010). Accuracy and completeness of publication and citation records in the Web of Science, PsycINFO, and Google Scholar: A case study for the computation of h indices in Psychology. *Journal of the American Society for Information Science and Technology, 61*(10), 2070–2085. doi:10.1002/asi.21372.

Garfield, E. (1972). Citation analysis as a tool in journal evaluation: Journals can be ranked by frequency and impact of citations for science policy studies. *Science, 178*(4060), 471–479.

Garfield, E. (1979). *Citation indexing: Its theory and application in science, technology, and humanities*. New York: Wiley.

Garfield, E. (2006). The history and meaning of the Journal Impact Factor. *Journal of the American Medical Association, 295*(1), 90–93.

Glänzel, W., Debackere, K., Thijs, B., & Schubert, A. (2006). A concise review on the role of author self-citations in information science, bibliometrics and science policy. *Scientometrics, 67*(2), 263–277.

Glänzel, W., Thijs, B., Schubert, A., & Debackere, K. (2009). Subfield-specific normalized relative indicators and a new generation of relational charts: Methodological foundations illustrated on the assessment of institutional research performance. *Scientometrics, 78*(1), 165–188.

Hemlin, S. (1996). Research on research evaluations. *Social Epistemology, 10*(2), 209–250.

Hirsch, J. E. (2005). An index to quantify an individual's scientific research output. *Proceedings of the National Academy of Sciences of the United States of America, 102*(46), 16569–16572. doi:10.1073/pnas.0507655102.

Huang, M.-H., Lin, C.-S., & Chen, D.-Z. (2011). Counting methods, country rank changes, and counting inflation in the assessment of national research productivity and impact. *Journal of the American Society for Information Science and Technology, 62*(12), 2427–2436. doi:10.1002/asi.21625.

Jacsó, P. (2009). Google Scholar's ghost authors. *Library Journal, 134*(18), 26–27.

Jacsó, P. (2010). Metadata mega mess in Google Scholar. *Online Information Review, 34*(1), 175–191. doi:10.1108/14684521011024191.

Jann, B. (2005). Tabulation of multiple response. *Stata Journal, 5*(1), 92–122.

Larivière, V., & Gingras, Y. (2011). Averages of ratios vs. ratios of averages: An empirical analysis of four levels of aggregation. *Journal of Informetrics, 5*(3), 392–399. doi:10.1016/j.joi.2011.02.001.

Lehmann, S., Jackson, A., & Lautrup, B. (2008). A quantitative analysis of indicators of scientific performance. *Scientometrics, 76*(2), 369–390. doi:10.1007/s11192-007-1868-8.

Leydesdorff, L., Bornmann, L., Mutz, R., & Opthof, T. (2011). Turning the tables in citation analysis one more time: Principles for comparing sets of documents. *Journal of the American Society for Information Science and Technology, 62*(7), 1370–1381.

Martin, B. R., & Irvine, J. (1983). Assessing basic research: Some partial indicators of scientific progress in radio astronomy. *Research Policy, 12*(2), 61–90.

Marx, W. (2011). Special features of historical papers from the viewpoint of bibliometrics. *Journal of the American Society for Information Science and Technology, 62*(3), 433–439. doi:10.1002/asi.21479

Marx, W., Schier, H., & Andersen, O. K. (2006). Using time-dependent citation rates (sales curves) for comparing scientific impacts. Retrieved from http://arxiv.org/ftp/physics/papers/0611/0611284.pdf.

Meho, L. I., & Spurgin, K. M. (2005). Ranking the research productivity of library and information science faculty and schools: An evaluation of data sources and research methods. *Journal of the American Society for Information Science and Technology, 56*(12), 1314–1331.

Moed, H. F. (2005). *Citation analysis in research evaluation*. Dordrecht, The Netherlands: Springer.

Moed, H. F., Van Leeuwen, T. N., & Reedijk, J. (1996). A critical analysis of the journal impact factors of *Angewandte Chemie* and the *Journal of the American Chemical Society*: Inaccuracies in published impact factors based on overall citations only. *Scientometrics, 37*(1), 105–116.

NORIA-net. (2011). *Comparing research at Nordic universities using bibliometric indicators: A publication from the NORIA-net "Bibliometric Indicators for the Nordic Universities."* Oslo, Norway: NordForsk.

Pendlebury, D. A. (2008). *Using bibliometrics in evaluating research*. Philadelphia: Research Department, Thomson Scientific.

Radicchi, F., & Castellano, C. (2011). Rescaling citations of publications in physics. *Physical Review E: Statistical, Nonlinear, and Soft Matter Physics, 83*(4). doi:10.1103/PhysRevE.83.046116.

Radicchi, F., Fortunato, S., & Castellano, C. (2008). Universality of citation distributions: Toward an objective measure of scientific impact. *Proceedings of the National Academy of Sciences of the United States of America, 105*(45), 17268–17272. doi:10.1073/pnas.0806977105.

Randić, M. (2009). Citations versus limitations of citations: Beyond Hirsch index. *Scientometrics, 80*(3), 809–818.

Rehn, C., Kronman, U., & Wadskog, D. (2007). *Bibliometric indicators—definitions and usage at Karolinska Institutet*. Stockholm, Sweden: Karolinska Institutet University Library.

Rousseau, R. (2012). Basic properties of both percentile rank scores and the I3 indicator. *Journal of the American Society for Information Science and Technology, 63*(2), 416–420. doi:10.1002/asi.21684

Ruegg, R., & Feller, I. (2003). *A toolkit for evaluating public R&D investment: Models, methods, and findings from ATP's first decade*. Gaithersburg, MD: National Institute of Standards and Technology.

Schreiber, M. (2012a). Inconsistencies of recently proposed citation impact indicators and how to avoid them. Retrieved from http://arxiv.org/abs/1202.3861.

Schreiber, M. (2012b). Uncertainties and ambiguities in percentiles and how to avoid them. Retrieved from http://arxiv.org/abs/1205.3588.

Seglen, P. O. (1992). The skewness of science. *Journal of the American Society for Information Science, 43*(9), 628–638.

Sheskin, D. (2007). *Handbook of parametric and nonparametric statistical procedures* (4th ed.). Boca Raton, FL: Chapman & Hall/CRC.

Snyder, H., & Bonzi, S. (1998). Patterns of self-citation across disciplines (1980–1989). *Journal of Information Science, 24*(6), 431–435.

StataCorp. (2011). *Stata statistical software: release 12*. College Station, TX: Stata Corporation.

Strotmann, A., & Zhao, D. (2010). Combining commercial citation indexes and open-access bibliographic databases to delimit highly interdisciplinary research fields for citation analysis. *Journal of Informetrics, 4*(2), 194–200. doi:10.1016/j.joi.2009.12.001.

Tijssen, R., & Van Leeuwen, T. (2006). Centres of research excellence and science indicators: Can "excellence" be captured in numbers? In W. Glänzel (Ed.), *Ninth International Conference on Science and Technology Indicators* (pp. 146–147). Leuven, Belgium: Katholieke Universiteit Leuven.

Tijssen, R., Visser, M., & Van Leeuwen, T. (2002). Benchmarking international scientific excellence: Are highly cited research papers an appropriate frame of reference? *Scientometrics, 54*(3), 381–397.

Vinkler, P. (2010). *The evaluation of research by scientometric indicators*. Oxford: Chandos.

Wright, M., & Armstrong, J. S. (2007). Verification of citations: Fawlty towers of knowledge? Retrieved from http://mpra.ub.uni-muenchen.de/4149.

12 Identifying and Quantifying Research Strengths Using Market Segmentation

Kevin W. Boyack and Richard Klavans

Introduction

There has been significant progress over the past 50 years in our ability to estimate research strengths using data on scientific publications. Prior to the 1960s, simple counts of articles and books using a library classification system were sufficient. Major improvements in measuring research strengths were achieved in the 1960s using citation counts (to weight the strength of each publication) and journal-based disciplinary classification systems (an improvement over library classification systems). Normalization schemes to account for differing publication and citation rates in different disciplines were introduced somewhat later, and are in common use today.

There are, however, significant shortcomings in the use of disciplines as the basis for evaluating research strengths. In today's world, most institutional research units (and thus potential strengths) are not disciplinary; they are either multidisciplinary or subdisciplinary. No single academic department, nor any single program at a funding agency, produces research that covers the entire footprint of a discipline. Researchers collaborate within and between institutions and departments on groups of research topics that are far smaller than disciplines.

Is there a reasonable alternative to the disciplinary approach? The answer to that question is best addressed by considering two perspectives from the field of management. One perspective, commonly adopted by economists, is to use standardized categories to track strengths. The second perspective, commonly adopted by managers, is to develop idiosyncratic categories. Industry and science both apply the economic perspective—the Standard Industrial Classification (SIC) codes in industry, and journal-based disciplinary categories in science. This is the top-down approach using predefined categories for reporting output and impact.

Industry applies the managerial approach using a technique known as market segmentation, which allows the manager to develop a unique definition of market segments to demonstrate competitive advantage. These self-defined segments are, in essence, the self-defined strengths or competencies of the institution, and they are promoted as such (e.g., ". . . is recognized as the world leader in stent design . . ."). Market segmentation is considered best practice in the business world and is used to overcome the limitations imposed by the economic, top-down, standardized category approach to measuring leadership. The world of science, however, has historically not used the practice of market segmentation.

In this chapter we apply the idea of market segmentation to bibliographic data and present an approach that accurately identifies and quantifies the research strengths of an institution. We compare results from this new approach with those obtained from traditional discipline-based approaches and explore a detailed example. The chapter closes by examining potential drawbacks of the new approach and suggesting research directions that can build and improve on the method.

Traditional Approach

The weaknesses of the traditional approach to research evaluation can be demonstrated using two examples. The biannual *Science and Engineering Indicators* (SEI) reports published by the U.S. National Science Board (2012, chap. 5 and related tables) provide publication counts by country for a set of 13 broad fields. These 13 broad fields (and accompanying 125 subfields or disciplines) cover a set of 5,085 internationally recognized, peer-reviewed journals. In addition to reporting publication counts by country, year, and field for the set of journals indicated, quantities are also reported for the top 1% of highly cited articles, giving an additional indicator for the highest-quality research.

Unfortunately, this technique does not identify the current strengths of a nation. Counting citations or counting the top *n*% of highly cited articles measures historical impact rather than current strengths. In addition, 125 disciplines is an extremely broad classification system for science. Rarely, if ever, will nations with smaller research budgets and known strengths in science appear to be a leader when the classifications are this large. For example, at this level of analysis Germany is not a research leader in any chemistry, physics, or engineering area (Klavans & Boyack, 2010), even though these are traditional areas of national strength that are still apparent in the scientific literature. In general, highly aggregated categories tend to overstate the strengths of the largest nation and correspondingly understate the strengths of smaller nations.

The Centre for Science and Technology Studies (CWTS) at Leiden University employs field normalization to rank universities using a variety of indicators. The so-called crown indicator (Moed, DeBruin, & Van Leeuwen, 1995) that was used for many years has now been superseded by the mean normalized citation score (MNCS, similar to the crown indicator) and $PP_{top10\%}$ (Waltman et al., 2012), which reports the fraction of an institution's articles that are among the top 10% highly cited by field. Fields or disciplines are defined using the Thomson Reuters Web of Science journal subject categories, a set of over 200 categories to which journals are multiply assigned. The normalization scheme used calculates an expected number of citations for each article based on its age and the subject category to which it belongs.

This approach is quite reasonable if one uses the economic perspective. However, the underlying motive of the economic perspective as applied to industry was to model profit-seeking behavior and to correspondingly identify potentially anticompetitive behavior. Industry categories were developed for regulatory purposes, to use as evidence in a court of law that a firm is a monopolist or that a proposed merger has anticompetitive effects. Industry categories were not developed by managers to measure a firm's strengths.

Disciplinary categories in science were developed for retrieval purposes, and not for evaluation. They share the same shortcomings as industry categories. The strengths of smaller universities are likely to be understated because the disciplinary categories are far too large. The strengths of any university that is following a multidisciplinary strategy will be understated. In addition, it is rare that managers in the scientific enterprise plan research efforts that are based on disciplinary categories (as defined by a few hundred clusters of journals). Rather, they plan research efforts based on their current resources (people, infrastructure, etc.), known "hot" topics, and the resources (e.g., funding) that are available. These factors do not often line up with traditional disciplinary structures.

Market Segment Approach

Several years ago we began a project to identify the research strengths at the University of California at San Diego (UCSD). Knowing that the National Science Foundation (NSF) journal classification (used in the SEI reports) had only 125 categories, we first decided to create a new journal classification system with a larger number of categories. We assumed that if we had more categories, UCSD might be ranked highly in a sufficient number of categories to use them as proxies for research strengths. We spent a great deal of time creating the so-called UCSD classification system and map of science (Börner et al., 2012), which has 554 categories, only to find that UCSD was ranked #1 in only one category internationally, and in only five categories when compared with

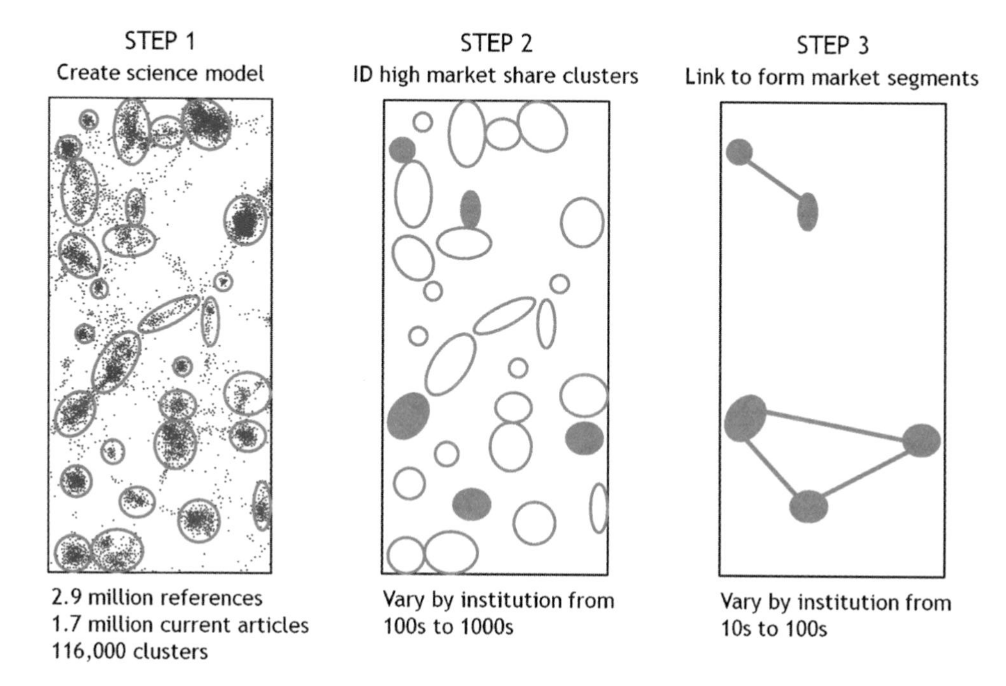

Figure 12.1
Visual representation of the steps used to identify research strengths (market segments) for an institution.

institutions in the United States. Increasing the number of journal-based disciplinary categories by a factor of four did not increase our ability to identify research strengths at this university in a meaningful way (Klavans & Boyack, 2010).

We thus decided to develop a new approach based on market segmentation, to identify and quantify the segments in which an institution is a research leader. Conceptually, this new approach consists of three steps (figure 12.1). Although the detailed methodology behind this approach is described in Klavans and Boyack (2010), we discuss it briefly here. First, a highly detailed model of the scientific literature is needed, one that will allow assembly of market segments of different sizes. We accomplish this step by generating a classification system at the article level rather than at the journal level using cocitation analysis (Boyack & Klavans, 2010). Using a single year of Scopus data, we identify a set of cited reference papers and then cluster those reference papers using a combination of force-directed layout and single-link clustering. Once the reference papers have been clustered, the current-year papers are fractionally assigned to the clusters using their reference lists. For the 2010 publication year, our model contains over 2.9 million references, 1.7 million current-year articles, and 116,000 clusters. We also fractionally assign articles from the previous four years (in this case 5.9 million articles from 2006 to 2009) to the clusters so that recent temporal trends can be derived for each cluster. Note that although this model is highly detailed and its creation is computationally intensive, it is only calculated once and can then be used for a variety of analyses.

Steps 2 and 3 in our approach for identifying and quantifying market segments for an institution are based on the clusters and their contents, and can be carried out separately for each institution to be analyzed. In step 2, we identify clusters in which the institution of interest has a high relative publication (market) share, where the relative share is calculated as the publication share of the institution divided by the publication share of the largest competitive institution. Clusters in which the institution has a high relative share are the topics on which the institution focuses. Large institutions, such as the largest universities, tend to have 1,000 or more (out of the 116,000) clusters in which they have a high relative market share. Smaller institutions tend to have hundreds of high market share clusters. However, a document cluster is not equivalent to a market segment. Rather, a market segment is a group of clusters that are strongly linked together by the institution. Thus, our third step is to link clusters using only the publications by the institution of interest. In other words, we group clusters into market segments when the authors themselves have explicitly linked those topics together through their citation patterns. Since most institutions tend to excel at the things in which they choose to specialize, these segments are areas in which the institution typically has high impact as well as high activity. Thus, we can consider these market segments to be the research strengths of the institution.

Quantifying Research Strengths

Once research strengths have been identified, they are described and quantified using several metrics that correspond to the concepts of publication leadership, reference leadership, and thought (or innovation) leadership (Klavans & Boyack, 2010). Each of these metrics is simple and has a different purpose. Publication leadership is meant to address current activity. An institution is a publication leader in its research strength if it has a high relative article share (RAS), which is calculated as the number of articles published by the institution divided by the number of articles published by the top competing institution. For example, in a case where the institution of interest published 50 articles in the research strength and the 2nd-ranked institution published 40 articles, the RAS value would be 1.25.

Reference leadership is intended to show recent activity combined with current impact. An institution is a reference leader in its research strength if it has a high relative reference share (RRS), which is calculated as the number of its reference articles published within the past five years divided by the number of reference articles published by the top competing institution over the same time period. For example, in the case where the institution of interest published 27 highly cited reference articles, and the top competing institution published 30, the RRS value would be 0.90. Note that this metric only considers references that are recent, and that were highly cited enough to have been included in the model (only about 12% of all references) generated in step 1 of our approach. Thus, recent papers with little impact are not considered and do not contribute to the concept of research leadership.

Thought or innovation leadership is not count-based, but is rather intended to reflect on the ability of the institution to quickly build on recent discoveries in the area of the research strength. The premise behind this indicator is that the more innovative institutions will rapidly build on advances in their areas of strength. Thought leadership is measured using an indicator called state of the art (SOA), which is calculated as the average median reference age for all of the articles published by all institutions in the research strength minus the average median reference age for the articles published by the institution of interest. Median reference age is calculated for each article and then averaged over articles. If the SOA value is positive, then the institution of interest is referencing more recent material than the world as a whole in the area of the research strength. Thought leadership is a very important type of research leadership that complements the other two types. For example, an institution can be the publication leader in an area of research strength, but if it is not building on recent research and does not have a high SOA value, it will not be taken as seriously by the rest of the world. In contrast, if an institution is one of the publication leaders in an area but is not ranked #1 in RAS, it is still highly regarded if it is working near the state of the art and has a high SOA value. In a sense, thought leadership is a measure of the quality of current activity.

Validation

We spent a considerable amount of time validating the market segment–based approach by conducting card-sorting exercises with over 60 researchers at two universities (UCSD and the University of Utah). After calculating research strengths for these two universities, we identified key researchers in the research strengths and asked them to participate in a card-sorting exercise. This exercise was designed to determine if the structure and content of the calculated research strength were consistent with the researcher's perception of his or her area of expertise and linkages to other topics and researchers at the university.

To prepare for each card-sorting exercise, we identified (a) the clusters in the research strength of interest, and (b) the clusters in which the researcher publishes. This mixture of clusters was intended to include topics in which the researcher was a recognized expert (whether or not they were part of the research strength), and the linked topics in the research strength that the researcher would likely be highly aware of but not working in personally. In most cases, this was limited to a set of 50 clusters. We then printed a set of postcard-sized cards, one card for each cluster. Each card contained a list of 10 high mutual information bigrams (phrases) extracted from the titles and abstracts of the articles in the cluster, a list of the 5 most highly cited recent articles in the cluster, and a list of the 5 most active researchers in the cluster. We found that this mix of information was highly recognizable by the researchers we interviewed. Using the combination of terms, articles, and authors, researchers were able to easily (and in most cases, instantly) recognize what each cluster was about, and were able to explain to us the fine distinctions between one cluster and another.

Each card-sorting exercise consisted of two parts. In the first part, we asked the researcher to separate the cards into three categories based on level of expertise (expert, familiar with, not aware of). The cards in the third category (not aware of) were put away, and the expert was then asked to group the remaining cards into higher-level sets of topics, to provide a short label or descriptor for each of these higher-level groupings, and to then describe the linkages (if any) between these higher-level groupings. This validation study produced several findings:

- Researchers understood and were able to differentiate between the fine-grained clusters in our model of science.
- Researchers were able to easily self-identify their areas of expertise using the information provided.
- Researchers were highly aware of the larger context in which their research was conducted. When considering first- and second-order linkages (linked topics in the research strength that were one or two linkages away from their own area of expertise), experts were intimately aware of over 90% of these linkages, and stated that in most cases these linkages were not accidental but were part of a larger research strategy.
- Researchers were typically not aware of linked topics in the research strength that were three or more linkages away from their own area of expertise.

In summary, although the full high-level structures of the calculated research strengths were only reproduced by researchers in about 25% of the cases, the correlation between the researchers' perceptions of their local networks (within a radius of two linkages of topic groupings) and the calculated research strengths was very high. In other words, the researchers understood and appreciated the idea of market segments as the broader strategic context within which they work. In addition, we note that this idea of market segment–based planning and evaluation of science is gaining traction; Elsevier's SciVal® Spotlight tool is based on this approach and is being used by institutions throughout the world.

Comparison of Approaches

It is instructive to compare the results from the different types of research evaluation approaches mentioned above. The Leiden Ranking 2011[1] provides metrics for 500 of the largest universities worldwide in terms of numbers of publications indexed in the Web of Science database. Sizes range from 33,511 articles indexed (Harvard) to 1,262 (Lille 2)—a substantial range in publication activity. This list of 500 universities is a very suitable set for comparing results of evaluation approaches.

Table 12.1 lists the size rank, number of articles (P), and Leiden's preferred metric, the proportion of top 10% publications ($PP_{top10\%}$) from the most recent Leiden Ranking for a sample of 21 institutions. $PP_{top10\%}$ is the fraction of a university's publications from 2005 to 2009 that are within the top 10% most frequently cited by discipline, publication year, and document type. A value of greater than 10% indicates that the university does better than the world at large at producing highly cited work across all fields. This single number does not, however, distinguish between disciplines—it does not show which disciplines are the strongest contributors to the whole. Yet it should be noted that, although not widely reported, in order to calculate $PP_{top10\%}$ the Leiden methodology does generate these values by discipline. $PP_{top10\%}$ ranges between 25.2% (MIT) and 2.0% (St. Petersburg State University) for the 500 universities in the sample.

Table 12.1 also shows results of a traditional set of rankings by discipline. Using the 2005–2009 publication years of Scopus, we calculated the numbers of publications (and citations) by discipline for the more than 4,000 institutions in our university database using the 554 journal categories in the UCSD journal classification system (Börner et al., 2012). For each of the 500 universities in the Leiden list, the number of disciplines in which the university was ranked in the top 10 by publication counts ($Disc_{top10}$) was noted. Although not reported here, we also calculated the number of disciplines in which each university was ranked in the top 10 by total citation counts. The correlation between these two values for the set of 500 universities is sufficiently high ($R^2 = 0.858$) that they do not need to be considered independently. Values of $Disc_{top10}$ range between 177 (Harvard) and zero (87 different universities). Nearly half (245) of the 500

Table 12.1
Comparison of results from research evaluation approaches

Rank $_P$	University	P	$PP_{top10\%}$	$Disc_{top10}$	STR
1	Harvard Univ	33,511	22.54%	177	661
6	Johns Hopkins Univ	16,343	16.45%	70	352
31	Univ Calif Berkeley	11,713	21.01%	37	220
56	Univ Texas Austin	8,743	16.68%	26	191
81	Univ Maryland	7,277	15.72%	8	179
106	Monash Univ	6,797	10.45%	12	201
131	Chinese Univ Hong Kong	6,029	10.13%	0	177
156	Univ Colorado Boulder	5,439	17.34%	5	132
181	Univ Ottawa	4,790	11.32%	4	167
206	Univ Torino	4,448	9.51%	0	141
231	Univ Buenos Aires	4,072	5.45%	3	117
256	Fed Univ Rio Grande Sul	3,796	4.93%	3	130
281	Univ Delaware	3,522	13.15%	3	100
306	Univ Regensburg	3,177	11.46%	1	109
331	Gazi Univ	2,952	4.07%	4	87
356	Kansas State Univ	2,676	11.06%	0	90
381	Giessen Univ	2,530	10.20%	0	80
406	Beijing Normal Univ	2,381	7.73%	5	77
431	Univ Rostock	2,222	9.45%	0	85
456	Univ Ioannina	2,031	8.71%	0	89
481	Univ Ferrara	1,822	8.78%	0	68

universities in our sample are ranked among the top 10 in two or fewer disciplines. Does this mean that these institutions, with few or no high disciplinary rankings, have few research strengths? Or does it simply show that disciplines are the wrong basis to use for identifying research strengths?

To help answer this question, our approach has been used to calculate the number of research strengths for each of the 500 institutions in our sample set. Table 12.1 shows that there are significant numbers of research strengths (STR) for institutions that do not fare well in traditional rankings. Numbers of strengths range from a high of 661 (Harvard) to 9; there are only two institutions in the list of 500 with fewer than 40 research strengths. This suggests that the discipline, embodied in most analyses as a group of journals, is not suitable as a basis for identifying the research strengths of an institution.

Table 12.1 also shows that the number of research strengths seems to be a function of the number of articles published by an institution. Figure 12.2, which shows the

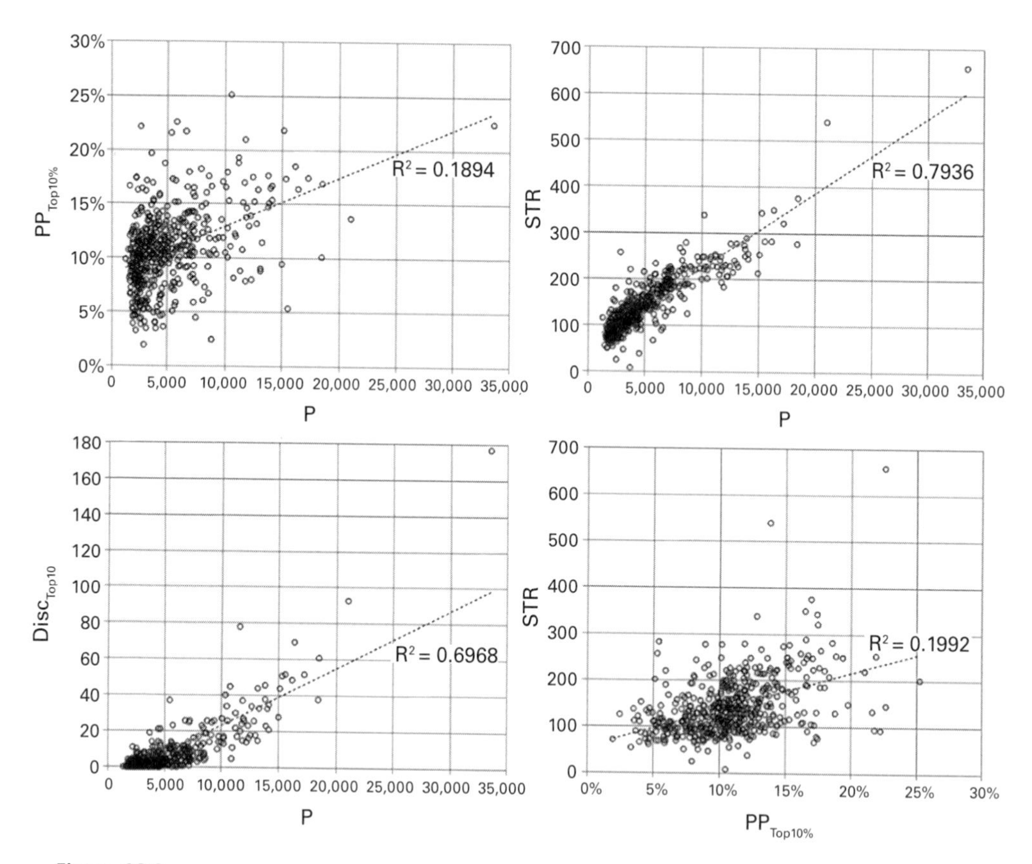

Figure 12.2
Correlations between metrics associated with different evaluation approaches.

numbers of strengths for all 500 universities as a function of size (upper right), affirms that this is true, and that the correlation between numbers of strengths and size is very high (R^2 = 0.794). Yet, it is also true that the number of disciplines in which an institution ranks in the top 10 is also strongly correlated with size (figure 12.2, lower left). $PP_{top10\%}$, which specifically attempts to normalize for size effects (Waltman et al., 2012), has a much lower, yet still measurable, correlation with size (figure 12.2, upper left). A major difference between the traditional rankings by discipline and the identification of research strengths using the new approach, even though both are strongly correlated with size, is that small institutions are shown to have significant numbers of research strengths despite their small size, while they have few, if any, high disciplinary rankings. The main reason for this is that areas defined as research strengths are much smaller than disciplines; the average size of the research strengths for the 500 universities in our sample is around 300 articles. The correlation between number of strengths and PPtop10% is also shown in figure 12.2 (lower right) to highlight the fact that it is relatively weak. Institutions with lower fractions of highly cited articles can still have research strengths in their areas of concentration.

Detailed Example

We have shown how highly specific research strengths can be identified for an institution, and have compared the numbers of research strengths identified using our new approach with more traditional metrics. Yet, this does not tell the real story. The purpose of identifying research strengths is not so much to know how many there are, but rather to know what they are. For this reason it is critical not only to identify these strengths, but to characterize them. We do this using the leadership metrics mentioned above, and also by looking at the actual contents of these strengths as exemplified by features such as authors, titles, terms, journal names, and discipline names.

Visualization can be a powerful entry point into these data. To that end, we have created a visual layout and template referred to as a circle of science. Briefly, the 554 disciplines from the UCSD classification system have been grouped into 13 major fields. These 13 fields, along with the 554 disciplines and the 116,000 clusters in our model of science, have been placed in order around a circle using multiple factor analyses (Börner et al., 2012), resulting in the circular structure shown in figure 12.3. This map orders fields from mathematics and physics at the top, clockwise through the physical sciences and engineering, through the medical, social, and computer sciences, and then back to mathematics. This continuous structure is important and is not arbitrary; it emerges naturally at a high level from nearly every comprehensive map of science generated over many decades using many different datasets and mapping approaches (Klavans & Boyack, 2009; Rafols, Porter, & Leydesdorff, 2010).

The map in figure 12.3 shows the research strengths (called competencies in the figure) of an *Example University* and their relative positions within the circle of science. Although the name of the institution is not given here, this is an actual example of a U.S. university that publishes around 2,500 articles per year. In terms of annual article counts, this university ranks in the 350–400 range internationally. Despite its modest size, this university has 101 measurable areas of research strength. Each strength is represented by a gray circle, sized to reflect the publication counts at the university within the strength, and located at the average position of the article clusters that are linked together in the strength. Strengths at the edge of the circle are more single-disciplinary, while those that are away from the edge of the circle are more interdisciplinary. The colored rays within each circle show the fields associated with the clusters that were

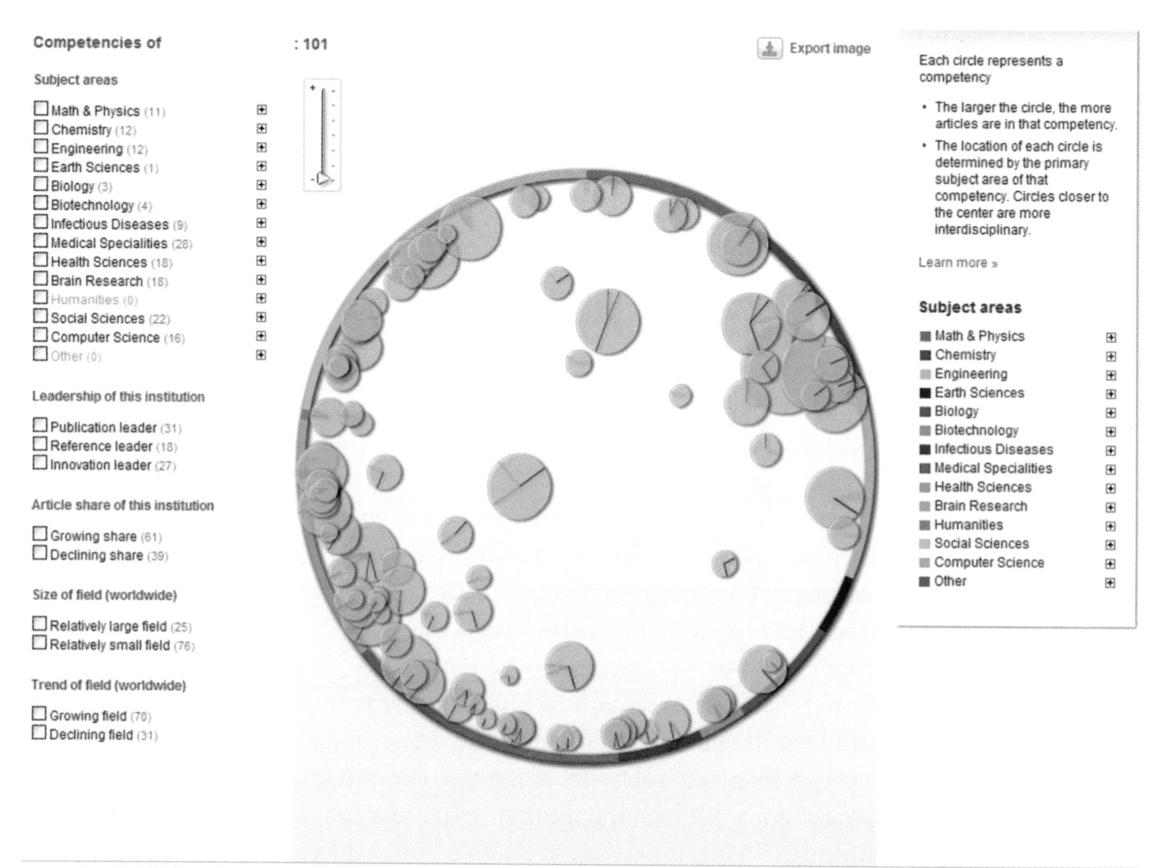

Figure 12.3
Visual representation of the research strengths (market segments) for an *Example University*. Graphic courtesy of Elsevier's SciVal® Spotlight.

linked to form the research strength. For example, the strength just to the left of the center of the circle contains orange, pink, and blue lines; topics from brain research, computer science, and chemistry all contribute to this strength. This is an example of a highly interdisciplinary research strength, one that could not have been identified using discipline- or journal-based approaches.

The distribution of research strengths around the map is fairly uniform, showing that this university has strengths in most of the major fields of science, with some concentration in the health sciences (orange) and chemistry (blue). This distribution is not surprising. While universities that specialize (e.g., medical or technical universities)

Table 12.2
Characterization of 10 of the larger research strengths of *Example University*

ID	Descriptors	P_{Tot}	P_{Univ}	RAS	RRS	SOA (yr)
6	powders, carbides, elastic moduli, nanoindentation	1,075	81	1.44	1.97	0.72
14	sky surveys, galaxies, quasars	1,250	61	0.56	0.27	0.49
19	sensors, resonance, natural frequencies	922	57	2.87	4	1.35
2	tissue engineering, mechanical properties, scaffolds	1,792	50	1.92	0.76	1.3
46	brain, near infrared spectroscopy, hemodynamics	602	49	4.7	4.33	1.41
18	capacitance, carbon, electrodes	985	49	1.02	1.07	0.97
16	plasmas, surface coating technology, endoscopy	808	44	2.1	1.28	1.09
1	complex regional pain syndromes, ketamine	1,721	44	0.93	0.86	2.36
8	axons, neurons, microtubules	1,797	43	1.9	1.25	1.35
22	nanopores, bacteria, DNA	983	39	1.52	0.57	0.21

tend to have research strengths in only one-half of the map, universities that do not specialize in that way tend to have strengths throughout the map.

Some details about 10 of the research strengths from figure 12.3 are given in table 12.2, including some descriptive terms. Several interesting combinations of terms appear, confirming that many of these strengths are interdisciplinary, and yet at the same time much smaller than a discipline. For example, strength #16 combines plasma physics and surface coating technology with a medical domain—endoscopy. The indicators are also important and show where this university is a clear leader in its self-defined market segments. For instance, the university has large (substantially greater than 1.0) RAS and RRS numbers in five of the ten strengths listed in table 12.2. This means that not only is the university publishing more research (RAS) in these areas than any other institution in the world, but that it has the largest number of highly cited articles in that area (RRS) over the previous five years as well. Thus its leadership in this segment is not simply based on the current year, but is built on a base of work lasting several years. The SOA values in these five areas range from 0.72 to 1.41 years; researchers at this university are building on current work in these areas more quickly than the world as a whole. These five strengths are thus robust in that they exhibit all three types of leadership: publication, reference, and thought leadership.

Two more examples are also very interesting. When looking at the numbers, it is clear that the university is not the overall leader in strength #14. The RAS value of 0.56

shows that it is only publishing half as much as the leader in the area, and the RRS value of 0.27 shows that others are generating far more of the highly cited literature on which this area is based. Strength #1 is intriguing in that the university is very close to being the leader in RAS (0.93) and RRS (0.86) and has a very large positive SOA value (2.36 years). Given that this university is rapidly incorporating key work, it is very possible that its leadership in this area will increase in the near future.

The data associated with research strengths can be used in other strategic ways as well. One of these is in identifying the level of collaboration with other institutions and how that overlaps with and supports the strengths. One can see which universities seem to be strategic partners across a range of strengths. Such information can be used as input to strategic planning. In addition, where levels of collaboration are low, the data show other researchers and institutions doing similar work; these are potential collaborators.

Data associated with research strengths can also enable a university to gain a better understanding of its key researchers and their contributions. Who exactly are the key researchers at a university? Some are very recognizable, while others may be less so. For example, researchers with tenure and high *h*-indexes or who run large laboratories are very well known to the university administrators (e.g., provosts, vice presidents for research, etc.) who often preside over internal funding decisions. Perhaps less well known are the younger, less established researchers with lower *h*-indexes who may, nevertheless, be key players in the research strengths of the university. Figure 12.4 shows a plot of *h*-indexes of the top 100 publishing researchers at our example university. Each of these researchers published 20 or more articles between 2007 and 2011. The highest *h*-index researchers (those at the upper right) are heavily involved in the research strengths of the university. Yet, there are also a few relatively high *h*-index researchers who do little or no publishing in the research strengths of the university. More importantly, there are a number of low *h*-index researchers who have significant numbers of publications in the research strengths of the university. These researchers have the potential to be the rising stars of the future. Analysis such as this provides a way to identify them.

Discussion

No approach to identifying research strengths is without its shortcomings. The major shortcoming of our new approach is its relative instability compared with traditional discipline-based approaches. A comparison of research strengths from one year to the next shows that while large research strengths tend to persist from year to year, smaller

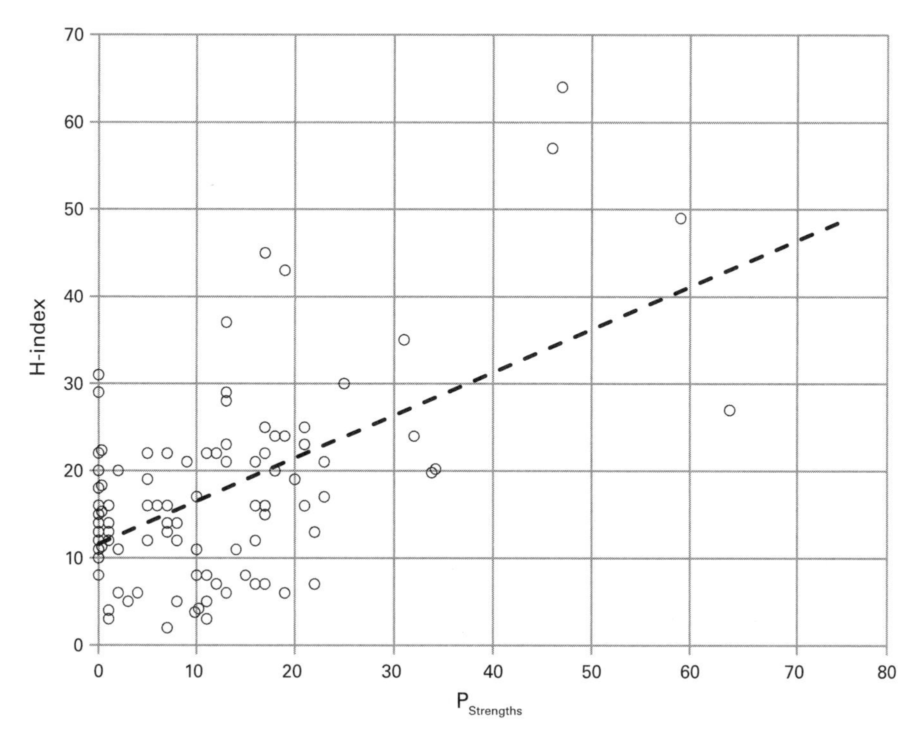

Figure 12.4
H-index values and numbers of articles in strengths for the top 100 publishing researchers at *Example University*.

research strengths are far less stable. Small research strengths may be there one year and gone the next.

This instability is a natural outcome of partitioning science into a large number of very small categories. It is well known that researchers constantly move between research problems. As they solve one problem they move to the next. When they try something that does not work, they reevaluate and then try to solve things in a different way. This is the essence of scientific research; there is a high degree of flux or instability at the bench level. This instability is naturally smoothed out as one aggregates the work to higher levels, such as journal-based disciplines. The birth and death rates of disciplines are extremely low; there is high stability in science at the discipline level. By contrast, the birth and death rates of topics or research problems is very high. This is the major source of the instability in the small research strengths identified by our new approach. An additional source of instability are the thresholds that are applied to identify high market share document clusters. Additional research is planned to seek ways to increase the stability of the smaller research strengths.

We are also very interested in the linkages between research strengths and both prior funding and future funding opportunities. Although these data are not yet linked in a systematic and comprehensive way, we envisage the day when those data will be linked and the effects of funding can be studied in a context that will provide for highly accurate results.

Conclusion

In this chapter we argue that a journal-based disciplinary approach, such as the one used in the SEI reports, is fundamentally flawed. The problem is not with the act of counting; this is a quite reasonable surrogate for strength. Rather, the problem is with the categories (i.e., disciplines) that are used to characterize different areas of science. The problem is simple: disciplines do not capture the unique multidisciplinary or subdisciplinary activities of sets of researchers. Researchers at a university or located in a region (state or nation) tend to self-organize around sets of multidisciplinary research problems. An assessment of strengths must take this into account.

We have introduced an approach modeled on the concept of market segmentation—an established business practice—to identify the research strengths of an institution from the literature. Using a detailed example, we have shown how this approach identifies a large number of research strengths for a university of modest size, and have further shown how these strengths can be quantified. Detailed information associated with research strengths can be used for a variety of strategic purposes.

This new approach is not intended to replace the traditional research evaluation approach, but rather to complement it. If single-numbered university or department comparisons (such as rankings) are needed, the traditional approach should be used. However, if detailed information about actual strengths is desired, this new market segment–based approach is well suited to providing such information.

Acknowledgment

We thank Mehul Pandya at Elsevier for supplying the data on research strengths that appear in table 12.1 and figure 12.2.

Note

1. http://www.leidenranking.com/ranking

References

Börner, K., Klavans, R., Patek, M., Zoss, A. M., Biberstine, J. R., Light, R. P., et al. (2012). Design and update of a classification system: The UCSD map of science. *PLoS ONE, 7*(7), e39464.

Boyack, K. W., & Klavans, R. (2010). Co-citation analysis, bibliographic coupling, and direct citation: Which citation approach represents the research front most accurately? *Journal of the American Society for Information Science and Technology, 61*(12), 2389–2404.

Klavans, R., & Boyack, K. W. (2009). Toward a consensus map of science. *Journal of the American Society for Information Science and Technology, 60*(3), 455–476.

Klavans, R., & Boyack, K. W. (2010). Toward an objective, reliable and accurate method for measuring research leadership. *Scientometrics, 82*(3), 539–553.

Moed, H., DeBruin, R. E., & Van Leeuwen, T. N. (1995). New bibliometric tools for the assessment of National Research Performance: Database description, overview of indicators and first applications. *Scientometrics, 33*(3), 381–422.

National Science Board. (2012). *Science and Engineering Indicators 2012* (NSB 12–01 and NSB 12–01A). Arlington, VA: National Science Foundation.

Rafols, I., Porter, A. L., & Leydesdorff, L. (2010). Science overlay maps: A new tool for research policy and library management. *Journal of the American Society for Information Science and Technology, 61*(9), 1871–1887.

Waltman, L., Calero-Medina, C., Kosten, J., Noyons, E. C. M., Tijssen, R. J. W., Van Eck, N. J., et al. (2012). The Leiden Ranking 2011/2012: Data collection, indicators, and interpretation. *arXiv:1202.3941v1*.

13 Finding and Recommending Scholarly Articles

Michael J. Kurtz and Edwin A. Henneken

Introduction

Communication is a two-way street; someone speaks, and someone else listens. Scholarly communication is essentially the same, but given the growing cacophony of so many voices, to whom should one listen?

The rate at which scholarly literature is being produced, similar to the rate at which all things of economic value (as measured by world gross domestic product or GDP) are being produced, has been increasing at approximately 3.5% per year for decades (estimated using data from the World Bank and the Web of Science [WoS][1]). This means that during a typical 40-year career the amount of new literature produced annually increases by a factor of four. The overall growth of available digital information is even more dramatic: in his IDC White Paper, Gantz (2008) observed that "the amount of digital information produced in the year [2001] should equal nearly 1,800 exabytes, or 10 times that produced in 2006" (p. 3). Over this short period of time people are not learning to read four times faster, nor are they becoming four times smarter. The methods scholars use to discover relevant literature must change. Just like everybody else involved in information discovery, scholars are confronted with information overload. Information overload is known to lead to decreased productivity and therefore has a negative economic impact, both to organizations and society as a whole. Spira (2010) calculated that in 2010 information overload costs the U.S. economy a minimum of $997 billion. The opposite is true as well, of course. Kurtz et al. (2005) estimated that the Smithsonian/NASA Astrophysics Data System (hereafter ADS), by making the process of finding information more efficient, saved the astronomy community about $250 million annually.

Two decades ago, this discovery process essentially consisted of paging through abstract books, talking to colleagues and librarians, and browsing journals. This was a time-consuming process that could take even longer if materials had to be shipped from elsewhere. Now, much of this discovery process is mediated by online scholarly information systems such as WoS, Scopus,[2] SciFinder,[3] ADS,[4] inSPIRE,[5] ACM-DL,[6] PubMed,[7] MathSciNet,[8] Microsoft Academic Search,[9] Google Scholar,[10] SSRN,[11] RePEc,[12] INSPEC,[13] and numerous others, large and small.

All of these systems are relatively new, and all are still changing. They all share a common goal: to provide their users with access to literature relevant to their specific needs. To achieve this, each system responds to actions by the user by displaying articles that the system judges relevant to the user's current needs. In plain English, the system recommends articles for the user's consideration.

Recently, search systems that use particularly sophisticated methodologies to recommend a few specific papers to the user have been called "recommender systems." Examples of successful recommender systems are Eigenfactor Recommends[14] which analyzes the citation network, and the Bx system (Bollen & Van de Sompel, 2006), which analyzes coreadership graphs. Google Scholar has implemented a recommender based on citation and authorship graphs. These methods are in line with the current use of the term "recommender system" in computer science (e.g., Lü et al., 2012; Jannach, Zanker, Felferning, & Friedrich, 2010); typically these methods are used to recommend products, books, movies, people (for dates), music, etc. In the domain of scholarly articles, this sort of system is primarily intended to provide browsing capability.

We do not adopt this definition; rather, we view systems like these as components of a larger whole, which are presented by the scholarly information systems themselves. In what follows, we view the recommender system as an aspect of the entire information system, one that combines the massive memory capacities of the machine with the cognitive abilities of the human user in order to achieve a human-machine synergy.

Recommendation and Search

From the point of view of a user of an information service "search," "recommendation," and "browse" are different functions. "Search" might be defined as returning answers to user-supplied specific questions, the equivalent of a person retrieving a fact from memory, while "recommendation" could be defined as returning specific answers to unstated or general questions, the equivalent of having a fact simply pop into one's head or by asking a question of an expert. In this sense, a recommender system really is a technological proxy for a social process. Users normally think of browsing as a serendipitous process, like finding a good book on a used-book table, or a relevant paper in a table of contents.

From the point of view of a scholarly information service, these functions are fundamentally the same. The goal of search systems, browse systems, and recommender systems is to provide the user with articles relevant to his or her current needs. Optimal results are very much user dependent; it is clear that lists of articles that would best serve a beginning student would not be the same as the lists that would best serve an established expert. The increasing sophistication of search systems has fully blurred the semantic difference between search and recommendation, a change captured by the popular aphorism "Recommendation is the new search."[15]

Modern information systems present the user with an information environment specific to each user. It is the task of the scholarly information system to orchestrate a continual transmutation of the user's intellectual environment to both anticipate and respond to that user's unique and ever-changing wants and needs. As extenders and enhancers of human thought, it is obvious that the abilities and capabilities of these machines will play a crucial role in the future development of humans.

There is a vast literature on search and recommendation systems (Lü et al., 2012) and they figure prominently in the list of the world's most valuable industrial corporations. For a good collection of reviews, see Ricci, Rokach, Shapira, and Kantor (2010). In addition, the yearly conference on recommendation systems of the Association for Computing Machinery[16] provides a snapshot of the current state of the art.

Here we make no attempt at a comprehensive literature review; rather, we look at the general issues and techniques for recommending scholarly articles, taking examples as needed primarily from our own work with the ADS.

The Unique Problem of Scholarly Recommendation

Recommendations form an important part of everyday life, providing suggestions for what restaurant to visit, or which book to read, or which play to watch, or which car to buy, etc. They are an important feature of most newspapers. Books and magazines such as the *Guide Michelin*[17] or *Consumer Reports*[18] have a long history. Blurring the distinction between recommending and advertising is a common and profitable practice, as the success of Google[19] and Amazon[20] shows.

Scholarly articles are a substantially denser and more subtly varied corpus than most commercial or entertainment domains. For example, the ADS has five times as many articles about a single subject, cosmology (a subfield of astrophysics, itself a subfield of physics, a substantially smaller discipline than chemistry or medicine), as there are films listed in the Videohound's Golden Movie Retriever[21] or available through Netflix.[22]

The usage pattern of scholarly articles also provides challenges. The maximum readership rate for an article is on the day it is published or posted online, before any use information is available. Use declines rapidly; the median download rate for a 10-year-old article from the *Astrophysical Journal*, the largest and most prestigious journal of astronomy, is once per month.

Citations provide another measure that can be used for developing recommendations (e.g., Küçüktunç, Saule, Kaya, & Çatalyürek, 2012), but they build up slowly. The most cited article from the *Astrophysical Journal* one year past publication has 50–60 citations; the median article has 5.

The users of scholarly publications tend to be highly discerning individuals, typically with doctorates in the subject matter of the publications being read. This increases the opportunities for interaction between the user and the search/recommendation system, because the user often has a detailed knowledge of what exactly she or he is seeking. This also makes the task of pure recommendation harder, since many of the apparently "best" articles to recommend are ones the user is fully aware of and has likely already read.

The Scholarly Information Garden

Bates (1989) has famously compared information seeking to berry picking, with the information system tasked with providing a bush filled with delicious berries. Today's (and tomorrow's) information systems must do more than provide berries; they must create complex, interactive virtual gardens. Besides a bush with delicious berries (e.g., a list of papers returned from a query), there are also multiple (intellectual) pathways and trail markers. In this metaphor, the search system provides the berry bush, the user interface provides the trail markers, and the recommender systems provide the pathways. The difference between this garden and a real garden is that the topology of the information garden changes as the information system's perception of the needs of the user changes. This continual transmutation of the garden makes the entire system highly interactive, a true human + machine collaboration.

Because the human is the master in the master-servant relation with the machine, there are substantial constraints on the recommender function, the portion of the system most under machine control. When the master asks for something the servant should fetch it, not return a set of objects that the servant thinks are best. However, the servant could be allowed to return items that other masters think are relevant (where "other masters" could be people very close in the information space, with perhaps a number of additional criteria).

As an example, a typical complex user query could be: "I want to see the paper by Dressler, et al. which is referenced in the recent paper on weak lensing by Kurtz, et al." This query can be asked, and the paper retrieved, within about 30 seconds (using the default settings of ADS, WoS, or SCOPUS). Using a recommender-based system such as Google Scholar takes substantially longer. The most valuable scholarly commodity is the time of our top scholars; this places strong constraints on recommendation systems.

An Overview of Recommendation in the ADS

Recommendation techniques have been built into the ADS system since its inception in 1992; in this section we discuss how they are implemented in the ADS, using as the model our new, streamlined (but still in beta test) user interface system. In the next section we discuss some algorithmic details of our implementation.

In addition to the recommenders built into the search system, the ADS provides a weekly, custom (for each user), stand-alone recommendation/notification service: myADS. Using a profile of the user, along with use and citation statistics, myADS provides a view of what the user should read/have read based on the papers released in the previous week. There are several versions of myADS, all based on input literature; the first author's version, based on data from the arXiv[23] e-print server (Ginsparg, 2011), can be found at http://adsabs.harvard.edu/myADS/cache/267336764_PRE.html. The page contains eight short lists of articles, each a form of recommendation. As an example, the upper-left list on the first author's page shows the most recent papers that have cited a paper where MJK is an author. This page changes every Friday; about 50% of all working astronomers subscribe to myADS.

There are two basic types of recommendations: ones the user specifically requests (a query), and those the user does not specifically request. The difference between the two is not algorithmic, but rather lies in how they are exposed through the user interface. The core of the ADS user interface, like many similar systems, consists of three main pages: a query page, a results/list page, and a document/abstract page. On each of these pages the system has different information concerning the current needs of the user, and the options and recommenders available on each page reflect this. Other web service designs are clearly possible; we use the three pages as a concrete example of how recommender techniques are incorporated into a scholarly information system.

The Query Page

The ADS main query page contains three major elements: the query input box, the query type toggles, and the recommender pane. We describe each in turn.

Queries to the ADS system can be free form, or entered into the query box using a highly structured query language, or a combination of the two. Hints are provided for the most common query constructs, and an auto-complete feature provides possible queries once the user begins to fill out the box (note that this is itself a type of recommendation).

Under the query box are seven toggles, which are used to direct how the system responds to the query. Choosing one of these options, and filling out the query box, initiates a user-specified recommendation. The seven query types are divided into two groups: four simple sorts, and three more complex to explore the field queries.

The four sort options arrange the results of a query by some value. Because the result of a query can be quite large (for example, the query "redshift survey" returns nearly 24,000 documents), the sort functions exactly as a recommender: the results on top are the ones you should look at.

1. "Most Recent" sorts the returned documents on publication date. This is the default query for the ADS, as well as for most other scholarly information systems (such as PubMed or WoS).
2. "Most Relevant" sorts on a combination of several indicators, including date, position of the query words in the document, position of the author in the author list, citation statistics, and usage statistics. This type of query is popular with commercial search engines such as Google or Bing.[24]
3. "Most Cited" simply sorts the returned documents on their citation count, the most cited being on top.
4. "Most Popular" sorts on the number of recent downloads.
5. The three "Explore the Field" options differ from the simple sorts in that the documents returned do not necessarily answer the query directly; rather, they use second-order operators (described in detail in the next section) on truncated, sorted query lists in order to form lists of recommended articles.
6. "What People Are Reading" returns a list of papers currently being heavily read by people in the subfield defined by the query.
7. "What Experts Are Citing" shows those papers that are most heavily referenced by the most relevant papers in the field defined by the query. These papers are often not about the query at all, but rather concern methods used in the subfield defined by the query.
8. "Reviews and Introductory Papers" returns just that, on the subfield defined by the query. This is achieved by finding articles that cite many highly cited articles on the desired topic.

Together, these options provide knowledgeable users with the ability to direct the actions of the machine assistant. The resulting lists either lead the user to desired articles or provide the basis for a more complex human-machine interaction.

The Recommender Pane shows the user lists of articles that the user did not specifically request, based on the machine's knowledge of the user and his or her recent actions. There are three different lists; the user determines which to see via a tab interface.

Perhaps the most useful list shows the day's release of papers from the arXiv e-print service, sorted according to the user's interest profile: the daily myADS-arXiv. This provides a quite complete (for astrophysics and other fields where arXiv is a fully integrated part), up-to-date answer to the question "What's new and of interest to me today?"

The other two lists are based on the user's recent search history. One is simply the papers the user most recently viewed (essentially a short-term memory). The second shows papers that are similar (via text similarity) to papers recently viewed, recently released, and very popular.

The Result/List Page

Lists of returned items are common to nearly all search and retrieval systems (such as Bing.com, Buy.com, Kayak.com, Scirus.com, Data.gov). These pages are the means by which the user can have a detailed interaction with the system. It is in these list pages that selection criteria can be modified and refined, and where items can be chosen. In the information garden metaphor, here is where the user either chooses some berries, decides to go further down a marked path, or asks the system to create a new garden.

A decade ago almost all result list pages were simply that, lists of results. Now it is common for such pages to have two or three columns, with one column devoted to the list of results. In the two-column format, a column (normally on the left) often provides lists of possible ways to filter/modify the list (facets); in the three-column format, the additional column (normally on the right) lists suggestions or recommendations that do not alter the list. The "recommendations" sections of Google's, Bing's, and Yahoo!'s[25] result pages consist of advertising. In addition to allowing filtering of the publications in the result list, the facets add a layer of useful information: Who are the most prolific authors in this field? Who has somebody coauthored with the most? Is this field still actively being publicized? What's the ratio of refereed to unrefereed publications?

All three designs are currently in use by digital libraries: the one-panel (MathSciNet), two-panel (ACM Digital Library), and three-panel (PubMed) versions. The ADS uses the two panel model, which does not allow unrequested article recommendations to be shown. This design decision resulted from the master/slave view of the literature discovery process. The master (user) has requested that a specific list of articles be presented; any additional lists of articles would necessarily take space away from the display of what the user has requested. For very large displays this may not be a problem, but the current trend is toward smaller displays (e.g., laptops, tablets, phones).

The list page is where the user has the most control over the recommendations returned by the system. Except for the most specific queries (e.g., "get *JASIST* 56, 36"), system responses always return documents that are ranked according to some desired criteria. These lists may be filtered, truncated, or edited by the user; the combined properties of the documents in the (edited) list can then be used to further extract information from the database.

These often-complex human-machine interactions can be understood in terms of operations on multipartite graphs (Kurtz, 2011) or in terms of operations on lists of attributes (Kurtz, Eichhorn, Accomazzi, Grant, & Murray, 2002). These operators use the combined properties of the articles in the list to extract information from the entire database. A typical example of this is:

- Return all papers containing the phrase "weak lensing" (done on the query page).
- Filter these papers to only include papers concerning the cluster of galaxies Abell 383.
- Filter these papers to only include papers that are based on data from the Hubble Space Telescope (HST).
- Sort these papers by date.
- Truncate the list to include only papers less than one year old.
- Find all the users who read one or more of these papers in the last three months.
- Filter these users to only include persons classified by their usage patterns as probable scientists.
- Find all the papers read by these scientist-users in the last three months.
- Sort this list by the number of these users who read each paper and return the sorted list to the user.

The list of papers resulting from this query will show what is currently noteworthy to scientists interested in HST measurements of weak lensing near the cluster of galaxies Abell 383. By constructing this query, an interested scientist can discover what is in the collective knowledge of astronomers with similar interests. This is obviously a custom recommender that relies heavily on the synergy of human and machine. In terms of the actual implementation, the user here must type in the original phrase ("weak lensing"), click on two buttons (Abell 383 and HST), slide the date slider, and click on the "Most Co-Read" button.

"Most Co-Read" is one of the recommender functions available from the list page; the "What People Are Reading" query, on the query page, is implemented using the "Most Co-Read" function. The two other "Explore the Field" options on the query page are implemented with two other recommender functions: the "Get Reference Lists" function is used in the "What Experts Are Citing" option, and the "Get Citation Lists" function is used in the "Reviews and Introductory Papers" option. "Get Reference Lists" simply collates all the reference lists of the papers in the original list and returns the collated list, sorted by the number of papers in the original list that referenced each paper. This gives the papers that were most cited by the articles in the original list. "Get Citation Lists" collates all the lists of citing papers for each of the papers in the original list and returns the collated list sorted by the number of papers in the original list that are cited by the citing paper. This gives the papers that have the most extensive discussions on the topic of the original list. Another function takes all the articles in the list, combines them, and treats them as a single document. It then returns the papers closest to this combination sorted by nearness according to a text-similarity metric.

A final recommender function is different from the others in that it is not intended to be used in combination with a (potentially filtered and edited) query list. The ADS allows users to create, edit, save, and recall lists. This capability is often used by authors to prepare bibliographies for their manuscripts. The "Citation Helper" function suggests papers that are not in the list but are near neighbors of these papers in the citation-reference network. This is intended to help find missed references.

The facets or filters can also be viewed as suggestions or recommendations to the user for possible additional restrictions that the user might want to make on the list. Which are the most useful possible restrictions is clearly a context-specific issue. As an example, typical subject matter queries can return thousands of articles with thousands of individual authors. Which ones should be on the top of the "Authors" facet? Simple, context-sensitive metrics allow the facets for a "show me the most recent papers on weak lensing" query to be different from the facets for a "show me the most popular/cited/etc. papers on weak lensing" query.

The Document/Abstract Page

The principal function of the "Abstract" page is to provide the user with a concentrated description of an article, sufficient for the user to decide whether to download and read it. By coming to an abstract page, the user has informed the system of his or her current interest, which permits useful recommendations to be made. In the ADS instantiation, this page includes the standard metadata of title, author, journal name, and abstract, but it also includes links to additional information and two sets of recommenders, one user-directed and one automated.

The four user-specified recommendations are nearly identical to the four functions on the list page (the only difference with the result obtained from a one-article list is the sort order), but are likely more familiar than their extensions. Each of these four buttons returns a list of recommended articles. The "References" button returns the list of papers referenced by the article, the "Citations" button returns the list of papers that cite the article, the "Co-Read" button returns the list of papers most read recently by scientists who read the article, and the "Similar Articles" button returns the list of articles most similar to the article by weighted text similarity. This "Abstract" page also includes a panel with eight recommended articles. Clicking on any of these returns the abstract page for the recommended article. These papers are each computed by a different algorithm (Henneken et al., 2011), but each recommendation is based on a list of articles that were recently published in major journals and are near the original article in a vector space that we create from the index terms attached to the papers in the reference lists of all the recent major journal articles, as well as their readership pattern among professional astronomy researchers.

Taking the list of papers near the original paper in the vector space (the near list), the eight recommended papers are: the paper in the near list closest to the original paper; the paper that was read by the largest number of scientists who also read a paper in the near list in the last three months; the paper that was most frequently read by a scientist immediately following that person reading a paper in the near list; the paper that was most frequently read by a scientist immediately before that person read a paper in the near list; the most recent paper, among the 30 most read papers by scientists who read a paper in the near list in the last three months; the paper that the papers in the near list cite most frequently; the paper that cites the largest number of papers in the near list; and finally, the paper that refers by name to the largest number of astronomical object (stars, galaxies, . . .) that are also referred to by papers in the near list. This last function uses data from the SIMBAD database of the Strasbourg (France) Data Center[26] (Wenger et al., 2000).

Implementation Techniques and Details

The actual implementation of the search and discovery tools in the ADS is very simple. We distinguish between "first-order" queries and "second-order" queries (Kurtz, 1992; Kurtz et al., 2002). First-order queries involve search terms and a sort order. An example might read as "bring me the papers that contain the phrase 'weak lensing' sorted by citation count." The facets allow these queries to be modified by further filtering, such as: "restrict the list to refereed papers by Hoekstra that use either data from the Hubble Space Telescope or the European Southern Observatory and use data from the Hubble Deep Field." The art here is not in how the queries are executed, which is pretty standard, but in making these capabilities easily available to the user.

Second-order queries take lists of articles as input and return different lists according to attributes of the entries in the original list. When the parameters defining the original list are well chosen, and the attributes chosen for the second-order query make sense, this methodology can result in highly specific results. Lists can either be created or edited by the user, as in the "Get Co-Read," "Reference," and "Citation" list queries on the list page, or else taken directly from a first-order query, as are the three "Explore the Field" options on the query page. The Abell 383 example above demonstrates the degree of sophistication possible with a user-intermediated query.

The "Reviews and Introductory Articles" option on the query page is an example of an automated chaining of a first- and second-order query. For example, taking "weak lensing" as the original query, the procedure is:

- Return the articles that contain the phrase "weak lensing."
- Sort this list by number of citations.
- Truncate this list to get the 200 most cited articles that contain the phrase "weak lensing."
- For each of these 200 articles retrieve the list of articles that cite it.
- Collate these lists and sort the collated list by frequency of appearance so the article that cites the largest number of the 200 articles is on top.

This returns articles that have extensive discussion sections on the desired topic (in this case "weak lensing"); normally these are review articles. The ADS has been providing this function since 1996.

In addition to the ones we expose to the users, other second-order operators are possible; the procedures used in the recommender pane of the abstract page list several of them. As the multipartite network of data objects and their attributes becomes both richer and broader, many more are likely to be developed. As an example, authors are data objects that have attributes; one can imagine queries that make use of these attributes. "Show me the most popular papers among researchers who read recent papers by young authors at Japanese research institutes" is a query that some might find useful. The ADS already contains all the information necessary to answer this query, save for the age or birth year of the authors.

The ADS also uses partial match word technologies in determining the distance between documents by text similarity. Currently we are, with the kind collaboration of essentially every publisher of physics and astronomy technical literature, establishing a complete collection of the full text of nearly every article ever published in those fields. We expect that these full-text data will form the basis for our article similarity measures in the future; thus, the ones currently implemented will become obsolete.

Basically, we now use two different methods. The first involves weighted word counts, modified by an extensive, subject-specific synonym list (Kurtz et al., 1993), a technique that has been standard since the SMART system (Salton & Lesk, 1965). The second method is an eigenvector method that uses as input the index terms associated with references in articles combined with the usage patterns of heavy users (Kurtz, Accomazzi, Henneken, Di Milia, & Grant, 2010); these techniques are derived from Kurtz (1993), which is based on the work of Ossorio (1966). We expect, in the near future, to build an article-similarity system based on a Latent Dirichlet Allocation (Blei, Ng, & Jordan, 2003) of the full text of recent articles in astronomy and physics. If successful, this will supplant both methods.

Measuring Effectiveness

Scholarly communication is a substantially different domain from the normal fields of commerce where not-so-highly-interactive recommender systems are found. Amazon.com can build a metric based on sales, Match.com on marriage rates, Netflix.com on user rankings, and Google Ads on clicks; what metric could measure the effectiveness of a highly interactive scholarly information system?

Certainly there are components of an information system that can be measured by a click rate. For example, one can easily imagine that different recommender algorithms could be compared in the recommender panel on the abstract page. These suggested articles are, however, a very small component of the total service. Ranking queries by relevance is also highly problematic. What is relevant to one person is not to another; the ADS alone has seven different types of rankings on its query page. There is no reason to believe that these seven are complete or unique.

Even when a particular query type can be measured, it is not obvious what to do, because the query itself is embedded in a larger context. For example, it is reasonable to expect that a measure of "importance" can be defined, and that network-based methods, such as Eigenfactor (West, Bergstrom, & Bergstrom, 2009; see also West and Vilhena, chapter 8, this volume), could show significantly better performance, albeit not hugely better than simple citation counts (Davis, 2008).

Were this a stand-alone query, the discussion would end with the measurements; however, it is not. The query is part of a system that includes the user; the user must decide to invoke this query sorting, and may choose to use the result in a further query. It is of crucial importance that the user understand what the query is doing. Even for the ADS's heavy users, essentially all of whom have doctorates in physics, it is much easier to understand a ranking based on citations than a ranking based on the eigenvalues of the principal eigenvector of the citation connectivity matrix (or some other similar technique).

The goal of scholarly information systems is to improve the quality of research. Few would argue that these systems have not improved the research effort, but by how much have they improved it?

Kurtz et al. (2000) devised a method to measure the impact of the (then new) digital literature system compared with the old paper system. They assumed that in the paper era a researcher did not waste her or his time (on average) by going to the library and reading an article. Since the electronic paper is available essentially instantly, they suggested that the no-longer-required overhead of going to the library, photocopying the document, etc., could be counted as additional research time obtained through use of the digital technology. Kurtz et al. (2005) used this technique to find that the use of electronic documents through the ADS contributed 736 additional full-time equivalent researcher-years to the worldwide astrophysics research effort, about 7% of the world's astronomers. Lesk (2011) has suggested that this could be an underestimate, because it does not take into account the improved mean quality of the papers read, which is a result of the improved search techniques.

Scholarly information systems do not just recommend articles. As an example, the Harvard Catalyst[27] is an information system covering medical research. As one of its features it recommends possible people as collaborators. This is likely a very useful function, but, considering privacy concerns, it is difficult to imagine, even in principle, a robust methodology for determining the quality of these recommendations apart from anecdotes.

As these systems mature there will likely be comparisons of the differences between similar techniques, with better performers replacing poorer performers. These systems are still quite new and rapidly changing. Their capabilities and usefulness depend far more on the vision of their respective creators and on the quality of the offered data than on any optimization techniques.

Conclusions and Future Directions

Recommendations are central to the effective use of our new digital environment. A decade ago, recommendation techniques were similar across fields. The ADS, for example, offered the capacity to find papers that "cited this (these) paper(s) also cited" in 1996; somewhat later, Amazon.com introduced their "people who bought this book also bought" feature. Now, however, the path of research literature has diverged from the mainstream.

In terms of information density, frequency of use, and user expectations, the problem of recommendation of research literature is substantially different from the more familiar commercial applications. It is simply not possible to solve this problem by algorithmic means alone; the information is too dense, the use too infrequent, and the expectations of the researcher/user too high. Scholarly recommender systems must be close collaborations of humans with machines.

This collaboration is becoming hugely powerful. With the aid of a machine-assisted memory astrophysicists can "remember" every single word of every one of the 125,000 articles that contain the word *redshift*, but which ones do they need to read now to facilitate their research? It is the ongoing task of scholarly information systems, and, more specifically scholarly recommendation systems, to develop the means by which users can search and find the berries that they need in the vast memory garden now at their disposal.

Amazing progress has already been made. It is now a simple matter to learn the answer to the question "What is everyone else in this field reading that I should be reading?" Automatically accessing the collective knowledge of relevant subgroups of scientists smacks of science fiction (Kurtz, 2011), but it is now happening routinely. The field is still very young, and major changes are certain to occur. No one over the age of 40 learned to use any of the major modern scholarly information systems as an undergraduate, because they did not exist then. Nearly every principal investigator on every major grant, and every full professor at every major university, learned to use the research literature on paper. In the future this will no longer be true. Information systems are becoming vastly richer, more complex, and more closely interconnected and interoperative. Articles, people, research objects, datasets, instruments, organizations, etc., are now data objects, with properties and attributes that can be combined and used in myriad ways. It will be the ongoing responsibility of the recommender aspect of these systems to permit researchers to make full use of these capabilities.

Acknowledgments

We especially thank the ADS team, led by Alberto Accomazzi. Whenever in the text we use the term "we" to refer to the ADS we actually mean the ADS team. We acknowledge conversations with Rudi Scheiber-Kurtz, Paul Ginsparg, Herbert Van de Sompel, Carl

Bergstrom, Johan Bollen, Jim Gray, Geoff Shaw, and Peter Ossorio. The ADS is funded by NASA grant NNX12AG54G.

Notes

1. webofknowledge.org
2. scopus.com
3. www.cas.org/products/scifinder
4. adsabs.org
5. inspirehep.net
6. dl.acm.org
7. pubmed.gov
8. ams.org/mathscinet/
9. academic.research.microsoft.com
10. scholar.google.com
11. ssrn.com
12. repec.org
13. www.theiet.org/inspec
14. http://mas.eigenfactor.org/recommendation.php
15. Remark made by Stephen Green, during a talk about Minion at Harvard University in 2008.
16. recsys.acm.org
17. viamichelin.com
18. consumerreports.org
19. google.com
20. amazon.com
21. movieretriever.com
22. netflix.com
23. arxiv.org
24. bing.com
25. yahoo.com

26. simbad.u-strasbg.fr

27. catalyst.harvard.edu

References

Bates, M. J. (1989). The design of browsing and berrypicking techniques for the online search interface. *Online Review, 13*, 407–424.

Blei, D. M., Ng, A. Y., & Jordan, M. I. (2003). Latent Dirichlet allocation. *Journal of Machine Learning Research, 3*, 993–1022.

Bollen, J., & Van de Sompel, H. (2006). An architecture for the aggregation and analysis of scholarly usage data. In *Proceedings of the 6th ACM/IEEE-CS Joint Conference on Digital Libraries* (pp. 298–307). New York: ACM.

Davis, P. M. (2008). Eigenfactor: Does the principle of repeated improvement result in better journal impact estimates than raw citation counts? *Journal of the American Society for Information Science and Technology, 59*, 2186–2188.

Gantz, J. F. (2008). *The diverse and exploding digital universe* [White paper]. Retrieved from http://www.emc.com/collateral/analyst-reports/diverse-exploding-digital-universe.pdf.

Ginsparg, P. (2011). ArXiv at 20. *Nature, 476*, 145–147.

Henneken, E. A., Kurtz, M. J., Accomazzi, A., Grant, C., Thompson, D., Bohlen, E., et al. (2011). Finding your literature match—a recommender system. *Astrophysics and Space Science Proceedings, 1*, 125–134.

Jannach, D., Zanker, M., Felferning, A., & Friedrich, G. (2010). *Recommender systems: An introduction.* Cambridge: Cambridge University Press.

Küçüktunç, O., Saule, E., Kaya, K., & Çatalyürek, Ü. V. (2012). Recommendation on academic networks using direction aware citation analysis. *arXiv:1205.1143*. Retrieved from http://arxiv.org/abs/1205.1143.

Kurtz, M. J. (1992). Second order knowledge: Information retrieval in the terabyte era. In A. Heck & F. Murtagh (Eds.), *Astronomy from large databases II. European Southern Observatory Conference and Workshop Proceedings, 43*, 85–97.

Kurtz, M. J. (1993). Advice from the oracle: Really intelligent information retrieval. In A. Heck & F. Murtagh (Eds.), *Intelligent information retrieval: The case of astronomy and related space sciences. Astrophysics and Space Science Library, 182*, 21–28. doi:10.1007/978-0-585-33110-2_3.

Kurtz, M. J. (2011). The emerging scholarly brain. In A. Accomazzi (Ed.), *Future professional communication in astronomy II. Astrophysics and Space Science Proceedings, 1*, 23–35. doi:10.1007/978-1-4419-8369-5_3

Kurtz, M. J., Accomazzi, A., Henneken, E., Di Milia, G., & Grant, C. S. (2010). Using multipartite graphs for recommendation and discovery. In Y. Mizumoto, K.-I. Morita, & M. Ohishi (Eds.),

Astronomical data analysis software and systems XIX. Astronomical Society of the Pacific Conference Proceedings, 434, 155–158.

Kurtz, M. J., Eichhorn, G., Accomazzi, A., Grant, C. S., Demleitner, M., & Murray, S. S. (2005). Worldwide use and impact of the NASA astrophysics data system digital library. *Journal of the American Society for Information Science and Technology, 56*, 36–45. doi:10.1002/asi.20095

Kurtz, M. J., Eichhorn, G., Accomazzi, A., Grant, C. S., & Murray, S. S. (2002). Second order bibliometric operators in the Astrophysics Data System. In J.-L. Starck & F. Murtagh (Eds.), *Astronomical data analysis II. Society of Photo-Optical Instrumentation Engineers (SPIE) Conference Series, 4847*, 238–245.

Kurtz, M. J., Eichhorn, G., Accomazzi, A., Grant, C. S., Murray, S. S., & Watson, J. M. (2000). The NASA astrophysics data system: Overview. *Astronomy & Astrophysics, Supplement Series, 143*, 41–59.

Kurtz, M. J., Karakashian, T., Grant, C. S., Eichhorn, G., Murray, S. S., & Watson, J. M., et al. (1993). Intelligent text retrieval in the NASA astrophysics data system. In R. J. Hanisch, R. J. V. Brissenden, & J. Barnes (Eds.), *Astronomical data analysis software and systems II. Astronomical Society of the Pacific Conference Proceedings, 52*, 132–136.

Lesk, M. E. (2011). Encouraging scientific data use—Michael Lesk [Blog post]. Retrieved from http://www.scilogs.com/the_fourth_paradigm/encouraging-scientific-data-use-michael.

Lü, L., Medo, M., Yeung, C. H., Zhang, Y.-C., Zhang, Z.-K., & Zhou, T. (2012). Recommender systems. ArXiv e-prints:1202.1112. Retrieved from http://arxiv.org/abs/1202.1112.

Ossorio, P. G. (1966). Classification space—a multivariate procedure for automatic document indexing and retrieval. *Multivariate Behavioral Research, 1*, 479–524.

Ricci, F., Rokach, L., Shapira, B., & Kantor, P. B. (Eds.). (2010). *Recommender systems handbook*. New York: Springer Verlag.

Salton, G., & Lesk, M. E. (1965). SMART automatic document retrieval system—an illustration. *Communications of the ACM, 6*, 391–398.

Spira, J. (2010). Information overload now $997 billion: What was changed? [Blog post]. Retrieved from http://www.basexblog.com/2010/12/16/io-997.

Wenger, M., Ochsenbein, F., Egret, D., Dubois, P., Bonnarel, F., Borde, S., et al. (2000). The SIMBAD astronomical database: The CDS reference database for astronomical objects. *Astronomy & Astrophysics, Supplement Series, 143*, 9–22. doi:10.1051/aas:2000332.

West, J., Bergstrom, T., & Bergstrom, C. (2009). Big Macs and Eigenfactor scores: Don't let correlation coefficients fool you. *Journal of the American Society for Information Science and Technology, 61*, 1800–1807.

Part IV Alternative Metrics

14 Altmetrics

Jason Priem

Introduction

This chapter discusses altmetrics (short for "alternative metrics"), an approach to uncovering previously invisible traces of scholarly impact by observing activity in online tools and systems. I argue that citations, while useful, miss many important kinds of impacts, and that the increasing scholarly use of online tools like Mendeley, Twitter, and blogs may allow us to measure these hidden impacts. Next, I define altmetrics and discuss research on altmetric sources—both research mapping the growth of these sources, and scientometric research measuring activity on them. Following a discussion of the potential uses of altmetrics, I consider the limitations of altmetrics and recommend areas ripe for future research.

What Are Altmetrics?

The Problem: Ideas Do Not Leave Good Tracks

It is beyond the scope of this chapter to trace the intellectual and historical lineage of citation mining; interested readers are encouraged to peruse De Bellis's chapter in this volume (chapter 2) for an overview. For present purposes, it is sufficient to assert a fact that is common knowledge: ideas, while invisible, are not untraceable. They leave tracks. For 40 years, bibliometricians have diligently hunted, followed, cataloged, and analyzed a particular kind of track: that humble but powerful "pellet of peer recognition" (Merton, 1988, p. 620), the citation.

As they have done so, though, recognition has grown both inside and outside the bibliometrics community that these tracks do not constitute a comprehensive source of data. Many important users of research do not cite; by some estimates, "Only about 15 to 20% of scientists in the United States have authored a refereed article" (King & Tenopir, 2004, para. 13). Many important artifacts are commonly not cited (MacRoberts & MacRoberts, 2010), notably datasets, an increasingly important scientific product. Eugene Garfield himself pointed out that "citation frequency reflects a journal's value and the use made of it, but there are undoubtedly highly useful journals that are not cited frequently" (Garfield, 1972, p. 535). Citations are often used for purposes not accounted for in Merton's normative conception of science, as social constructivists have argued (Bornmann & Daniel, 2008). Most importantly, citations are products of a slow, rigid *formal* communication system, while scientific ideas themselves are born, nursed, and raised in messy, fast-moving *informal* invisible colleges (Price & Beaver, 1966). The heart of scientific communication is neither formal publications nor citations but "visits, personal contacts, and letters" (Bernal, 1944, p. 303). Because of this, citations are typically late visitors to the impact "crime scene" (De Bellis, 2009, p. 103).

Solution: Observe New Tracks

The growing pervasiveness of the Web is creating an environment in which scholars and other users create new kinds of tracks that reveal once-invisible scholarly activities. Inexorably, the daily work of scholars, like other knowledge workers, is moving online. As it does, the background of scholarship—the dog-eared manuscripts and hallway conversations—is pushed out onto the stage. Many online tools are enjoying dramatic growth in usage—growth that seems likely to continue as a "born-digital" generation moves into tenured positions. Social media tools like reference managers, microblogging, and bookmarking services are becoming increasingly important in scholars' workflows, as several recent studies have made clear. Thirteen percent of UK academics frequently use Web 2.0 in novel forms of scholarly communications (Procter et al., 2010), and 80% of scholars have social media accounts (Tinti-Kane, Seaman, & Levy, 2010). A UK study reports that 10% of doctoral students "use and value" Twitter for research, and that social media tools are affecting the scholarly workflow (Carpenter, Wetheridge, Smith, Goodman, & Struijvé, 2010).

Importantly, these tools are facilitating scholarly practice but do not seem to be remaking it. Historically, a researcher might have read an article, liked it, saved it in a box or file, discussed it over lunch with her colleagues, cited it in her next paper, and even formally endorsed or recommended it at a conference. Today she might download it from an online journal, save it in her reference manager, discuss it with peers on Twitter and blogs, and recommend it on Faculty of 1000. The difference is that each of these latter activities leaves traces that can be measured.

Of course, an interest in alternative ways to track the flow of ideas is not new. Many alternatives and complements to citations have been proposed and successfully used in the past. Scientometricians have followed and analyzed all sorts of novel impact traces, including acknowledgments (Cronin & Overfelt, 1994), patents (Pavitt, 1985), mentorships (Marchionini, Solomon, Davis, & Russell, 2006), news articles (Lewison, 2002), readings in syllabi (Kousha & Thelwall, 2008), and many others, separately and in various combinations (Martin & Irvine, 1983). Each of these methods has strengths and weaknesses, and all have produced interesting findings. However, they also share a common weakness—they are difficult and time consuming to harvest, especially when compared with citations gathered using commercial indexes like Thomson Reuters' Web of Science (WoS) and Elsevier's Scopus.

Indeed, the idea of measuring impact online is not itself a new one. Practitioners of webometrics have been pursuing digital tracks for some time (Almind & Ingwersen, 1997). However, scalability has been a concern. Although lauded as a solution to the narrowness and slowness of citation metrics (Cronin, Snyder, Rosenbaum, Martinson, & Callahan, 1998), this method has in many ways failed to live up to early expectations (Thelwall, 2010). This is in no small measure due to the difficulty of collecting data—a process commercial search engines have seen little reason to ease, given their heavy investment in proprietary indexes, as well as the relatively tiny number of users interested in these kinds of searches.

The latest sources of online impact differ from earlier citation alternatives, both offline and online, in a key respect: they are far easier to gather automatically, since they are based on *tools* and *environments* with clearly defined borders, data types, and (in many cases) structured Application Programming Interfaces or web APIs to allow access (Priem & Hemminger, 2010). This allows them to be used at scale, in ways to date uncommon except for citation (and perhaps patent) data. Building on these open APIs, a number of web-based software tools have emerged to aggregate, process, and present data from multiple APIs. Early tools included ReaderMeter, CitedIn, ScienceCard, and *Public Library of Science* Article Level Metrics (*PLoS* ALM); of these, only *PLoS* ALM seems to remain under heavy development at the time of this writing, and is geared more toward use by journals than individual researchers. Recently, the Webometric Analyst package has added Mendeley references to the data it gathers. Two aggregators of particular note for scientometricians are total-impact[1] and Altmetric.[2] The former aggregates a dozen different altmetrics (including *Wikipedia* citation, Mendeley use, and tweets) on multiple scholarly products (including slides, datasets, software, and articles); it is open-source and free. The latter is a commercial product that offers particularly good coverage of Twitter, as well as Mendeley, mainstream news outlets, Reddit, and more. These tools are under rapid development.

Altmetrics

Altmetrics (jasonpriem, 2010; Priem, Taraborelli, Groth, & Neylon, 2010) is the study and use of scholarly impact measures based on activity in online tools and environments. Unlike the case of "bibliometrics" or "scientometrics," the term is used to describe both the field of study and the metrics themselves (one might advocate the use of a particular "altmetric"). Since altmetrics is concerned with measuring scholarly activity, it is a subset of scientometrics, except when tracking nonscience scholarship. Since it is also concerned with measuring activity on the Web, altmetrics is a proper subset of webometrics. It is distinct in that it focuses more narrowly on online *tools and environments* than on the Web as a whole. Although in theory any web tool or environment could support altmetrics, in practice, altmetrics researchers have opportunistically sought out rich data pastures, watered by open APIs. They have also focused on tools with large and growing scholarly use. To date, these have included *Wikipedia*, social reference managers like Mendeley and CiteULike, blogs, Twitter, and several others discussed in the next section.

Altmetrics Sources: A Summary of Research

Several sources in the mid to late aughts identified the potential of altmetrics and called for further research (Jensen, 2007; Neylon & Wu, 2009; Taraborelli, 2008). Since then, research has proceeded apace. However, it is important to note that altmetrics is still in its infancy. Since "it took approximately a generation (20 years) for bibliographic citation analysis to achieve acceptability as a measure of academic impact" (Vaughan & Shaw, 2005, p.1315), we should expect nothing different from other new metrics. Although altmetrics research is at a very early stage, several methods have already shown themselves useful in its study. Each of these has been employed in the validation of citation metrics, as is shown in the next section.

Methods

Correlation and Prediction with Established Metrics Garfield (1979) used a correlational and predictive approach to justify citation counts as measures of individual impact, showing how they predicted future Nobel Prize recipients. Narin summarized early efforts to connect citations and esteem measures (Narin, 1976). This has been a common approach in altmetrics research to date, but should be used with caution. We should not expect or even desire perfect correlation between new metrics and traditional metrics (Sugimoto, Russell, Meho, & Marchionini, 2008); part of the value of altmetrics is the ability to measure forms of impact partly or wholly unrelated to what citation captures. Eysenbach (2012), for example, has pursued predictive work with Twitter, while several studies have examined correlation between citation and reference manager inclusion (Haustein & Siebenlist, 2011; Li, Thelwall, & Giustini, 2011), or between multiple altmetrics and citations (Priem, Piwowar, & Hemminger, 2012; Yan & Gerstein, 2011).

Content Analysis There has been a long tradition of context-analytic studies of citation, beginning with Moravcsik and Murugesan's (1975) influential categorization of citations from 30 high-energy physics articles (for a review, see Bornmann and Daniel, 2008). Cronin et al. (1998) analyzed web mentions of scholars, and Thelwall and others (Kousha & Thelwall, 2006; Thelwall, 2003) analyzed the context of scholarly hyperlinks. Similar studies have investigated altmetrics, particularly tweets (Letierce, Passant, Breslin, & Decker, 2010; Priem & Costello, 2010; Ross, Terras, Warwick, & Welsh, 2011). Techniques for automatically extracting contextual information can be used to isolate and describe contextual information in tweets (Stankovic, Rowe, & Laublet, 2010).

Creator Feedback Creator feedback studies are also known as "citer motivation" or "citer behavior" studies (Borgman & Furner, 2002). Here, researchers use interviews or surveys to investigate authors' reasons for creating certain types of records. Bornmann and Daniel (2008) review many creator feedback studies of traditional citation beginning with Brooks' (1986) interview-based and Vinkler's (1987) survey-based investigations. More recently, Cronin and Overfelt (1994) used surveys to establish authors' motivations for creating acknowledgments, and Priem and Costello (2010) used interviews to investigate scholars' motivations for citing on Twitter.

Prevalence Studies Although citation practices vary between disciplines, the practice is ubiquitous. The distribution of altmetrics, on the other hand, depends on the uptake of scholarly tools being examined. Consequently, many studies useful to altmetrics have focused less on metrics and more on simply describing the scholarly use of a given environment. Some of these studies are discussed below.

Notable Altmetrics Findings by Source

An important property of altmetrics is the ability to track impact on broad or general audiences, as well as on scholars. This is reflected in table 14.1, which separates impact types by audience. This section reviews findings relevant to altmetrics, discussing each cell in table 14.1 in turn. I highlight findings that relate specifically to measuring

Table 14.1
Altmetrics sources by type and audience

	General users	Scholarly users
Recommendation	Web-based mainstream media	Faculty of 1000
Citation	*Wikipedia*	Citation from within peer-reviewed literature
Conversation	Twitter, Facebook, blogs	Scholarly blogs, article comments, tweets from scholars
Reference	Social bookmarking	Social reference managers
Reading	HTML views	PDF downloads

scholarly impact, but also select studies indicating the prevalence and uptake of the different communication channels and media. Most studies discussed in this section focus on one source. Two notable ones that discuss multiple sources use sets of articles from the open-access publisher *Public Library of Science* (*PLoS*) (Priem et al., 2012; Yan & Gerstein, 2011), thanks to *PLoS's* Article-Level Metrics (ALM) API, which presents lists of altmetrics data by article, making this a useful test set for early altmetrics investigations. I begin by discussing indicators of impact on the general public. These impacts are vital and not just for research with an immediate public benefit (e.g., medical research). Since the public indirectly funds most research, public awareness and outreach are important, especially in lean economic times.

Public

Mainstream Media Lewison (2002) tracked mass media mentions of scholarly articles, finding little relation between journal citation rates and citedness by major news outlets, suggesting that this is a legitimately distinct form of impact. However, other work shows that these mentions seem to affect citation rates, implying that the two are not completely unconnected (Kiernan, 2003). Since then, many major news outlets like the *Guardian* and the *New York Times* have launched open APIs that make searching their text much easier, or at least have Really Simple Syndication (RSS) feeds that can be automatically crawled. Altmetric.com provides a convenient interface for many of these; this should encourage more work tracking this form of impact.

Wikipedia For much of the world, especially students (Head & Eisenberg, 2010; Schweitzer, 2008), *Wikipedia* is the first stop for information. Influencing *Wikipedia*, therefore, means influencing the world in a profound way. Citation on *Wikipedia* could be considered a public parallel to scholarly citation (a discussion of the latter is out of scope for this chapter, but it is included in table 14.1 as a reference point). Nielsen (2007) has shown that citations in *Wikipedia* correlate well with data from the Journal Citation Reports, establishing a relationship between impact on *Wikipedia* and in more traditional contexts. Priem et al. (2012) report that around 5% of *PLoS* articles are cited in *Wikipedia*, and report correlations between 0.1 and 0.4 between normalized *Wikipedia* citations and traditional citations, depending on the journal. Wikimedia Research's Cite-o-Meter tool shows an up-to-date league table of which academic publishers are most cited on *Wikipedia*.

Conversation (Twitter) Twitter presents a rich source of data, but a difficult one for investigators of scholarly impact. Unlike many tools discussed in this chapter, Twitter does not make apparent the audience interacting with a research product. Twitter is heavily used by some scholars, as we will see later, but nonscholars make up the bulk of Twitter users. Given this, we assume tweets come from layreaders unless given evidence to the contrary (such contrary evidence could come from biographical information, follower lists, or contents of tweets; Altmetric.com uses an algorithm to decide). This assumption is supported by the near-zero correlation between tweets and citations found in the *PLoS* dataset (Priem et al., 2012), suggesting that the high visibility of *PLoS* attracts many nonscholarly readers. On the other hand, Eysenbach's (2012) study of *Journal of Medical Internet Research* (JMIR) papers found a distinct link between tweeting and citation: articles in the top quartile of tweets were 11 times more likely to be in the top quartile of citations two years later. This may be because *JMIR* tweeters are more likely to be the same scholars who might cite the work. More research is needed that delineates the identity of Twitter users.

Conversation (Social News) Social news includes sites like Reddit, Digg, and Slashdot. Lerman and Galstyan (2008) demonstrated that early "Diggs" predict the later importance of news stories, which encourages inquiry into similar predictive validity for scholarly publications. However, while Norman suggests using a "Slashdot index" to measure scholarly impact (Cheverie, Boettcher, & Buschman, 2009), no research has yet emerged tracking scholarly articles' mentions on recommendation sites like these. Reddit is a particularly interesting case, since users can create focused "subreddits." Although these can appear, mutate, and disappear rapidly, at the time of writing, there were subreddits built around requesting paywalled scholarly papers, discussing the latest science research (reporting nearly 2 million readers), and others, making subreddits an interesting subject for future altmetrics research.

Conversation (Bookmarking) The small body of research into social bookmarking of scholarly resources has focused on Delicious[3] (Lund, Hammond, Flack, & Hannay, 2005). Ding, Jacob, Caverlee, Fried, and Zhang (2009) found that "scientific domains, such as bioinformatics, biology, and ecology, are also among the most frequently occurring tags" (p. 2398), suggesting at least some scholars use the service. Priem et al. (2012) report that about 10% of *PLoS* articles are bookmarked in Delicious.

HTML Views With the growth of the open-access (OA) movement, it has become possible for us to talk meaningfully about numbers of nonscholarly readers. HTML downloads, when reported separately from PDF downloads, may be useful for identifying nonscholarly readers. Priem et al. (2012) observe that HTML views tend to cluster with other metrics of public impact. This is probably because the general public is more likely to read only the abstract, or to skim the text of the article quickly, while academics are more likely to download and print the paper. Further research investigating the ratio between HTML views and PDF downloads could uncover interesting findings about how the public interacts with the open-access research literature.

Scholars

In addition to tracking scholarly impacts on traditionally invisible audiences, altmetrics holds potential for tracking previously hidden scholarly impacts.

Faculty of 1000 Faculty of 1000 (F1000) is a service publishing reviews of important articles, as adjudged by a core "faculty" of selected scholars. Wets, Weedon, and Velterop (2003) argue that F1000 is valuable because it assesses impact at the article level, and adds a human-level assessment that statistical indicators lack. Others disagree ("Revolutionizing Peer Review?," 2005), pointing to a very strong correlation ($r = 0.93$) between F1000 score and Journal Impact Factor. This said, the service has clearly demonstrated some value, because over two-thirds of the world's top research institutions pay the annual subscription fee to use F1000 (Wets et al., 2003). Moreover, F1000 has been to shown to spot valuable articles that "sole reliance on bibliometric indicators would have led [researchers] to miss" (Allen, Jones, Dolby, Lynn, & Walport, 2009, p. 1). In the *PLoS* dataset, F1000 recommendations were not closely associated with citation or other altmetrics counts and formed their own factor in factor analysis, suggesting they track a relatively distinct sort of impact.

Conversation (Scholarly Blogging) In this context, "scholarly blogging" is distinguished from its popular counterpart by the expertise and qualifications of the blogger.

While a useful distinction, this is inevitably an imprecise one. One approach has been to limit the investigation to science-only aggregators like ResearchBlogging (Groth & Gurney, 2010; Shema & Bar-Ilan, 2011). Academic blogging has grown steadily in visibility; academics have blogged their dissertations (Efimova, 2009), and the ranks of academic bloggers contain several Fields Medalists, Nobel laureates, and other eminent scholars (Nielsen, 2009). Economist and Nobel laureate Paul Krugman, himself a blogger, argues that blogs are replacing the working-paper culture that has in turn already replaced economics journals as distribution tools (Krugman, 2012). Given its importance, there have been surprisingly few altmetrics studies of scholarly blogging. Extant research, however, has shown that blogging shares many of the characteristics of more formal communication, including a long-tail distribution of cited articles (Groth & Gurney, 2010; Shema & Bar-Ilan, 2011). Although science bloggers can write anonymously, most blog under their real names (Shema & Bar-Ilan, 2011).

Conversation (Twitter) Scholars on Twitter use the service to support different activities, including teaching (Dunlap & Lowenthal, 2009; Junco, Heiberger, & Loken, 2011), participating in conferences (Junco et al., 2011; Letierce et al., 2010; Ross et al., 2011), citing scholarly articles (Priem & Costello, 2010; Weller, Dröge, & Puschmann, 2011), and engaging in informal communication (Ross et al., 2011; Zhao & Rosson, 2009). Citations from Twitter are a particularly interesting data source, since they capture the sort of informal discussion that accompanies early important work. There is, encouragingly, evidence that tweeting scholars take citations from Twitter seriously, both in creating and reading them (Priem & Costello, 2010). The number of scholars on Twitter is growing steadily, as shown in figure 14.1. The same study found that, in a sample of around 10,000 doctoral students and faculty members at five representative universities, 1 in 40 scholars had an active Twitter account. Although some have suggested that Twitter is only used by younger scholars, rank was not found to significantly associate with Twitter use and faculty members' tweets were twice as likely to discuss their and others' scholarly work.

Conversation (Article Commenting) Following the lead of blogs and other social media platforms, many journals added article-level commenting to their online platforms in the middle of the last decade. In theory, the discussion taking place in these threads is another valuable lens into the early impacts of scientific ideas. In practice, however, many commenting systems are virtual ghost towns. In a sample of top medical journals, fully half had commenting systems lying idle, completely unused (Schriger, Chehrazi, Merchant, & Altman, 2011). But commenting was far from universally

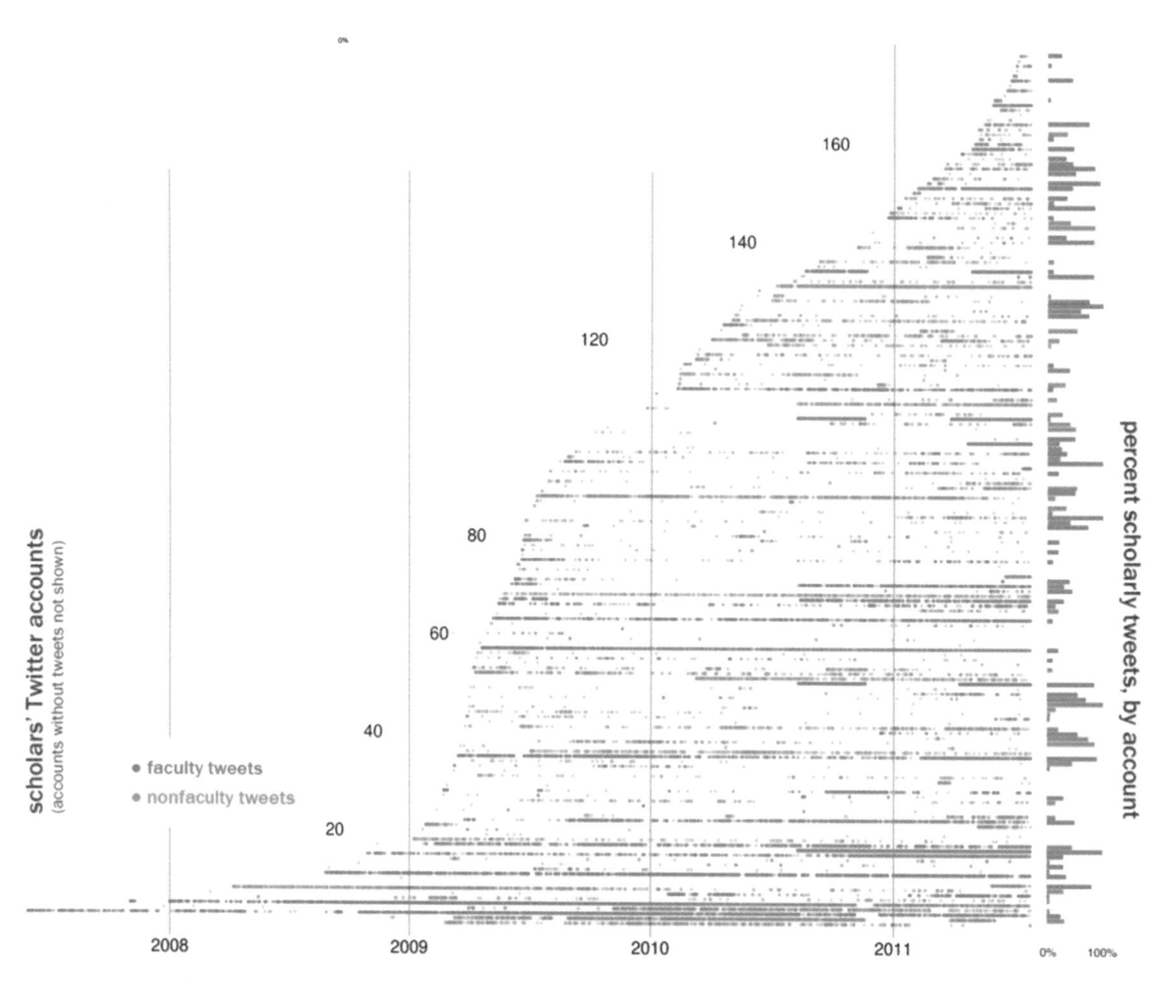

Figure 14.1
Growth in number of scholars actively using Twitter, from Priem et al. (2011).

unsuccessful; several journals had comments on 50% to 76% of their articles. In a sample from the *British Medical Journal*, articles had, on average, nearly five comments each (Gotzsche, Delamothe, Godlee, & Lundh, 2010). Additionally, many articles may accumulate comments in other environments; a growing number of external comment sites allow users to post comments on articles published elsewhere. These have tended to appear and disappear quickly over the last few years. Neylon (2010) argues that online article commenting is thriving, particularly for controversial papers, but that "people are much more comfortable commenting in their own spaces" (para. 5) (i.e., on their blogs and on Twitter).

Reference Managers Reference managers like Mendeley and CiteULike are very useful sources of altmetrics data and are currently among the most studied. Although scholars have used electronic reference managers for some time, this latest generation offers scientometricians the chance to query their datasets, offering a compelling glimpse into scholars' libraries. It is worth summarizing three main points, though. First, the most important social reference managers are CiteULike and Mendeley. Another popular reference manager, Zotero, has received less study (Lucas, 2008). Papers and ReadCube are newer, smaller reference managers. Connotea and 2Collab both dealt poorly with spam; both have now closed. Second, the usage base of social reference managers—particularly Mendeley—is large and growing rapidly. Mendeley's coverage, in particular, rivals that of commercial databases like Scopus and Web of Science (WoS) (Bar-Ilan et al., 2012; Haustein & Siebenlist, 2011; Li et al., 2011; Priem et al., 2012). Finally, inclusion in reference managers correlates with citation more strongly than most other altmetrics. Working with various datasets, researchers have reported correlations of 0.46 (Bar-Ilan, 2012), 0.56 (Li et al., 2011), and 0.5 (Priem et al., 2012) between inclusion in users' Mendeley libraries and WoS citations. This closer relationship is likely because of the importance of reference managers in the citation workflow. However, the lack of perfect or even strong correlation suggests that this altmetric, too, captures influence not reflected in the citation record. There has been particular interest in using social bookmarking for recommendations (Bogers & van den Bosch, 2008; Jiang, He, & Ni, 2011).

PDF Downloads As discussed earlier, most research on downloads today does not isolate HTML views in PDF downloads. However there is a substantial and growing body of research investigating article downloads and their relation to later citation. Several researchers have found that downloads predict or correlate with later citation (Perneger, 2004; Brody, Harnad, &Carr, 2006). The MESUR project is the largest of these studies to date, and employed linked usage events to create a novel map of the connections between disciplines as well as analyses of potential metrics using download and citation data in novel ways (Bollen, et al., 2009a). Shuai, Pepe, and Bollen (2012) show that downloads and Twitter citations interact, with Twitter likely driving traffic to new papers, and also reflecting reader interest.

Uses, Limitations, and Future Research

Uses

Several uses of altmetrics have been proposed, which aim to capitalize on their speed, breadth, and diversity, including in evaluation, analysis, and prediction.

Evaluation The breadth of altmetrics could support more holistic evaluation efforts; a range of altmetrics may help solve the reliability problems of individual measures by triangulating scores from easily accessible "converging partial indicators" (Martin & Irvine, 1983, p. 1). Altmetrics could also support the evaluation of increasingly important, nontraditional scholarly products like datasets and software, which are currently underrepresented in the citation record (Howison & Herbsleb, 2011; Sieber & Trumbo, 1995). Research that impacts wider audiences could also be better rewarded; Neylon (2012) relates a compelling example of how tweets reveal clinical use of a research paper—use that would otherwise go undiscovered and unrewarded. The speed of altmetrics could also be useful in evaluation, particularly for younger scholars whose research has not yet accumulated many citations. Most importantly, altmetrics could help open a window on scholars' "scientific 'street cred'" (Cronin, 2001, p. 6), helping reward researchers whose subtle influences—in conversations, teaching, methods expertise, and so on—influence their colleagues without perturbing the citation record. Of course, potential evaluators must be aware that while uncritical application of any metric is dangerous, this is doubly so with altmetrics, whose research base is not yet adequate to support high-stakes decisions.

Analysis Altmetrics could greatly benefit students of science, supplying a set of instruments possessing unprecedented scope and granularity. Some have suggested combining multiple metrics using regression analysis (Harnad, 2009) or combining metrics using Principal Component Analysis (Bollen, Van de Sompel, Hagberg, & Chute, 2009) or time-series analysis (Kurtz et al., 2005) to determine "underlying" forms of impact—the impact equivalent of Spearman's *g*. However, in many ways this approach moves in the wrong direction. Impact invariably begs the question, "Impact on *what?*"

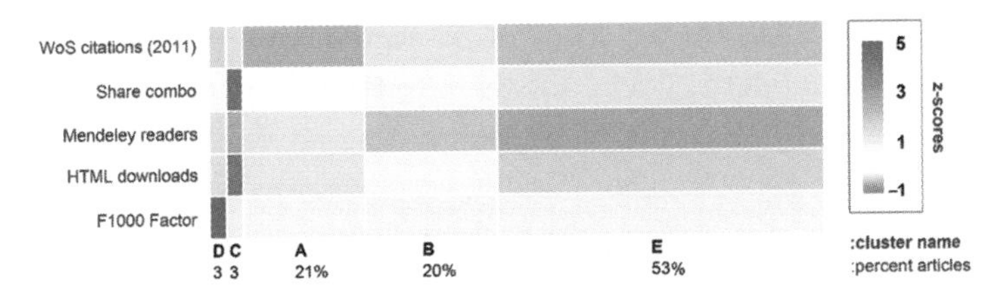

Figure 14.2
Article clusters based on altmetrics event types. Columns show the centers of clusters; rows represent altmetrics. Bluer cells indicate cluster centers on a relatively high standardized value for the given metric (Priem et al., 2012).

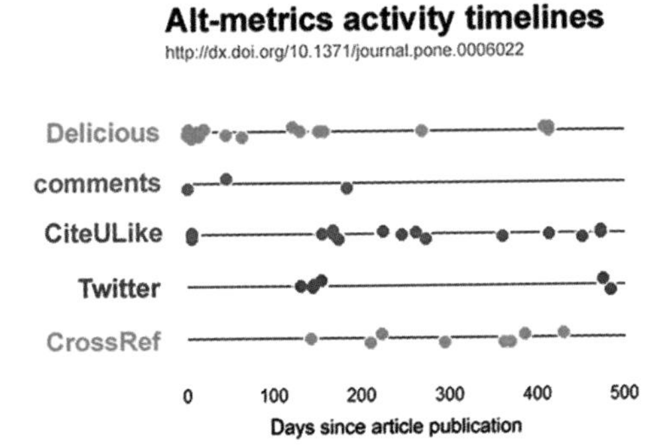

Figure 14.3
Events relating to a single article, by time since the article's publication. Each dot represents one event (Priem et al., 2012).

As long as there are multiple answers to these questions, reductive approaches will founder. Instead, it may be more useful to think not in terms of underlying impact, but, rather, of different kinds or flavors of impact, based on different audiences, products, and goals (Piwowar, 2012). Priem et al. (2012) found evidence of five such impact types from a cluster analysis of 25,000 *PLoS* articles (figure 14.2), including "popular hits" (cluster C in the figure), and heavily referenced but relatively uncited papers (cluster B).

The temporal resolution of altmetrics has the potential to help scientometricians better understand how knowledge moves through the varyingly dense media of different communication systems. Figure 14.3 gives an example for a single article. It shows that Delicious, CiteULike, and *PLoS* comments are active immediately after the article's publication. There follows a second wave of activity four or so months later, possibly initiated by readers alerted to the article by its first external citation.

Prediction Altmetrics, in most cases, are fast. They often begin accumulating a few days after publication (Kurtz et al., 2005; Priem, Costello, & Dzuba, 2011) and hold potential as leading indicators of other types of use supporting an "embryology of learned inquiry" (Harnad & Carr, 2000, p. 629). Data like these shown in figure 14.3 could be aggregated across thousands or millions of articles to produce powerful predictive models.

Forecasting based on social media has already proved surprisingly effective in diverse areas, including predicting stock prices, election results, and movie box-office returns (Asur & Huberman, 2010; Bollen, Mao, & Zeng, 2011; Tumasjan, Sprenger, Sandner, & Welpe, 2010). Altmetrics could support the extension of the same techniques to academic trends and research fronts. Wang, Wang, and Xu (2012) show the potential of this approach, identifying emerging research areas using download counts and keyword data from the journal *Scientometrics*. Here the speed of altmetrics compared with traditional citation measures is of tremendous value. Trend detection using traditional citations is rather like predicting the weather after the first few raindrops; altmetrics could be a radar-sized leap forward in our predictive power.

Recommendation Although article recommendation systems based on citations have been proposed before (McNee, 2006; see also Kurtz and Henneken, chapter 13, this volume), they have not seen wide usage, largely because of the lag inherent in citation tracking. By the time the citation record has caught up, researchers' interests have often evolved. The speed of altmetrics could be valuable in this area. One could imagine receiving daily, emailed recommendations based on the aggregated judgments of one's trusted peers, helping to repair the "filter failure" (Shirky, 2008) plaguing the modern research reader. A system like this, if widely used, would supplement the peer review system, harnessing reader-centered teams of reviewers by tracking their interactions with recently published literature (Neylon & Wu, 2009). Such a "soft peer review" (Taraborelli, 2008) system could aggregate and compile readership metrics and user-generated comments and reviews; this information could help readers judge the article's importance and relevance. While this is unlikely to replace traditional peer review, it could support faster-publishing "postpublication peer review" journals, or add an additional layer of open quality control to traditional publications.

A more radical use would be to move away from traditional peer review entirely. Given a sufficiently advanced recommender system, one must wonder, why would we need peer review at all? It could be that the aggregated readership, conversation, and access to the collection of one's trusted peers is enough to separate the gold from the dross. Many have suggested "deconstructed" (Smith, 2003) or "decoupled" journals to take advantage of this. Extending the open-access model, such journals would offer paying authors a shopping cart of copyediting, review, translation, and other services. On publication, readers could use their own robust filtering systems, built around the aggregated readership, conversation, and collection judgments of their peers to uncover the most promising material to read. By leveraging collective usage information, a decentralized peer review system might filter the literature in a way similar to how Google filters the Web based on aggregated link formation.

Limitations

Despite the potential of altmetrics, it has provoked criticism. The concerns have tended to focus on three areas: a lack of theory, ease of gaming, and bias.

It is certainly true that altmetrics lacks a cohesive body of theory; this lack should be remedied before altmetrics can be widely employed. However, it should also be acknowledged that lack of theory, while important to address, has not kept other metrics from being useful in practice. One need look no further than bibliometrics itself for examples of this. As late as 1979, Garfield acknowledged gaps in our theoretical understanding of citation, but argued that this should not preclude the use of bibliometrics:

> We still know very little about how sociological factors affect citation rates. . . . On the other hand, we know that citation rates say something about the contribution made by an individual's work, at least in terms of the utility and interest the rest of the scientific community finds in it. (Garfield, 1979, p. 372)

The same can be said for today's altmetrics. The introduction of new sources for impact metrics such as patents (Oppenheim, 2000), acknowledgments (Cronin et al., 1998), mentorships (Marchionini et al., 2006; see also Sugimoto, chapter 19, this volume), and scholarly hyperlinks (Ingwersen, 1998) has typically preceded robust theoretical underpinnings; indeed, it is difficult to imagine otherwise. A theory of altmetrics should be a priority for the new field, but not a prerequisite.

A second concern is the ease with which altmetrics counts can be manipulated. Again, this is a legitimate concern—but we should not imagine that extant metrics are free from it, either. Any metric will spawn attempts to exploit it (Espeland & Sauder, 2007). The Journal Impact Factor is a noteworthy example of a heavily gamed metric. The increasing importance of this measure has resulted in a range of tips and tricks for artificially boosting citation scores (Falagas & Alexiou, 2008), including, recently, the formation of "citation cartels" (Davis, 2012; Franck, 1999) in which journals collude to cite one another. In an extreme example, a single scholar's questionable editorial practices were enough to catapult the University of Alexandria—"not even the best university in Alexandria"—into the *Times Higher Education* top 200 rankings (Guttenplan, 2010, para. 6; see also the discussion in Gingras, chapter 6, this volume). For every extreme case like this, there are likely many more going undetected.

That said, it is certainly true that the relative ease of creating and using social media profiles seems to make generating false data easier for malfeasants. Indeed, artificial inflation of social media metrics is already a well-established practice outside academia. However, successful businesses and tools have evolved immune systems to combat this kind of gaming, in the form of antispam and antigaming measures. Perhaps the highest-profile of these is Google; with millions of dollars in traffic at stake, advertisers have assaulted Google search results with "black-hat" search engine optimization (Malaga, 2010). While these have not been entirely unsuccessful, they have not significantly undermined users' trust in the value of Google results. Google uses a variety of constantly evolving algorithms to differentiate spam sites from legitimate ones; similar statistical techniques, or "algorithmic forensics," can help control social media gaming as well. For instance, the automated WikiScanner tool[4] exposed and helped correct corporate tampering with *Wikipedia* articles (Borland, 2007). For Twitter users, Twitteraudit scans lists of followers, using tuned algorithms to spot bots. Moving to academia, the SSRN preprint server measures and reports download statistics for articles posted there. These statistics have become important for evaluation purposes in several disciplines, and have consequently attracted gaming (Edelman & Larkin, 2009). The Social Science Research Network (SSRN) has used algorithmic forensics to detect fraudulent downloads based on the observed properties of millions of legitimate ones. One particular virtue of an approach examining multiple social media ecosystems is that data from different sources can be cross-calibrated, exposing suspicious patterns invisible in a single source; *PLoS*'s "DataTrust" uses this approach to spot fraudulent altmetrics counts. While additional work in this area is certainly needed, there is evidence to suggest that social metrics, properly filtered and cautiously interpreted, could be relatively robust despite attempts to game them.

A third concern with altmetrics is that they will be systematically biased—in particular, toward younger or more fad-embracing researchers or outlets. Will altmetrics create a science dominated by shallow, fame-seeking narcissists, constantly chasing the next trend? While a glib "How would that be different from now?" is needlessly cynical, it contains an element of truth: scientists constantly promote themselves and their ideas across many venues, from conferences to mailing lists to press releases to the popular press. If the new metrics reward scientists who make the best use of available technology to provoke conversation among their peers and capture the imagination of the public, surely this is a bias we can accept. It is far from clear that efficient and successful use of available communication technologies to forcefully advocate one's ideas is something we should recoil from rewarding. Additionally, there is little evidence confirming naive assumptions about the demographics of researchers who are adopting new communication technologies. Priem et al. (2011), for example, find that doctoral students are no more likely to be active on Twitter than university faculty members. This said, we should take pains in using altmetrics to compare like with like, just as we do with citations; a field like digital humanities, where anecdotal reports suggest a majority of scholars use Twitter, will of course generate different numbers of tweets compared with a more technologically conservative discipline.

Future Research

The newness of altmetrics presents many opportunities for continued research. Most pressing is the need for environmental and content analysis surveys to build a fuller picture of how scholars use the tools and environment from which altmetrics draws. What does inclusion in a reference manager mean? Why and when do researchers bookmark content? How does linking to supporting papers from a blog post differ from traditional citation within the literature? These and similar questions need to be answered more fully before altmetrics can be used in any but the most cautious ways.

It will also be important to add more of a network perspective to altmetrics. Historically, bibliometricians have made surprisingly little use of network-analytic approaches. Things are changing in the citation literature, and this welcome trend should extend to altmetrics as well. Network-aware measurement will be particularly vital in informing recommendation and assessment algorithms. As Priego (2012) observes, the *number* of altmetrics events—tweets, for example—is less important than *who* they came from. Search engines faced similar problems in the early days of indexing the Web; simply counting links was not sufficient, since some are more important than others. Approaches like Google's PageRank and related approaches use the authority of linking pages—recursively calculated by the authority of other linking pages—to weight links. Similar approaches will be important for altmetrics; a tweet from a Nobel laureate is much more meaningful for many purposes than a tweet from the general public (as well as, arguably, more meaningful than a traditional citation).

Visualization holds promise for altmetrics as well; the multidimensionality and complexity of altmetrics make visualization more challenging, but also more important. The creation of composite metrics is another valuable research project. We might care less about the absolute numbers of, say, Mendeley and Facebook mentions than we do about the ratio between them. Very low Facebook likes could be an indication of minimal popular interest that could make a high number of Mendeley bookmarks more compelling as a sign of scholarly value—a negative term in an academic impact regression equation. Similar relationships may exist for other measures, or for more complex combinations of measures. Finally, altmetrics researchers should continue to expand the number and variety of data sources they investigate, examining scholarly impact from YouTube videos, Slideshare presentations, VIVO profiles, SSRN and Academia.edu downloads, Stack Overflow reputation, and more.

Altmetrics is a young, but growing, field. For now, there is perhaps more promise than results, but the promise is sufficient to justify further research. The results we have seen give us good reason to believe that scholars, like other information workers, will continue to move their work into online tools and environments. To the extent that these tools allow us to peer into processes once hidden—and it seems this extent is significant—altmetrics will become increasingly important as a way to understand the hidden stories of scholarly impact, by more directly observing the tracks of scholarly ideas.

Notes

1. total-impact.org
2. http://altmetric.com/
3. http://delicious.com
4. http://wikiscanner.virgil.gr

References

Allen, L., Jones, C., Dolby, K., Lynn, D., & Walport, M. (2009). Looking for landmarks: The role of expert review and bibliometric analysis in evaluating scientific publication outputs. *PLoS ONE, 4*(6), e5910.

Almind, T. C., & Ingwersen, P. (1997). Informetric analyses on the World Wide Web: Methodological approaches to "webometrics." *Journal of Documentation, 53*(4), 404–426.

Asur, S., & Huberman, B. A. (2010). Predicting the future with social media. *arXiv:1003.5699*. Retrieved from http://arxiv.org/abs/1003.5699.

Bar-Ilan, J. (2012). *JASIST@mendeley*. Presented at altmetrics12: An ACM Web Science Conference 2012 Workshop, Evanston, IL.

Bar-Ilan, J., Haustein, S., Peters, I., Priem, J., Shema, H., & Terliesner, J. (2012). Beyond citations: Scholars' visibility on the social Web. *arXiv:1205.5611*. Retrieved from http://arxiv.org/abs/1205.5611.

Bernal, J. D. (1944). *The social function of science*. London: Routledge.

Bogers, T., & Van den Bosch, A. (2008). Recommending scientific articles using citeulike. In *Proceedings of the ACM 2008 Conference on Recommender Systems* (pp. 287–290). New York: ACM.

Bollen, J., Mao, H., & Zeng, X. (2011). Twitter mood predicts the stock market. *Journal of Computational Science*, *2*(1), 1–8.

Bollen, J., Van de Sompel, H., Hagberg, A., Bettencourt, L., Chute, R., Rodriguez, M. A., et al. (2009a). Clickstream data yields high-resolution maps of science. *PLoS ONE*, *4*(3), e4803.

Bollen, J., Van de Sompel, H., Hagberg, A., & Chute, R. (2009b). A principal component analysis of 39 scientific impact measures. *PLoS ONE*, *4*(6): e6022.

Borgman, C. L., & Furner, J. (2002). Scholarly communication and bibliometrics. *Annual Review of Information Science & Technology*, *36*(1), 2–72.

Borland, J. (2007, August 8). See who's editing Wikipedia—Diebold, the CIA, a campaign. *Wired*. Retrieved from http://www.wired.com/politics/onlinerights/news/2007/08/wiki_tracker?currentPage=1.

Bornmann, L., & Daniel, H. D. (2008). What do citation counts measure? A review of studies on citing behavior. *Journal of Documentation*, *64*(1), 45–80.

Brooks, T. A. (1986). Evidence of complex citer motivations. *Journal of the American Society for Information Science*, *37*(1), 34–36.

Carpenter, J., Wetheridge, L., Smith, N., Goodman, M., & Struijvé, O. (2010). *Researchers of tomorrow annual report, 2009–2010*. JISC/British Library. Retrieved from http://explorationforchange.net/index.php/rot-home.html.

Cheverie, J. F., Boettcher, J., & Buschman, J. (2009). Digital scholarship in the university tenure and promotion process: A report on the sixth scholarly communication symposium at Georgetown University Library. *Journal of Scholarly Publishing*, *40*(3), 219–230.

Cronin, B. (2001). Bibliometrics and beyond: Some thoughts on web-based citation analysis. *Journal of Information Science*, *27*(1), 1–7.

Cronin, B., & Overfelt, K. (1994). The scholar's courtesy: A survey of acknowledgement behaviour. *Journal of Documentation*, *50*(3), 165–196.

Cronin, B., Snyder, H. W., Rosenbaum, H., Martinson, A., & Callahan, E. (1998). Invoked on the Web. *Journal of the American Society for Information Science*, *49*(14), 1319–1328.

Davis, P. (2012, April 10). The emergence of a citation cartel. *The Scholarly Kitchen*. Retrieved from http://scholarlykitchen.sspnet.org/2012/04/10/emergence-of-a-citation-cartel.

De Bellis, N. (2009). *Bibliometrics and citation analysis: From the Science Citation Index to cybermetrics*. Lanham, MD: Scarecrow Press.

Ding, Y., Jacob, E. K., Caverlee, J., Fried, M., & Zhang, Z. (2009). Profiling social networks: A social tagging perspective. *D-Lib Magazine, 15*(3/4).

Dunlap, J., & Lowenthal, P. (2009). Tweeting the night away: Using Twitter to enhance social presence. *Journal of Information Systems Education, 20*(2), 129–135.

Edelman, B. G., & Larkin, I. (2009). Demographics, career concerns or social comparison: Who games SSRN download counts? *SSRN eLibrary*. Retrieved from http://papers.ssrn.com/sol3/papers.cfm?abstract_id=1346397.

Efimova, L. (2009). PhD—Mathemagenic. *Mathemagenic*. Retrieved from http://blog.mathemagenic.com/phd.

Espeland, W. N., & Sauder, M. (2007). Rankings and reactivity: How public measures recreate social worlds. *American Journal of Sociology, 113*(1), 1–40.

Eysenbach, G. (2012). Can tweets predict citations? Metrics of social impact based on Twitter and correlation with traditional metrics of scientific impact. *Journal of Medical Internet Research, 13*(4).

Falagas, M., & Alexiou, V. (2008). The top-ten in journal impact factor manipulation. *Archivum Immunologiae et Therapiae Experimentalis, 56*(4), 223–226.

Franck, G. (1999). Scientific communication—a vanity fair? *Science, 286*(5437), 53–55.

Garfield, E. (1972). Citation analysis as a tool in journal evaluation. *Science, 178*(4060), 471–479.

Garfield, E. (1979). Is citation analysis a legitimate evaluation tool? *Scientometrics, 1*(4), 359–375.

Gotzsche, P. C., Delamothe, T., Godlee, F., & Lundh, A. (2010). Adequacy of authors' replies to criticism raised in electronic letters to the editor: Cohort study. *BMJ, 341*: c3926.

Groth, P., & Gurney, T. (2010). *Studying scientific discourse on the Web using bibliometrics: A chemistry blogging case study*. Presented at WebSci10: Extending the Frontiers of Society On-Line, Raleigh, NC. Retrieved from http://journal.webscience.org/308.

Guttenplan, D. D. (2010, November 14). Questionable science behind academic rankings. *New York Times*. Retrieved from http://www.nytimes.com/2010/11/15/education/15iht-educLede15.html.

Harnad, S., & Carr, L. (2000). Integrating, navigating, and analysing open eprint archives through open citation linking (the OpCit project). *Current Science, 79*(5), 629–638.

Harnad, S. (2009). Open access scientometrics and the UK research assessment exercise. *Scientometrics, 79*(1), 147–156.

Haustein, S., & Siebenlist, T. (2011). Applying social bookmarking data to evaluate journal usage. *Journal of Informetrics, 5*(3), 446–457.

Head, A. J., & Eisenberg, M. B. (2010). How today's college students use Wikipedia for course-related research. *First Monday, 15*(3). Retrieved from http://firstmonday.org/htbin/cgiwrap/bin/ojs/index.php/fm/article/view/2830/2476.

Howison, J., & Herbsleb, J. D. (2011). Scientific software production: Incentives and collaboration. In *Proceedings of the ACM 2011 Conference on Computer Supported Cooperative Work* (pp. 513–522). New York: ACM.

Ingwersen, P. (1998). The calculation of Web impact factors. *Journal of Documentation, 54*(2), 236–243.

jasonpriem. (2010, September 28). I like the term #articlelevelmetrics, but it fails to imply *diversity* of measures. Lately, I'm liking #altmetrics. Retrieved from https://twitter.com/#!/jasonpriem/status/25844968813.

Jensen, M. (2007, June 15). The New Metrics of Scholarly Authority. The Chronicle of Higher Education. Retrieved from http://www.lexisnexis.com/us/lnacademic/frame.do?tokenKey=rsh-20.918629.3985793049&target=results_listview_resultsNav&reloadEntirePage=true&rand=1254107715289&returnToKey=20_T7441126159&parent=docview.

Jiang, J., He, D., & Ni, C. (2011). Social reference: Aggregating online usage of scientific literature in CiteULike for clustering academic resources. *Proceeding of the 11th Annual International ACM/IEEE Joint Conference on Digital Libraries* (pp. 401–402). Ottawa, Canada: ACM.

Junco, R., Heiberger, G., & Loken, E. (2011). The effect of Twitter on college student engagement and grades. *Journal of Computer Assisted Learning, 27*(2), 119–132.

Kiernan, V. (2003). Diffusion of news about research. *Science Communication, 25*(1), 3–13.

King, D., & Tenopir, C. (2004). An evidence-based assessment of the "author pays" model. *Nature web focus: Access to the literature*. Retrieved from http://www.nature.com/nature/focus/accessdebate/26.html.

Kousha, K., & Thelwall, M. (2006). Motivations for URL citations to open access library and information science articles. *Scientometrics, 68*(3), 501–517.

Kousha, K., & Thelwall, M. (2008). Assessing the impact of disciplinary research on teaching: An automatic analysis of online syllabuses. *Journal of the American Society for Information Science and Technology, 59*(13), 2060–2069.

Krugman, P. (2012). Open science and the econoblogosphere. *Paul Krugman Blog*. Retrieved from http://krugman.blogs.nytimes.com/2012/01/17/open-science-and-the-econoblogosphere.

Kurtz, M. J., Eichhorn, G., Accomazzi, A., Grant, C. S., Demleitner, M., Murray, S. S., et al. (2005). The bibliometric properties of article readership information. *Journal of the American Society for Information Science, 56*(2), 111–128.

Lerman, K., & Galstyan, A. (2008). Analysis of social voting patterns on digg. In *Proceedings of the First Workshop on Online Social Networks* (pp. 7–12). New York: ACM.

Letierce, J., Passant, A., Breslin, J., & Decker, S. (2010). Understanding how Twitter is used to spread scientific messages. Presented at the Web Science Conference, Raleigh, NC. Retrieved from http://journal.webscience.org/314/2/websci10_submission_79.pdf.

Lewison, G. (2002). From biomedical research to health improvement. *Scientometrics, 54*(2), 179–192.

Li, X., Thelwall, M., & Giustini, D. (2011). Validating online reference managers for scholarly impact measurement. *Scientometrics, 91*(2), 461–471.

Lucas, D. V. (2008). A product review of Zotero. Master's thesis, University of North Carolina at Chapel Hill. Retrieved from http://ils.unc.edu/MSpapers/3388.pdf.

Lund, B., Hammond, T., Flack, M., & Hannay, T. (2005). Social bookmarking tools (II). *D-Lib Magazine, 11*(4), 1082–9873.

MacRoberts, M. H., & MacRoberts, B. R. (2010). Problems of citation analysis: A study of uncited and seldom-cited influences. *Journal of the American Society for Information Science and Technology, 61*(1), 1–12.

Malaga, R. A. (2010). Search engine optimization—black and white hat approaches. *Advances in Computers: Improving the Web 78*, 1–39.

Marchionini, G., Solomon, P., Davis, C., & Russell, T. (2006). Information and library science MPACT: A preliminary analysis. *Library & Information Science Research, 28*(4), 480–500.

Martin, B. R., & Irvine, J. (1983). Assessing basic research: Some partial indicators of scientific progress in radio astronomy. *Research Policy, 12*(2), 61–90.

McNee, S. M. (2006). Meeting user information needs in recommender systems. Unpublished doctoral dissertation, University of Minnesota.

Merton, R. K. (1988). The Matthew effect in science, II. *Isis, 79*, 606–623.

Moravcsik, M. J., & Murugesan, P. (1975). Some results on the function and quality of citations. *Social Studies of Science, 5*(1), 86–92.

Narin, F. (1976). *Evaluative bibliometrics: The use of publication and citation analysis in the evaluation of scientific activity*. Cherry Hill, NJ: Computer Horizons.

Neylon, C. (2010, December 6). Forward linking and keeping context in the scholarly literature. *Science in the Open*. Retrieved from http://cameronneylon.net/blog/forward-linking-and-keeping-context-in-the-scholarly-literature.

Neylon, C. (2012, August 10). *Research assessment to support research impact*. University of Cape Town, Cape Town: South Africa. Retrieved from http://www.slideshare.net/CameronNeylon/research-assessment-to-support-research-impact.

Neylon, C., & Wu, S. (2009). Article-level metrics and the evolution of scientific impact. *PLoS Biology, 7*(11), e1000242.

Nielsen, F. (2007). Scientific citations in Wikipedia. *First Monday, 12*(8). Retrieved from http://firstmonday.org/ojs/index.php/fm/article/view/1997/1872.

Nielsen, M. (2009). Doing science online. *Michael Nielsen Blog*. Retrieved from http://michaelnielsen.org/blog/doing-science-online.

Oppenheim, C. (2000). Do patent citations count? In B. Cronin & H. B. Atkins (Eds.), *The web of knowledge: A festschrift in honor of Eugene Garfield* (ASIS Monograph Series, pp. 405–434). Medford, NJ: ASIS.

Pavitt, K. (1985). Patent statistics as indicators of innovative activities: Possibilities and problems. *Scientometrics, 7*(1–2), 77–99. doi:10.1007/BF02020142.

Perneger, T. V. (2004). Relation between online "hit counts" and subsequent citations: prospective study of research papers in the *BMJ. BMJ (Clinical Research Ed.), 329*(7465), 546–547. doi:10.1136/bmj.329.7465.546.

Piwowar, H. (2012, January 31). 31 flavors of research impact through #altmetrics. *Research Remix*. Retrieved from http://researchremix.wordpress.com/2012/01/31/31-flavours.

Price, D. J. de S., & Beaver, D. (1966). Collaboration in an invisible college. *American Psychologist, 21*(11), 1011–1018.

Priego, E. (2012, August 24). "Altmetrics": Quality of engagement matters as much as retweets. *Guardian Higher Education Network*. Retrieved from http://www.guardian.co.uk/higher-education-network/blog/2012/aug/24/measuring-research-impact-altmetic.

Priem, J., & Costello, K. (2010). How and why scholars cite on Twitter. *Proceedings of the 73rd ASIS&T Annual Meeting*. Pittsburgh, PA: ASIS&T.

Priem, J., Costello, K., & Dzuba, T. (2011). *Prevalence and use of Twitter among scholars*. Presented at the Metrics 2011 Symposium on Informetric and Scientometric Research, New Orleans. Retrieved from http://jasonpriem.org/self-archived/5uni-poster.png.

Priem, J., & Hemminger, B. H. (2010). Scientometrics 2.0: Toward new metrics of scholarly impact on the social Web. *First Monday, 15*(7). Retrieved from http://firstmonday.org/htbin/cgiwrap/bin/ojs/index.php/fm/article/view/2874/2570.

Priem, J., Piwowar, H. A., & Hemminger, B. M. (2012). Altmetrics in the wild: Using social media to explore scholarly impact. *arXiv:1203.4745*. Retrieved from http://arxiv.org/abs/1203.4745.

Priem, J., Taraborelli, D., Groth, P., & Neylon, C. (2010). Altmetrics: A manifesto. Retrieved from http://altmetrics.org/manifesto.

Procter, R., Williams, R., Stewart, J., Poschen, M., Snee, H., Voss, A., & Asgari-Targhi, M. (2010). Adoption and use of Web 2.0 in scholarly communications. *Philosophical Transactions of the Royal Society A: Mathematical, Physical and Engineering Sciences, 368*(1926), 4039–4056. doi:10.1098/rsta.2010.0155.

Revolutionizing peer review? (2005). *Nature Neuroscience, 8*(4), 397.

Ross, C., Terras, M., Warwick, C., & Welsh, A. (2011). Enabled backchannel: Conference Twitter use by digital humanists. *Journal of Documentation, 67*(2), 214–237.

Schriger, D. L., Chehrazi, A. C., Merchant, R. M., & Altman, D. G. (2011). Use of the Internet by print medical journals in 2003 to 2009: A longitudinal observational study. *Annals of Emergency Medicine, 57*(2), 153–160.

Schweitzer, N. J. (2008). Wikipedia and psychology: Coverage of concepts and its use by undergraduate students. *Teaching of Psychology, 35*(2), 81–85.

Shema, H., & Bar-Ilan, J. (2011). *Characteristics of researchblogging.org science blogs and bloggers*. Presented at altmetrics11: Tracking scholarly impact on the social Web (an ACM Web Science Conference 2011 workshop), Koblenz, Germany. Retrieved from http://altmetrics.org/workshop2011/shema-v0.

Shirky, C. (2008). *It's not information overload: It's filter failure*. Presented at the Web 2.0 Expo, New York. Retrieved from http://www.youtube.com/watch?v=LabqeJEOQyI&feature=youtube_gdata.

Shuai, X., Pepe, A., & Bollen, J. (2012). How the scientific community reacts to newly submitted preprints: Article downloads, Twitter mentions, and citations. *arXiv:1202.2461*. Retrieved from http://arxiv.org/abs/1202.2461.

Sieber, J., & Trumbo, B. (1995). (Not) giving credit where credit is due: Citation of data sets. *Science and Engineering Ethics, 1*(1), 11–20.

Smith, J. W. T. (2003). The deconstructed journal revisited—a review of developments. *ICCC/IFIP Conference on Electronic Publishing—ElPub03–From Information to Knowledge* (pp. 2–88). http://elpub.scix.net/cgi-bin/works/show_id=280_elpub2008&sort=DEFAULT&search=Galina&hits=2/BrowseTree?field=series&separator=:&recurse=0&order=AZ&value=ELPUB%3a2003&first=0.

Stankovic, M., Rowe, M., & Laublet, P. (2010). Mapping tweets to conference talks: A goldmine for semantics. *The Third Social Data on the Web Workshop SDoW2010, collocated with ISWC2010, Shanghai, China*. Retrieved from http://ceur-ws.org/Vol-664.

Sugimoto, C. R., Russell, T. G., Meho, L. I., & Marchionini, G. (2008). MPACT and citation impact: Two sides of the same scholarly coin? *Library & Information Science Research, 30*(4), 273–281.

Taraborelli, D. (2008). Soft peer review: Social software and distributed scientific evaluation. In *Proceedings of the 8th International Conference on the Design of Cooperative Systems* [COOP '08]. Carry-Le-Rouet, France.

Thelwall, M. (2003). What is this link doing here? Beginning a fine-grained process of identifying reasons for academic hyperlink creation. *Information Research, 8*(3).

Thelwall, M. (2010). Webometrics: Emergent or doomed? *Information Research, 15*(4). Retrieved from http://informationr.net/ir/15-4/colis713.html.

Tinti-Kane, H., Seaman, J., & Levy, J. (2010). *Social media in higher education: The survey*. Retrieved from http://www.slideshare.net/PearsonLearningSolutions/pearson-socialmediasurvey2010.

Tumasjan, A., Sprenger, T. O., Sandner, P. G., & Welpe, I. M. (2010). Predicting elections with Twitter: What 140 characters reveal about political sentiment. *Proceeding of the Fourth International AAAI Conference on Weblogs and Social Media* (pp. 178–185). Menlo Park, CA: AAAI Press.

Vaughan, L., & Shaw, D. (2005). Web citation data for impact assessment: A comparison of four science disciplines. *Journal of the American Society for Information Science, 56*(10), 1075–1087.

Vinkler, P. (1987). A quasi-quantitative citation model. *Scientometrics, 12*(1–2), 47–72.

Wang, X., Wang, Z., & Xu, S. (2012). Tracing scientists' research trends realtimely. *arXiv:1208.1349*. Retrieved from http://arxiv.org/abs/1208.1349.

Weller, K., Dröge, E., & Puschmann, C. (2011). Citation analysis in Twitter: Approaches for defining and measuring information flows within tweets during scientific conferences. In *Proceedings of Making Sense of Microposts Workshop* (# MSM2011, pp. 1–12), co-located with Extended Semantic Web Conference, Crete, Greece.

Wets, K., Weedon, D., & Velterop, J. (2003). Post-publication filtering and evaluation: Faculty of 1000. *Learned Publishing, 16*, 249–258.

Yan, K.-K., & Gerstein, M. (2011). The spread of scientific information: Insights from the Web usage statistics in PLoS Article-Level Metrics. *PLoS ONE, 6*(5), e19917.

Zhao, D., & Rosson, M. B. (2009). How and why people Twitter: The role that micro-blogging plays in informal communication at work. In *Proceedings of the ACM 2009: International Conference on Supporting Group Work* (pp. 243–252). New York: ACM.

15 Web Impact Metrics for Research Assessment

Kayvan Kousha and Mike Thelwall

Introduction

Quantitative research impact assessment has relied primarily on analyses of citations between published academic documents (mostly journal articles) in conventional citation indexes, such as Scopus, Web of Science (WoS), or the Chinese Citation Index. For practical reasons these can only cover a limited subset of the scientific literature, though some have expanded in recent times from high-impact, mainly English-language journal articles to a wider variety of documents, including conference proceedings, books, and preprints. For example, Thomson Reuters introduced the "Web Citation Index" to include citation data from nearly 450,000 full-text web sources, including preprints, conference proceedings, technical reports, theses, and white papers (Thomson Scientific, 2006), and also offers a book citation index covering approximately 10,000 volumes per year (Thomson Reuters, 2012). These wider types of information sources, notably books and monographs, are typically needed for monitoring research impact in the social sciences and humanities (Nederhof, 2006). Other academic activities that are essential to the progress of science but rarely result in formal publications—such as teaching, discussions, and presentations—may sometimes cite research in ways that would be undetectable through traditional citation indexes.

Information scientists have argued that since web sources are increasingly used in research and scholarly communication, it may be possible to extend bibliometric methods beyond conventional citation indexes to the Web (e.g., Borgman & Furner, 2002; Cronin, 2001). This is because the web includes a vast range of digital objects (e.g., articles, course syllabi, presentations, blogs, scientific videos, and images) that could potentially be useful for assessing the impact of research. Figure 15.1 gives an overview of the extent to which web pages are cited in published articles for the

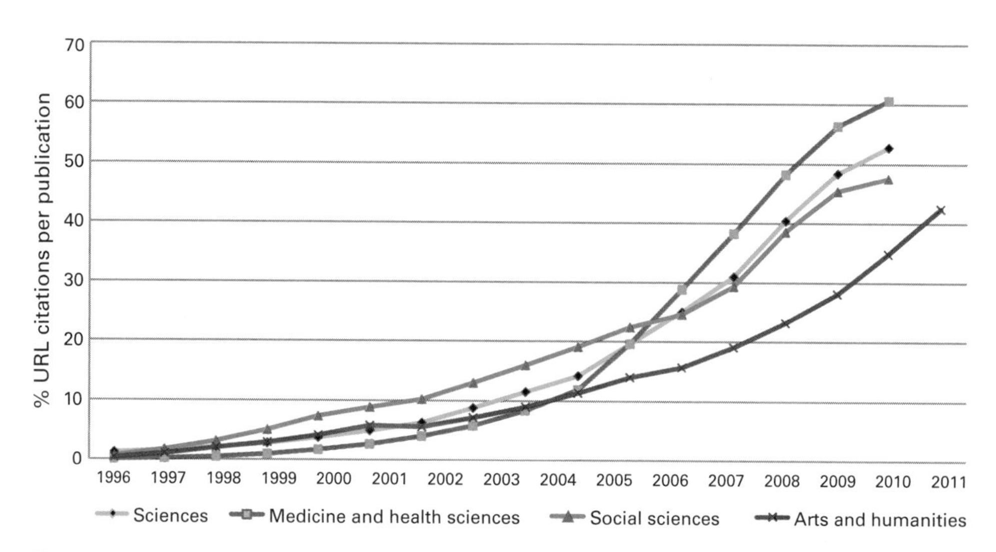

Figure 15.1
URLs cited in Scopus (1996–2011) in four broad subject areas (authors' own data).

period 1996 to 2011. It shows a constant growth in the number of web sources cited in scholarly publications indexed by Scopus in the four broad areas. The data suggest that the Web is of increasing importance to academic research.

Early web impact assessment was predicated on similarities between traditional bibliometric indicators (such as journal citations) and web links (e.g., Almind & Ingwersen, 1997; Ingwersen, 1998; Rousseau, 1997), web-extracted citations (e.g., Kousha & Thelwall, 2007a; Vaughan & Shaw, 2005), or download and usage statistics (e.g., Bollen & Van de Sompel, 2008; Brody, Harnad, & Carr, 2006). However, the emergence of social media has led to many alternative metrics (see Priem, chapter 14, this volume). "Scientometrics 2.0" type of analysis (e.g., Priem & Hemminger, 2010) includes Twitter (Eysenbach, 2011; Priem & Costello, 2010), blogs (Kjellberg, 2010; Shema, Bar-Ilan, & Thelwall, 2012), online reference managers (e.g., Jiang, He, & Ni, 2011; Li, Thelwall, & Giustini, 2012), and social bookmarking (Haustein & Siebenlist, 2011).

This chapter discusses the use of web metrics for assessing the impact of academic research—whether artifacts, articles, researchers, or institutions. We believe that web impact assessment should supplement conventional impact metrics by including new types of *sources* of impact (e.g., presentations or syllabi) and new types of *outputs* (e.g., online videos or science blogs).

General Web Impact Assessment

This section discusses different methods to assess web impact, including hyperlinks, text, and URL citations.

Web Hyperlinks

A web hyperlink, like a citation, is a virtual interdocument connection. Its source and target could be a department, an online resource, an individual academic, a business, or anything else with a web presence. Hyperlinks are more diverse than traditional citations and hence useful for wider types of citation analysis. Ingwersen (1998) first demonstrated how they could be used to calculate the web impact of an entire nation. Most early research attempted to validate hyperlinks as evidence of impact in a particular context by showing that hyperlink counts correlated with traditional citation metrics. The rationale behind this was that if hyperlinks reflected an aspect of scholarly impact, they should have a positive correlation with citation counts, which are established impact indicators. This should still be true even if they reflect different aspects of scholarly communication since it is unlikely that two aspects of scholarly communication would be completely unrelated. There have been other attempts to validate the use of hyperlinks for impact evaluation, such as content analyses of reasons for hyperlink creation (e.g., Bar-Ilan, 2004; Wilkinson, Harries, Thelwall, & Price, 2003).

The early years of the Web saw the emergence of the first online open-access journals and websites for print journals. A logical use for hyperlink metrics was to assess the online impact of entire journals, in parallel with Journal Impact Factors (JIFs). A series of investigations found that JIFs were not equivalent to metrics based on counts of hyperlinks to journal websites because of disciplinary differences in the amount of hyperlinking and differing amounts of content on journal websites (e.g., Vaughan & Hysen, 2002).

A second potential use of hyperlinks was to assess the online impact of entire universities, the purpose of which was to produce league tables of universities to help academics identify the status of institutions in various countries. In practice, most studies had a primarily methodological focus, testing different ways of gathering and counting hyperlinks. They correlated counts of hyperlinks with university websites, sometimes normalized for university size, with an external source of information about the university's research quality, choosing only countries for which such a source was available. The results were generally positive, with high correlations found between the online and offline metrics (e.g., Thelwall & Harries, 2004). The idea of ranking universities based on web data has flourished in the form of the Ranking Web of World Universities (Aguillo, Granadino, Ortega, & Prieto, 2006), which promotes open-access research.

From a methodological perspective, one clear finding from the hyperlink research was that it was preferable to count hyperlinking *sites* rather than hyperlinking *pages* because of the anomalies created by occasional large-scale replication of links within a site (Thelwall, 2002). The hyperlinks used were collected either by a web crawler or by queries to commercial search engines, including AltaVista and Yahoo!. The different sources gave results that were comparable but not identical. Commercial search engines had two advantages: they were easier to use and queries could include the whole Web rather than just counting links between a fixed set of websites. The withdrawal of all hyperlink searches from major search engines, still the case at the time of writing, has undermined hyperlink web impact research. However, as discussed below, URL citations and title mentions have been proposed as reasonable alternatives to hyperlink searches for web impact analysis (Thelwall, Sud, & Wilkinson, 2012).

Text: Web Citations and Web Mentions

A traditional citation gives enough details to uniquely identify the cited document. This is true for all standard style formats, including Chicago and Harvard. The Web contains many such citations as well as mentions of people and organizations, all of which can potentially be used for research evaluation.

The earliest attempt to investigate web mentions for impact assessment queried the names of highly cited academics through commercial search engines and classified the resulting web pages. Academics' names were invoked in a wide variety of scholarly contexts, such as conference pages, course reading lists, current awareness bulletins, resource guides, personal or institutional homepages, listservs, and tables of contents (Cronin, Snyder, Rosenbaum, Martinson, & Callahan, 1998). The continuous growth of the Web has made subsequent similar studies more difficult. One of us (Thelwall) has recently evaluated the web profiles of musicians in a set of European music schools, with the underlying hypothesis that a musician's fame was a good indicator of his or her scholarly contribution in the absence of an orientation toward published research.[1]

The term *web citation* has been used by Vaughan and Shaw (2003) to refer to the occurrence of an exact article title in a web page, although others have operationalized it through their web query strategy as also including additional bibliographic information (e.g., author name, source title, or publication year) (Kousha & Thelwall, 2008). Web citation counts can thus reflect how often a publication has been mentioned on web pages. Commercial search engines (e.g., Google, Bing) have been used to identify web citations for impact indicators. Most web citation analyses have compared web impact metrics with conventional counterparts (e.g., WoS citations) to assess whether these web metrics are relevant for research evaluation. For example, one study compared web citation counts with traditional citations to library and information science journal articles and found significant correlations between them (Vaughan & Shaw, 2003). This result was subsequently confirmed for other fields and journals (Kousha & Thelwall, 2007a; Vaughan & Shaw, 2005), giving preliminary quantitative evidence of a relationship between conventional and web-extracted research impact measures. Nevertheless, further qualitative studies in several science and social science disciplines showed that only 23% and 19%, respectively, of Google web citations were from references in online documents. In contrast, most web citations were created for navigation (e.g., scientific databases or online bibliographies) or self-publicity (i.e., personal or institutional curriculum vitae) (Kousha & Thelwall, 2007b). Because of these nonstandard reasons for citation creation, web citations should never be claimed to be *measuring* impact. This weakens but does not invalidate the value of the results, since web citations can still be used as *indicators* of research impact.

Practical applications of web citation/mention counting include calculating indicators for objects outside of traditional citation indexes, including new journals, preprints, and white papers.

URL Citations

URL citations are mentions of a specific URL in the text of a web page, whether hyperlinked or not (see below). Thus, URL citation counting is an alternative method to capture web citations to different web objects (e.g., open-access articles, online videos, blog posts, or other web documents). URL citation queries can be used in commercial search engines even if they do not permit hyperlink queries; this is an important advantage. Below is a Google query for URL citations to an open-access journal article. The -site: after the URL of a journal article excludes possible URL mentions created by the journal for navigational reasons (e.g., in a table of contents).

"informationr.net/ir/5–4/paper82.html" -site:informationr.net/ir

URL citations are different from both hyperlink and web citations and have some advantages and disadvantages for online impact assessment. One advantage of URL citation searches over web citation searches (e.g., mentions of exact article titles) is that URLs are unique and therefore URL citation counting can prevent false matches. Nevertheless, URL citation searches are only useful for counting citations to open-access articles or other digital objects, particularly where it would be reasonable to expect a URL to be explicitly mentioned in a citation. Another major restriction of URL citation searching is that it cannot mine hyperlinked citations embedded in images or titles of online documents.

A practical application of URL citation counting for individual articles is to calculate JIFs for journals and citation counts for articles and other digital artifacts beyond conventional citation indexes.

Hybrid Approaches

Web citations, URL citations, and hyperlinks all retrieve overlapping as well as unique results for online impact assessment. For instance, web citation queries may miss URL citations in some web documents, while URL citations will fail to capture hyperlinked citations embedded in titles or images. Consequently, an attempt has been made to use a combined data-gathering method, web/URL citation counting, to capture both sources (Kousha & Thelwall, 2007a).

Impact Assessment with Web Databases

Bibliographic databases have different policies for indexing publications and therefore their coverage can be considerably different from a comprehensive academic literature collection (e.g., in terms of language, source types). Consequently, impact metrics are limited by database coverage. Many have argued that selective coverage of the scientific literature by traditional citation indexes leads to unrepresentative impact measures (MacRoberts & MacRoberts, 1996; Moed, 2005), especially in the social sciences and humanities (Archambault, Vignola-Gagne, Cote, Larivière, & Gingras, 2006; Glänzel, 1996; Hicks, 1999). For instance, the number of publications in the Thomson Reuters Arts & Humanities Citation Index is 1,689 (Arts & Humanities Citation Index, 2012), whereas Elsevier's Scopus database indexes about 3,500 arts and humanities publications (Scopus Arts & Humanities Coverage, 2011).

The development of automatic web-based citation indexes and other document search tools has been a turning point that has enabled the extraction of web metrics of different types from those available from conventional citation databases. Consequently, many studies have used CiteSeer, Google Scholar, Google Books, and other online databases to investigate alternative impact metrics.

Specialist Citation Indexes

Early attempts to capture alternative citation metrics from online databases were based on specialist digital libraries. CiteSeer, for instance, is an autonomous citation index that primarily covers online publications (e.g., postprints or preprints) in computer science and information science. A comparison of citation counts from online computer science papers indexed in CiteSeer with WoS citations found more recent CiteSeer citations, especially from conference papers (Goodrum, McCain, Lawrence, & Giles, 2001). The same was true for the Extended Markup Language (XML) research field, which also attracted more citations from CiteSeer than from the WoS database (Zhao & Logan, 2002). It is not clear, however, whether the subsequent introduction of the Thomson Reuters' Conference Proceedings Citation Index has altered the situation.

Google Scholar

Since the advent of Google Scholar in 2004, many studies have used it as an online citation database for bibliometric analysis (e.g., Eysenbach, 2011; Meho & Yang, 2007). While Google Scholar seems to give access to a larger citation index than traditional offerings, like most sources of web impact it generates false matches and inflated citation counts (e.g., Jacsó, 2005, 2006), and lacks the quality control mechanism needed for robust bibliometric analysis (Aguillo, 2012; Delgado-López-Cózar & Cabezas-Clavijo, 2012). Nevertheless, with the ever-increasing number of open-access publications (including preprints and postprints) and Google Scholar's wide geographic and linguistic coverage of online publications, it seems reasonable to use Google Scholar citation metrics (citations counts, *h*-indexes, etc.) for small-scale impact calculations, especially in the social sciences, arts, and humanities, when traditional citation indexes (e.g., WoS) are not available or have insufficient coverage. However, for important large-scale research assessment exercises, Google Scholar should be used cautiously as an alternative to traditional citation databases due to its potential to be manipulated and its inclusion of non-peer-reviewed documents.

Google Books

While there is general agreement on the importance of books in research communication for the social sciences and humanities (for reviews, see Huang & Chang, 2008; Nederhof, 2006), book citation analysis has been little used in research evaluation. One reason is that the traditional journal-based citation databases have not been primarily designed for book citation analysis and therefore most citation analyses of books have been based on analyses of cited references from a selected number of nonbook publications, such as WoS-indexed articles (e.g., Bar-Ilan, 2010a; Butler & Visser, 2006). Moreover, the earlier absence of citations from books to books in traditional citation databases has been a problem for impact assessment in book-based disciplines. A practical issue has been how to assess the research performance of academics or departments in book-based disciplines other than via peer review. Books and monographs tend to be much longer than journal articles and this complicates large-scale peer review evaluation exercises. For instance, the proportion of books submitted to the 2008 UK Research Assessment Exercise in social sciences and humanities subject areas was 31%, compared with only 1.2% across the sciences (Kousha, Thelwall, & Rezaie, 2011). For impact assessment of book-based fields, bibliometricians need a database with large numbers of books (Garfield, 1996). In recognition of this need, Thomson Reuters finally released its "Book Citation Index" in 2011, but Google Books is a free web-based alternative and currently is more comprehensive for bibliometric analyses.

Google Books (http://books.google.com) includes a huge number of digitized books and allows full-text keyword searching. Although it does not contain a citation index, the Google Books full-text search capability makes it a uniquely large source to locate cited references from books, book chapters, and monographs, although it also generates some false matches and its content coverage is unknown. Kousha and Thelwall (2009) compared citations from Google Books searches with WoS citations to a sample of journal articles. Google Books citations were 31% to 212% as numerous as WoS citations in the social sciences and humanities but only 3% to 5% in the sciences, except for computing (46%; perhaps because of computing conference proceedings being frequently published as books). Google Book Search is thus valuable for research evaluation in book-oriented disciplines (e.g., assessment of departments or promotion and tenure), though it also needs minor manual data cleaning to cope with false matches. A subsequent study compared citations from Google Books searches to 1,000 books submitted to the 2008 UK Research Assessment Exercise with Scopus citations across seven book-oriented subject categories, and found that Google Books citations were 1.4 times more numerous than Scopus citations, which suggests Google Books citations can be a valuable tool to support peer review in research evaluation exercises, at least in the United Kingdom (Kousha et al., 2011).

In conclusion, Google Books is useful for monitoring the research performance of book-based disciplines. It is a practical tool for small-scale impact assessments of individual researchers or departments, where no alternative online data source is available, and it has superior coverage to the Thomson Reuters Book Citation Index.

New Types of Impact Assessment on the Web

One interesting feature of web impact analysis is its potential for generating impact metrics from types of academic documents not found in citation indexes (Barjak, 2006; Palmer, 2005). Prior to the Web, "About 90 per cent of the scientific results published in journal articles [were] previously disseminated in one of the channels of the informal communication domain" (Schubert, Zsindely, & Braun, 1983, p. 177). The subsequent partial shift of some informal academic information to the Web has motivated studies of new web sources that might be helpful for impact assessment, such as online course reading lists, presentations, and blogs.

Online Course Syllabi

An unknown number of university course syllabi and reading lists are available on the Web. Citations from online syllabi and reading lists reflect the educational impact of research, something that was previously impossible to assess using bibliometric methods. This is now possible with web searches. For example, Kousha and Thelwall (2008) identified citations from online syllabi to over 70,000 journal articles in 12 subject areas using automatically constructed queries submitted to web search engines. These citations were sufficiently numerous to provide evidence of the educational utility of research in a number of social science disciplines, including political science and information science.

Of course, no search engine indexes all online syllabi and many online academic syllabi are stored in password-protected databases. Nevertheless, researchers wishing to demonstrate the educational impact of their research can use this approach to generate quantitative evidence of impact.

Online Presentations

Conference and seminar presentations are important ways of disseminating research findings in many disciplines, including computer science (Bar-Ilan, 2010b). Although many scientific presentations are subsequently published as journal articles or in conference proceedings (e.g., Bird & Bird, 1999), some may never be formally published. In some subject areas less than one-third of scientific presentations are ever published (e.g., Fennewald, 2005; Miguel-Dasit, Marti-Bonmati, Aleixandre, Sanfeliu, & Valderrama, 2006; Montane & Vidal, 2007). Moreover, there might be a significant time lag between a presentation and the subsequent full-paper publication (e.g., two years in urology; Autorino, Quarto, Di Lorenzo, De Sio, & Damiano, 2007). Hence, in some fields where presentations are particularly important but rarely published in a timely manner, research is inadequately evaluated by traditional citation counts.

Thelwall and Kousha (2008) considered the value of citations in online presentations for impact assessment. Based on searches of PowerPoint files citing any article in a list of 1,807 ISI-indexed journals in 10 science and 10 social science disciplines, presentations were found to be useful for the impact assessment of whole journals but were not numerous enough for the impact assessment of individual articles (Thelwall

& Kousha, 2008). However, many scholarly presentations are also available in PDF or HTML format, and there is no practical search method to separate out these formats to assess presentation impact more widely. This may change if new online presentation formats (e.g., slideshare.net or slideshow.com) become popular and are searchable for citations.

Download Counts

The uptake of scientific publications has been commonly assessed using citation counts rather than readership information, partly because the latter has been impossible to accurately determine for individual papers (see Haustein, chapter 17, this volume). However, with the advent of the Web a variety of usage-related indicators can now be calculated. In particular, the number of times that an article has been downloaded is potentially a good indicator of its uptake, although downloading does not imply reading and readers may prefer a print version instead (for reviews, see Bollen & Van de Sompel, 2008; Bollen, Van de Sompel, Smith, & Luce, 2005). Attempts have been made to assess the impact of digital library documents based on downloads extracted from server logs (Bollen & Luce, 2002; Kaplan & Nelson, 2000). Later studies found evidence of a relationship between citations and downloads for research papers (Brody et al., 2006; Kurtz et al., 2005). At the level of whole journals, download metrics such as the journal download immediacy index (Wan, Hua, Rousseau, & Sun, 2010) and the usage impact factor and usage half-life (Schloegl & Gorraiz, 2010) are useful. An important issue to bear in mind when comparing download counts between journals is that if they are in different digital libraries, technical factors may invalidate the results.

Social Web Impact Metrics

A recent survey suggests that there are many scholars who publicize, share, or access research through social network tools, especially in the social sciences and humanities (Rowlands, Nicholas, Russell, Canty, & Watkinson, 2011). This supports the call to develop online metrics to assess the impact of scholarly research based on nonstandard academic sources and indicators (Priem & Hemminger, 2010). For instance, ImpactStory[2] gives readership-based statistics and metrics to assess the impact of scientific publications in terms of online readership.

Science blogs can be used for discussions of research results and "informal post-publication peer-review" (Shema et al., 2012; see also Bar-Ilan and colleagues, chapter 16, this volume). For instance, citations to major blogs *from* Scopus publications increased dramatically from 21 citations before 2003 to just under 5,000 in 2011 (see table 15.1). Nevertheless, based on blog citations *to* two library and information science journals,

Kousha, Thelwall, and Rezaie (2010) concluded that there were too few citations from blogs to be worth counting. This may not be true for all fields, however.

Altmetrics studies suggest that various other online sources can be useful for research impact assessment, including social bookmarks (Haustein & Siebenlist, 2011), tweets (Eysenbach, 2011; Priem & Costello, 2010), and online reference managers (Li et al., 2012). More research is needed to understand why social web citations occur, and how often they provide evidence of readership and use.

Beyond Academic Publishing: The Web Impact of Nontraditional Outputs

Although the above discussion mainly concerns the impact of traditional academic publications, scholars also engage in other activities that produce tangible outputs, such as videos, that could be assessed to determine their value. Table 15.1 gives an overview of how frequently a range of social websites have been formally cited, providing evidence of the scholarly value of their content. There has been a steady upward trend in the citing of different types of sites in publications indexed by Scopus. For instance, while there were only 9 and 34 citations to *Wikipedia* articles and blog posts in 2003, respectively, and 6 citations to online videos in 2006, the numbers rose dramatically to 7,176, 4,910, and 1,188, respectively. Table 15.1 gives some evidence that nonstandard outputs can be useful in science.

A number of studies have analyzed the impact of specific types of nonstandard scientific outputs, including videos. Online videos play an increasingly valuable role in science communication, for example by illustrating science experiments or recording academic presentations and lectures. These videos are cited in novel ways by authors to support research across subject areas, ranging from real-time scientific demonstrations and laboratory experiments in the sciences and medicine to artistic outputs (e.g., music and dance) in the arts and humanities (Kousha & Thelwall, 2012; Kousha, Thelwall, & Abdoli, 2012). In terms of impact measurements, a study of TED (Technology Entertainment Design, www.ted.com) talks showed that it was possible to develop a wide range of online impact indicators for videos (Sugimoto & Thelwall, 2013). This was not limited to viewer numbers but included demographics obtained from YouTube so that a much more in-depth impact analysis was possible.

Ideally, a scientist's overall contribution could be assessed by a single indicator that combined traditional citations as well as citations to nontraditional outputs. This seems unlikely, however, due to the variety of audiences that nontraditional outputs attract; it would not be fair to compare the download counts of a popular science video with those of a technical demonstration. The logical solution seems to be to use multiple,

Table 15.1
The number of Scopus publications citing social networking sites over the years* (authors' own data)

Year	*Wikipedia*	Blogs[a]	Video sharing[b]	File sharing[c]	General social networks[d]	Online reference managers[e]	Professional SNS[f]	Image sharing[g]	Total
Up to 2002	0	21	0	0	0	0	0	0	**21**
2003	9	34	0	0	0	0	0	0	**43**
2004	83	66	0	0	0	0	0	0	**149**
2005	359	184	0	0	0	0	0	2	**545**
2006	1,210	313	6	1	1	2	0	2	**1,535**
2007	2,143	609	53	6	6	2	0	11	**2,830**
2008	3,189	1,098	144	40	24	8	1	12	**4,516**
2009	4,844	2,252	431	220	71	37	5	31	**7,891**
2010	6,449	3,717	767	713	158	67	18	61	**11,950**
2011	7,176	4,910	1,188	1,195	285	157	26	81	**15,018**
Total	**25,462**	**13,204**	**2589**	**2175**	**545**	**273**	**50**	**200**	**44,498**

*Excluding general citations in the reference lists such as Wikipedia.org, Facebook.com, Twitter.com, YouTube.com

a. Citations to 68 major blogs such as WordPress.com and ScienceBlogs.com

b. Citations to 15 major video-sharing sites such as YouTube.com, Ted.com, Vimeo.com

c. Citations to 7 major file-sharing sites such as DropBox.com and Scribd.com

d. Citations to 3 major general social networking sites including Facebook.com, Twitter.com, and MySpace.com

e. Citations to 8 major online reference manger sites such as CiteULike.org and Mendeley.com

f. Citations to 2 professional social networking sites such as LinkedIn.com

g. Citations to 4 major image-sharing sites such as Flicker.com and PhotoBucket.com

separate indicators to fashion a narrative of a scientist's contribution rather than use a stand-alone statistic (Thelwall, Kousha, Weller, & Puschmann, 2012).

Conclusions

There are many ways in which research impact can be assessed using the Web. In most cases, web impact indicators do not require a database subscription. Their real advantage, however, is the wider coverage that they give compared with traditional citation indexes, although the latter now include more source types. Nevertheless, web impact indicators suffer from a generic lack of quality control compared with scholarly citations, and hence should be used cautiously in research evaluation. As the popularity of the altmetrics movement has shown, there is a niche that can be occupied by web impact metrics, both to supplement traditional indicators through the wider range of source types available on the Web and to provide early indicators of likely future citation impact. Using web impact statistics to provide evidence of the value of nontraditional outputs, such as videos, seems trickier, however, and it remains to be seen whether this approach is widely adopted. Finally, as the Web evolves, it is likely that new sources of web impact will emerge and old ones will fade, making web impact evaluation research an ongoing necessity.

Acknowledgment

This chapter was partially funded by the EU FP7 ACUMEN project (Grant agreement: 266632).

Notes

1. This work is unpublished. Please contact Mike Thelwall for more information on this research.

2. http://impactstory.org

References

Aguillo, I. F. (2012). Is Google Scholar useful for bibliometrics? A webometric analysis. *Scientometrics*, *91*(2), 343–351.

Aguillo, I. F., Granadino, B., Ortega, J. L., & Prieto, J. A. (2006). Scientific research activity and communication measured with cybermetrics indicators. *Journal of the American Society for Information Science and Technology*, *57*(10), 1296–1302.

Almind, T. C., & Ingwersen, P. (1997). Informetric analyses on the World Wide Web: Methodological approaches to "webometrics." *Journal of Documentation, 53*(4), 404–426.

Archambault, E., Vignola-Gagne, E., Cote, G., Larivière, V., & Gingras, Y. (2006). Benchmarking scientific output in the social sciences and humanities: The limits of existing databases. *Scientometrics, 68*(3), 329–342.

Arts & Humanities Citation Index. (2012). Retrieved from http://ip-science.thomsonreuters.com/cgi-bin/jrnlst/jlresults.cgi?PC=H.

Autorino, R., Quarto, G., Di Lorenzo, G., De Sio, M., & Damiano, R. (2007). Are abstracts presented at the EAU meeting followed by publication in peer-reviewed journals? A critical analysis. *European Urology, 51*(3), 833–840.

Bar-Ilan, J. (2004). A microscopic link analysis of academic institutions within a country—the case of Israel. *Scientometrics, 59*(3), 391–403.

Bar-Ilan, J. (2010a). Citations to the "Introduction to Informetrics" indexed by WoS, Scopus and Google Scholar. *Scientometrics, 82*(3), 495–506.

Bar-Ilan, J. (2010b). Web of Science with the Conference Proceedings Citation Indexes: The case of computer science. *Scientometrics, 83*(3), 809–824.

Barjak, F. (2006). The role of the Internet in informal scholarly communication. *Journal of the American Society for Information Science and Technology, 57*(10), 1350–1367.

Bird, J. E., & Bird, M. D. (1999). Do peer-reviewed journal papers result from meeting abstracts of the biennial conference on the biology of marine mammals? *Scientometrics, 46*(2), 287–297.

Bollen, J., & Luce, R. (2002). Evaluation of digital library impact and user communities by analysis of usage patterns. *D-Lib Magazine, 8*(6).

Bollen, J., & Van de Sompel, H. (2008). Usage impact factor: The effects of sample characteristics on usage-based impact metrics. *Journal of the American Society for Information Science and Technology, 59*(1), 136–149.

Bollen, J., Van de Sompel, H., Smith, J. A., & Luce, R. (2005). Toward alternative metrics of journal impact: A comparison of download and citation data. *Information Processing & Management, 41*(6), 1419–1440.

Borgman, C. L., & Furner, J. (2002). Scholarly communication and bibliometrics. *Annual Review of Information Science & Technology, 36*, 3–72.

Brody, T., Harnad, S., & Carr, L. (2006). Earlier web usage statistics as predictors of later citation impact. *Journal of the American Society for Information Science and Technology, 57*(8), 1060–1072.

Butler, L., & Visser, M. S. (2006). Extending citation analysis to non-source items. *Scientometrics, 66*(2), 327–343.

Cronin, B. (2001). Bibliometrics and beyond: Some thoughts on web-based citation analysis. *Journal of Information Science, 27*(1), 1–7.

Cronin, B., Snyder, H. W., Rosenbaum, H., Martinson, A., & Callahan, E. (1998). Invoked on the Web. *Journal of the American Society for Information Science, 49*(14), 1319–1328.

Delgado-López-Cózar, E., & Cabezas-Clavijo, Á. (2012). Google Scholar Metrics: An unreliable tool for assessing scientific journals. *El Profesional de la Información, 21*(4), 419–427.

Eysenbach, G. (2011). Can Tweets predict citations? Metrics of social impact based on Twitter and correlation with traditional metrics of scientific impact. *Journal of Medical Internet Research, 13*(4).

Fennewald, J. (2005). Perished or published: The fate of presentations from the Ninth ACRL Conference. *College & Research Libraries, 66*(6), 517–525.

Garfield, E. (1996). Citation indexes for retrieval and research evaluation. *Consensus Conference on the Theory and Practice of Research Assessment*, Capri. Retrieved from http://www.garfield.library.upenn.edu/papers/ciretreseval-capri2.pdf.

Glänzel, W. (1996). A bibliometric approach to social sciences, national research performances in 6 selected social science areas, 1990–1992. *Scientometrics, 35*(3), 291–307.

Goodrum, A. A., McCain, K. W., Lawrence, S., & Giles, C. L. (2001). Scholarly publishing in the Internet age: A citation analysis of computer science literature. *Information Processing & Management, 37*(5), 661–675.

Haustein, S., & Siebenlist, T. (2011). Applying social bookmarking data to evaluate journal usage. *Journal of Informetrics, 5*(3), 446–457.

Hicks, D. (1999). The difficulty of achieving full coverage of international social science literature and the bibliometric consequences. *Scientometrics, 44*(2), 193–215.

Huang, M.-H., & Chang, Y.-W. (2008). Characteristics of research output in social sciences and humanities: From a research evaluation perspective. *Journal of the American Society for Information Science and Technology, 59*(11), 1819–1828.

Ingwersen, P. (1998). The calculation of Web impact factors. *Journal of Documentation, 54*(2), 236–243.

Jacsó, P. (2005). Google Scholar: The pros and the cons. *Online Information Review, 29*(2), 208–214.

Jacsó, P. (2006). Deflated, inflated and phantom citation counts. *Online Information Review, 30*(3), 297–309.

Jiang, J., He, D., & Ni, C. (2011). Social reference: Aggregating online usage of scientific literature in CiteULike for clustering academic resources. In *Proceedings of the ACM/IEEE Joint Conference on Digital Libraries* (pp. 401–402). Ottawa, Ontario, June 13-17, 2011.

Kaplan, N. R., & Nelson, M. L. (2000). Determining the publication impact of a digital library. *Journal of the American Society for Information Science and Technology, 51*(4), 324–339.

Kjellberg, S. (2010). I am a blogging researcher: Motivations for blogging in a scholarly context. *First Monday, 15*(8).

Kousha, K., & Thelwall, M. (2007a). Google Scholar citations and Google Web/URL citations: A multi-discipline exploratory analysis. *Journal of the American Society for Information Science and Technology, 58*(7), 1055–1065.

Kousha, K., & Thelwall, M. (2007b). How is science cited on the web? A classification of Google unique web citations. *Journal of the American Society for Information Science and Technology, 58*(11), 1631–1644.

Kousha, K., & Thelwall, M. (2008). Assessing the impact of disciplinary research on teaching: An automatic analysis of online syllabuses. *Journal of the American Society for Information Science and Technology, 59*(13), 2060–2069.

Kousha, K., & Thelwall, M. (2009). Google Book search: Citation analysis for social science and the humanities. *Journal of the American Society for Information Science and Technology, 60*(8), 1537–1549.

Kousha, K., & Thelwall, M. (2012). Motivations for citing YouTube videos in the academic publications: A contextual analysis. In Proceedings of STI 2012 Montréal (pp. 488–497), Montréal, Québec, September 5-8, 2012.

Kousha, K., Thelwall, M., & Abdoli, M. (2012). The role of online videos in research communication: A content analysis of YouTube videos cited in academic publications. *Journal of the American Society for Information Science and Technology, 63*(9), 1710–1727.

Kousha, K., Thelwall, M., & Rezaie, S. (2010). Using the Web for research evaluation: The Integrated Online Impact indicator. *Journal of Informetrics, 4*(1), 124–135.

Kousha, K., Thelwall, M., & Rezaie, S. (2011). Assessing the citation impact of books: The role of Google Books, Google Scholar, and Scopus. *Journal of the American Society for Information Science and Technology, 62*(11), 2147–2164.

Kurtz, M. J., Eichhorn, G., Accomazzi, A., Grant, C., Demleitner, M., Murray, S. S., et al. (2005). The bibliometric properties of article readership information. *Journal of the American Society for Information Science and Technology, 56*(2), 111–128.

Li, X., Thelwall, M., & Giustini, D. (2012). Validating online reference managers for scholarly impact measurement. *Scientometrics, 91*(2), 461–471.

MacRoberts, M. H., & MacRoberts, B. R. (1996). Problems of citation analysis. *Scientometrics, 36*(3), 435–444.

Meho, L., & Yang, K. (2007). Impact of data sources on citation counts and rankings of LIS faculty: Web of Science vs. Scopus and Google Scholar. *Journal of the American Society for Information Science and Technology, 58*(13), 2105–2125.

Miguel-Dasit, A., Marti-Bonmati, L., Aleixandre, R., Sanfeliu, P., & Valderrama, J. C. (2006). Publications resulting from Spanish radiology meeting abstracts: Which, where and who. *Scientometrics, 66*(3), 467–480.

Moed, H. K. (2005). *Citation analysis in research evaluation*. New York: Springer.

Montane, E., & Vidal, X. (2007). Fate of the abstracts presented at three Spanish clinical pharmacology congresses and reasons for unpublished research. *European Journal of Clinical Pharmacology, 63*(2), 103–111.

Nederhof, A. J. (2006). Bibliometric monitoring of research performance in the social sciences and the humanities: A review. *Scientometrics, 66*(1), 81–100.

Palmer, C. L. (2005). Scholarly work and the shaping of digital access. *Journal of the American Society for Information Science and Technology, 56*(11), 1140–1153.

Priem, J., & Costello, K. L. (2010). How and why scholars cite on Twitter. *Proceedings of the American Society for Information Science and Technology, 47*(1), 1–4.

Priem, J., & Hemminger, B. M. (2010). Scientometrics 2.0: Toward new metrics of scholarly impact on the social Web. *First Monday, 15*(7).

Rousseau, R. (1997). Sitations: An exploratory study. *Cybermetrics,* 1(1).

Rowlands, I., Nicholas, D., Russell, B., Canty, N., & Watkinson, A. (2011). Social media use in the research workflow. *Learned Publishing, 24*(3), 183–195.

Schloegl, C., & Gorraiz, J. (2010). Comparison of citation and usage indicators: The case of oncology journals. *Scientometrics, 82*(3), 567–580.

Schubert, A., Zsindely, S., & Braun, T. (1983). Scientometric analysis of attendance at international scientific meetings. *Scientometrics, 5*(3), 177–187.

Scopus Arts & Humanities Coverage. (2011). Retrieved from http://www.info.sciverse.com/scopus/scopus-in-detail/arts-humanities.

Shema, H., Bar-Ilan, J., & Thelwall, M. (2012). Research blogs and the discussion of scholarly information. *PLoS ONE, 7*(5): e35869.

Sugimoto, C. R., & Thelwall, M. (2013. Scholars on soap boxes: Science communication and dissemination via TED videos. *Journal of the American Society for Information Science and Technology, 64*(4), 663–674.

Thelwall, M. (2002). Conceptualizing documentation on the Web: An evaluation of different heuristic-based models for counting links between university web sites. *Journal of the American Society for Information Science and Technology, 53*(12), 995–1005.

Thelwall, M., & Harries, G. (2004). Do the web sites of higher rated scholars have significantly more online impact? *Journal of the American Society for Information Science and Technology, 55*(2), 149–159.

Thelwall, M., & Kousha, K. (2008). Online presentations as a source of scientific impact?: An analysis of PowerPoint files citing academic journals. *Journal of the American Society for Information Science and Technology, 59*(5), 805–815.

Thelwall, M., Kousha, K., Weller, K., & Puschmann, C. (2012). Assessing the impact of online academic videos. In G. Widen-Wulff & K. Holmberg (Eds.), *Social information research* (pp. 195–213). Bradford, UK: Emerald.

Thelwall, M., Sud, P., & Wilkinson, D. (2012). Link and co-inlink network diagrams with URL citations or title mentions. *Journal of the American Society for Information Science and Technology, 63*(4), 805–816.

Thomson Reuters. (2012). Book Citation Index in Web of Science. Retrieved from http://wokinfo.com/media/pdf/bkci_fs_en.pdf.

Thomson Scientific. (2006). Web Citation Index. Retrieved from http://scientific.thomson.com/tutorials/wci/wci1tut1.html.

Vaughan, L., & Hysen, K. (2002). Relationship between links to journal web sites and impact factors. *Aslib Proceedings, 54*(6), 356–361.

Vaughan, L., & Shaw, D. (2003). Bibliographic and web citations: What is the difference? *Journal of the American Society for Information Science and Technology, 54*(14), 1313–1322.

Vaughan, L., & Shaw, D. (2005). Web citation data for impact assessment: A comparison of four science disciplines. *Journal of the American Society for Information Science and Technology, 56*(10), 1075–1087.

Wan, J. K., Hua, P. H., Rousseau, R., & Sun, X. K. (2010). The journal download immediacy index (DII): Experiences using a Chinese full-text database. *Scientometrics, 82*(3), 555–566.

Wilkinson, D., Harries, G., Thelwall, M., & Price, L. (2003). Motivations for academic web site interlinking: Evidence for the web as a novel source of information on informal scholarly communication. *Journal of Information Science, 29*(1), 49–56.

Zhao, D. Z., & Logan, E. (2002). Citation analysis using scientific publications on the Web as data source: A case study in the XML research area. *Scientometrics, 54*(3), 449–472.

16 Bibliographic References in Web 2.0

Judit Bar-Ilan, Hadas Shema, and Mike Thelwall

Introduction

The Internet and the World Wide Web (Web) have had a major influence on scientific communication. Even before Web 2.0 applications were introduced, scientists used email and discussion lists to exchange information, create homepages, access articles electronically, upload their publications to the Web, and receive information regarding conferences and funding opportunities. Web 2.0 initiatives enhanced interactivity, sharing, and collaboration on the Web and thus further influenced scholarly communication. With Web 2.0, scientists have platforms such as blogs, YouTube, and Facebook that enable them to disseminate their studies and views to the wider public, as well as to interact and have discussions with the public. Currently, with the increased emphasis on the societal impact of science and need for accountability about how taxpayers' money is spent, Web 2.0 platforms provide useful means for showcasing research.

Several interesting questions arise in relation to science and Web 2.0 platforms:

- What do the researchers think about the usefulness and effectiveness of these platforms?
- How can we measure the "impact" of these platforms on science and scientists?
- What is the relation between the "impact" of these platforms on science and traditional bibliometric measures?

This chapter provides a short review of studies discussing the first question. It then introduces a new branch of bibliometrics called altmetrics (Priem, Taraborelli, Groth, & Neylon, 2010; see also Priem, chapter 14, this volume), which proposes measures to study the impact of web-based platforms on science. Note that we intentionally enclosed the term "impact" in quotation marks. According to the Oxford English Dictionary, to impact is "to have a (pronounced) effect on" something or someone

("Impact," 2012). The use of Web 2.0 platforms for science is rather recent, and it remains to be seen whether they are going to have a pronounced effect on the way science is conducted. This chapter focuses on science blogs and open reference managers—sites where scholarship is simultaneously referenced according to the rules of scientific citation and exposed to alternative metrics.

Researchers and Web 2.0

This section reviews some studies exploring researchers' attitudes toward Web 2.0 applications. Many of these applications are new, and some of them are transient; thus, before suggesting metrics to assess the effect of these platforms, we should try to understand what scientific value researchers attach to them.

The Research Information Network (RIN)[1] conducted a series of case studies with various collaborators on the information needs and uses of scientists in various fields, including their use of Web 2.0 platforms. Humanities scholars still rely heavily on print sources, but they increasingly make use of contemporary technologies, publishing their work on personal websites or in repositories, maintaining blogs, and participating in other social media (RIN, 2011b). Some also use Zotero, an online reference manager. There are humanities scholars who use Twitter and Facebook in order to discuss and disseminate research. Some have doubts about quality assurance on these platforms and concerns about how to present their work to the general public. In contrast, physicists use wikis and email lists and rely heavily on repositories like arXiv and the NASA Astrophysics Data System (ADS) (the latter is discussed in Kurtz and Henneken, chapter 13, this volume). Chemists, at least the participants in this case study, make little use of Web 2.0, but the earth science case study showed the use of blogs as a means for communicating with the public. Scientists working with the Galaxy Zoo citizen science project were the only group in the physical sciences reporting heavy use of Web 2.0 services (RIN, 2011a). In an earlier study of the life sciences (RIN, 2009), Web 2.0 platforms, except for wikis for sharing information within the projects, were hardly mentioned—possibly because life scientists are less open to the use of Web 2.0 platforms or because this study preceded the other studies by two years.

Another set of case studies was conducted by the Center for Studies in Higher Education at the University of California, Berkeley (Harley, Acord, Earl-Novell, Lawrence, & King, 2010), in order to understand the needs and practices of faculty members regarding in-progress scholarly communication and archival publications. Dissemination practices were found to include submissions to preprint repositories or publishing on personal websites. Early public sharing of scientific results was rare, and in many fields, young scholars were especially concerned about exposing their ideas too soon. The conclusion of the study was that it was "premature to assume that Web 2.0 platforms geared toward early public exposure of research ideas or data are going to spread among scholars in the most competitive institutions" (Harley et al., 2010, p. 4). Social platforms were not cited as common ways in which the 160 interviewed scholars in different disciplines broadcast and received information. Most of the interviewees viewed maintaining blogs as a waste of time, due to the lack of peer review and quality control. However, a number of participants mentioned some "good" blogs, and some read blogs related to their research, but generally they complained about a lack of time to follow blogs. Gruzd, Staves, and Wilk (2011) conducted interviews with 51 members of the American Society for Information Science and Technology. They found that the use of online social media was not taken into account in tenure and promotion processes in most institutions. However, they surmised that this situation was likely to change in the future.

In addition to case studies and interviews, other means of learning about the value of Web 2.0 platforms for researchers include analyzing existing Web 2.0 pages and profiles (Bukvova, 2011) and carrying out surveys. Ponte and Simon (2011) conducted a survey and received 349 responses to questionnaires sent to mailing lists. The respondents were mainly from the social sciences and computer science disciplines. They were asked about different relevance criteria for evaluating researchers. Citations and publication quality ranked highest, while personal web pages, personal blogs, membership in professional networks, and user-generated tagging services were considered irrelevant by about 60%, 70%, 75%, and 85% of the respondents, respectively. Gruzd, Goertzen, and Mai (2012) surveyed 367 academics (mainly from the social sciences) and found "an increasing acceptance of blogs and microblogs as legitimate and trustworthy methods for gathering and disseminating scholarly information" (p. 3). Scholars also noted that participation in social media for professional purposes is time consuming and creates unnecessary distractions. Unlike the previously mentioned studies, the results of this survey show that academics like to read or comment on blog posts, although only a few maintained blogs. In addition to time constraints, they were wary of potential negative exposure. This study shows that academic social networking sites (e.g., academia.edu and researchgate) and online reference managers are becoming popular. Finally, 8% of the respondents stated that social media activities are taken into account in the tenure/promotion processes at their home institutions.

Nicholas and Rowlands (2011) conducted a large-scale survey with about 100,000 invitations to different lists. They received 2,414 replies from researchers: 1,923 (80%

of the respondents) used social media tools and 491 did not. Social networking tools were used in research by 27% of the respondents, blogging by 15% (it was not clear whether this related to reading or writing), microblogging by 9%, and social tagging and bookmarking by 9%. The highest use of social media in research was reported in earth science (95% of the respondents in this field), compared with 74% in business and management (note that 80% of the respondents were active social media users). In the research cycle, social media and blogging were mainly used for disseminating research findings, research collaboration, identifying research opportunities, and reviewing the literature. Microblogging was used mainly for disseminating research findings. Major perceived benefits were the ability to communicate internationally, faster dissemination, and the ability to connect with people outside the academy. The major barriers were lack of time and issues with authority and trust.

In summary, it seems that there is some use of social media tools in academia; the use is somewhat field dependent, but many view these tools as a waste of time. Nevertheless, there are indications that in the future social media tools will be used more widely for research, and thus it is reasonable to consider ways to measure their "impact."

Altmetrics

Altmetrics is a new branch of bibliometrics that aims to "expand our view of what impact looks like, but also of what's making the impact" (Priem et al., 2010, para. 5) and to measure the impact of online scholarly tools. Tools have been developed to provide alternative metrics for individual authors or for individual articles (e.g., ImpactStory—http://impactstory.org) and individual metrics have been proposed based upon blog citations, Twitter mentions, reference manager readers and other social media. Here we discuss two platforms: science blogs and reference managers.

Science Blogs

Science blogs have become popular with a section of the scholarly community. Respected scholarly media outlets such as *National Geographic*, the *Nature* Group, *Scientific American*, and the *PLoS* journals all have science blogging networks. A *Nature Medicine* editorial discussing blogs and peer review concluded that "online science blogs are a valuable forum for commenting on published research, but their present importance lies in complementing rather than replacing the current system of peer review" ("Perfecting Peer Review?", 2011, pp. 1–2). This chapter introduces a number of examples to illustrate the various functions of science blogs and their importance in scholarly communication.

Between Blogs and Formal Scientific Discourse

On December 2, 2010, *Science* published an online paper by NASA scientists claiming to have discovered bacteria using arsenic, rather than phosphorus, in their DNA (Wolfe-Simon et al., 2010). *Science* bloggers were deeply skeptical about the findings (Zivkovic, 2011), and scientists tweeted extensively about the subject. The criticism made its way into peer-reviewed journals (Katsnelson, 2010) and articles in mainstream media outlets (Grossman, 2010; Overbye, 2010), which quoted the blogs. Before *Science* had published any relevant technical comments (including one from Rosemary Redfield, a blogger and a microbiologist at the University of British Columbia), the scientific community had thoroughly commented on and criticized the paper online. In July 2012, *Science* published two articles (one with Redfield as a coauthor) contradicting the Wolfe-Simon article and showing that bacteria need phosphorus in their DNA (Schiermeier, 2012). This, and other examples, suggests that blogging can be particularly useful to discuss and critically evaluate high-profile research and may be able to identify flaws overlooked by the standard peer review process.

Social Change—the Elsevier Boycott

On January 21, 2012, Fields Medalist Timothy Gowers declared on his blog his intention to boycott the academic publisher Elsevier. He justified the decision by listing Elsevier's high prices, their support of legislation against Open Access, and aggressive selling techniques. One of Gowers' readers was quick to open a website called "The Cost of Knowledge" (http://thecostofknowledge.com), where people could make public their commitment to boycott Elsevier. The boycott, with credit to Gowers' original post, was widely covered in major peer-reviewed journals such as *Science* and *Nature* (De Vrieze, 2012; Whitfield, 2012), as well as in the *New York Times*, the *Guardian*, and other mainstream media (Lin, 2012; Jha, 2012). By August 2012, 12,600 people had made public their decision to boycott Elsevier.

Investigative Science Blogging

Another way in which blogs can provide a concrete service to the academic community is by investigating apparent malpractice within science. Phil Davis of *The Scholarly Kitchen* blog received a "tip from a concerned scientist" about anomalous citation practices (Davis, 2012, para. 13). Davis used Journal Citation Report (JCR) data and published evidence about the existence of a "citation cartel" (Davis, 2012, para. 3)—that is, groups of editors working together to improve the impact factors of their respective journals. In one case, the journal *Medical Science Monitor* published a 2010 review citing 490 articles, 445 of which were published in 2008–2009 in the journal *Cell Transplantation*. Three of the authors were *Cell Transplantation* editors. Thomson Reuters eventually suspended three out of the four journals from the JCR, and another blog, *Retraction Watch*, reported that two of the manipulating articles were retracted (Oransky, 2012).

Blogs as Metric Sources

Citations to academic journal articles from blogs can potentially be used as an alternative source of impact evidence (i.e., an altmetric). One study has shown that it is possible, at least on a small scale, by using Google Blog Search to calculate blog citations to a set of published articles (Kousha, Thelwall, & Rezaie, 2010). Although blog citations were found to be much rarer than academic citations, they could still be useful evidence of research impact on wider discussions, especially in the social sciences and humanities. Nevertheless, an evaluation of science blogs is necessary to determine their validity as metric sources.

Scientific journals and their articles are normally sustainable, with past issues preserved both by their publishers and by libraries. Journals can be continually published for years, decades, and even centuries. Their citations are, as Cronin (1981, p. 16) put it, "frozen footprints in the landscape of scholarly achievement; footprints which bear witness of the passage of ideas." In comparison with these "frozen footprints," the sustainability of blogs is questionable. In 2006, *Nature News* published a list of 50 popular science blogs written by scientists (Butler, 2006). By August 2012 (figure 16.1), two (4%) were unreachable, sixteen (32%) were inactive (had not posted for over a year or officially announced their closing down), and three (6%) were hibernating—for two of them it has been less than a year but more than six months since their last post, and one posted a hiatus announcement. Overall, 29 (58%) of the blogs were active: 16 at their 2006 address and 13 at a new address.

An interesting phenomenon occurring in science blogs is that some bloggers choose to cite their sources in a structured style, much like that of refereed journals. Researchblogging.org (RB) aggregates blog posts referring specifically to peer-reviewed research. It is a self-selecting aggregator, allowing bloggers to cite peer-reviewed research in an academic citation format. Bloggers discussing peer-reviewed research can register with the aggregator, and when they mark relevant posts in their blog, these posts appear on the aggregator site, giving one-stop access to a variety of research reviews from different authors. The site has human editors who ensure that blogs submitted to the aggregator

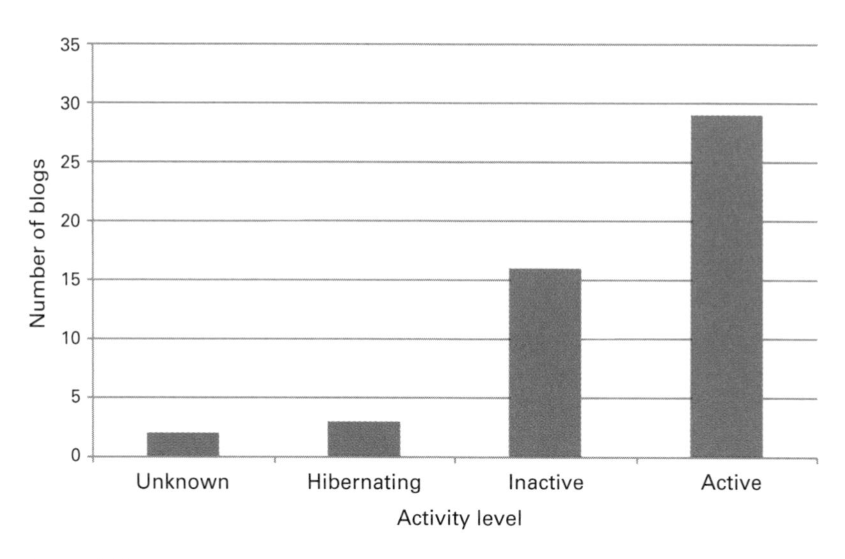

Figure 16.1
Popular 2006 science blogs and their August 2012 activity level (authors' own data).

follow the guidelines and are of appropriate quality. It also has an altmetric role, since it serves as one of the article-level metrics (ALM) displayed for each *PLoS* article.

The first Researchblogging.org study was conducted by Groth and Gurney (2010) and focused on 295 aggregated posts tagged "Chemistry." The literature cited in these posts was mostly up to date and came from top journals: 71% of the cited articles were from the top 20 journals in the field of chemistry, and 21% were from the 60 top publications across all disciplines. Another study (Shema, Bar-Ilan & Thelwall, 2012a) focused on established blogs and bloggers that had at least 20 posts aggregated in RB between January 1, 2010, and January 15, 2011. The chosen blogs were noncommercial and written by either one or two authors. The sample included 126 blogs and 135 bloggers. The most popular blog category was "Life Science" (39%), followed by "Psychology," "Psychiatry, Neurosciences & Behavioral Science," and "Medicine" (19%). Blogs about "Social Sciences & Humanities" and about "Computer Science & Engineering" were the least popular (5% and 1%, respectively). The study confirmed the preference for high-impact journals found by Groth and Gurney (2010). The journals most cited in blog posts were *Science*, *Nature*, and *PNAS*. The bloggers were highly educated (32% had a PhD; 27% were graduate students), and 59% of them were affiliated with a university or a research institution, which would explain their familiarity with the academic citation styles. Most (84%) of the bloggers wrote under their own name, which suggests that they see blogging as a career enhancer, or at least as career-neutral.

RB uses a tagging system that focuses mainly on the life and natural sciences. For example, "Astronomy" has 10 subtags, "Psychology" 21, and "Biology" 28. "History," "Economics," and "Sociology," on the other hand, are represented only as subtags of the "social science" tag. This tagging bias could be the result of the underrepresented disciplines' bloggers' lack of awareness or willingness to aggregate in RB.[2] The lack of tags in a specific field could also be a factor in a blogger's decision to aggregate in RB, creating a feedback loop. Another possibility is that the RB tagging system merely reflects a reality in which most blogging about peer-reviewed research occurs in certain fields. In fact, the number of life sciences doctorates awarded is rising (National Science Foundation, Division of Science Resources Statistics, 2010a), and more than two-thirds of the academic postdoctoral appointments in the United States were in the life and medical sciences (National Science Foundation, Division of Science Resources Statistics, 2010b). This could partially explain the dominance of life sciences blogs and life sciences papers in our sample. Moreover, according to Bora Zivkovic, *Scientific American*'s blog editor, blogs are "written by graduate students, postdocs and young faculty, a few by undergraduates and tenured faculty, several by science teachers, and just a few by professional journalists" (Bonetta, 2007, p. 443).

Technically speaking, RB has a rather slow interface, which by August 2012 had not been updated in quite some time. It is possible that it will be replaced in the future by its younger, more comprehensive sister site ScienceSeeker.Org. Similarly to RB, ScienceSeeker allows bloggers to create structured citations and aggregate them. In addition, it aggregates posts from science blogs that do not use a structured citation style, as long as the blogs have been approved by the editors. It also allows filtered searching for posts and blogs that cite refereed publications. ScienceSeeker integrates Twitter into its homepage, showing science blog posts that are popular on Twitter.

Self-Citing Blogs

The citing of one's own work is common practice in formal scientific communication and is well documented in the bibliometric literature. To discover whether science blogs share this trait, one study (Shema, Bar-Ilan & Thelwall, 2012b) investigated four RB categories: "Ecology/Conservation," "Computer Science/Engineering," "Mathematics," and "Philosophy." The small "Mathematics" category was studied in full (until April 30, 2012), and the other categories were sampled between January 1, 2010, and April 30, 2012. Only bloggers who used their real name and posts signed by them were included. Posts that belonged to several categories were counted in each category separately. A self-citing post belonging to several categories can be seen in figure 16.2.

Figure 16.2
A self-citing post belonging to several RB categories.

In comparison with our previous study (Shema et al., 2012a), among the self-citing bloggers there was a higher percentage of bloggers with PhDs (80% compared with 32% in the general sample), and more (82%) were affiliated with a research institute or university (59% in the general sample). The rate of self-citing posts was low overall but varied according to discipline, with "Mathematics" having the highest percentage of self-citing posts (10%), "Computer Science" and "Philosophy" having a slightly lower percentage (9%), and "Ecology" having the lowest (5%).

This low rate of self-citation can be explained in several ways: in the academic world, formal publishing is a necessity ("publish or perish"), while blogs are more of an extra-curricular activity. Even authors who are science bloggers can blog without referring to their own research, while academic publications often build on the authors' previous work because researchers tend to conduct related studies. Moreover, bloggers have the freedom to post anonymously, while academic authors do not. A science blogger does not have to be a published author in a peer-reviewed journal. Bloggers who are under-graduates or professional science writers might not have peer-reviewed publications to cite. It seems that bloggers who have earned a PhD and are affiliated with a research institute are likely to have authored more refereed publications than other science bloggers and therefore have more publications to cite. This is in line with the findings by Fowler and Aksnes (2007) and Costas, Leeuwen, and Bordons (2010) about the positive correlation between productivity and self-citations.

In conclusion, it seems that science blogs fill a gap in scientific communication. They share certain characteristics with the formal scholarly discourse, while allowing rapid, informal discussion about peer-reviewed literature and subjects of interest to the research community.

Reference Managers

Established reference managers like Endnote, Bibtex, and Refworks aim to help authors with the referencing process as they write, as well as to help in formatting the citations according to the appropriate citation style. Some of the newer software also allows reference sharing and storing of articles or linking to stored items.

Some reference managers, like BibSonomy,[3] CiteULike,[4] Connotea,[5] and Mendeley,[6] have additional features, such as reporting the number of users of the system who bookmarked a specific item. These services are sometimes called scholarly social bookmarking services. The users who bookmarked an item are called "readers" on Mendeley (and we use this term from this point on), and the number of readers indicates how many users of the system expressed interest in the specific item and saved its details to their library. Although bookmarkers are called readers on Mendeley, they may not have read a bookmarked article, and even if they did, their opinion of the item is unknown. This is quite unlike science blogs, where the bloggers review the cited papers; thus it seems safe to assume that they have read the paper being cited, and they typically express their opinion in the blog post on the paper. However, the reference managers have many users. The reader count is easily accessible on these systems and thus has naturally been proposed as an altmetric, although it is not clear what kind of impact it measures. On the other hand, it seems reasonable to assume that an item that has been bookmarked by hundreds of readers is probably of interest, assuming that it is not artificially generated spam. In this sense "readers" may be compared to "citers," and readership counts compared with citation counts. Later in this chapter we discuss several studies in this direction.

In addition to the number of readers, in systems like BibSonomy and CiteULike it is also possible to study the distribution of the free-text tags assigned by users to the bookmarked items. These tags supplement author- and indexer-provided keywords and can be useful for discovery (Kipp, 2011; Lee & Schleyer, 2010). Tags can help in clustering the items, and can point to "hot" topics and to items bookmarked by many users within these topics (Haustein & Siebenlist, 2011; see also Haustein, chapter 17, this volume). In one paper, Ke and Chen (2012) studied the tagging patterns of 1,600 Library and Information Science (LIS) papers on CiteULike. The purpose of the paper was to identify tag types, but the authors also provided a list of the most frequently occurring tags. The top tags in decreasing order of frequency were: information, information behavior, similarity, cocitation, and human factors—indicating the value of tags as a data source for classifying items.

Since social bookmarking services contain huge amounts of accessible information on scholarly articles saved by users, it is possible to carry out different informetric analyses on them. For example, Borrego and Fry (2012) studied a set of more than 370,000 records from BibSonomy created by about 3,000 users. Only records with a URL of the source were considered. Almost half of the bookmarked items were journal articles, followed by conference papers and books. Ma, Duhon, Hardy, and Börner (2009) also analyzed more than 250,000 records in BibSonomy. Similarly to Borrego and Fry (2012), they found that articles constitute about 50% of the bookmarked items. They created a visualization of the domains covered by the bookmarked records[7] and listed the most frequently bookmarked journals, with *Physica A* leading the list. Physics, economics, and statistics are particularly common among the top bookmarked journals.

Haustein and Siebenlist (2011) defined four indicators for studying journal usage on online reference managers: usage ratio, usage diffusion, article usage intensity, and journal usage intensity. Here we discuss two of these, usage ratio and article usage intensity. Usage ratio is the ratio of the number of articles bookmarked in the system out of the total number of articles published by a journal during the time period under consideration. Article usage intensity is the average number of readers per bookmarked article in a journal. These indicators were applied to articles from 45 solid-state physics journals published between 2004 and 2008. Data were collected from three online reference managers—BibSonomy, CiteULike, and Connotea—and bookmarks and readership counts were combined. Usage ratio measures the coverage of the reference manager systems; it was highest for the *Reviews of Modern Physics* (0.630), while the average usage ratio for all 45 journals was low at 0.035 (i.e., 3.5% of articles had at least one reader). Article usage intensity was 1.17 for all journals combined, and again it was highest for the *Reviews of Modern Physics* at 3.89. It should be noted that *Reviews of Modern Physics* is the journal with the highest impact factor among all physics journals: its 2011 impact factor is 43.933, and it is the journal with the 4th highest 2011 impact of all JCR journals.

Haustein and Siebenlist (2011) found, on average, fewer bookmarks than citations. However, Saeed, Afzal, Latif, Stocker, and Tochtermann (2008) found that papers from the 2006 WWW Conference revealed a different tendency, at least for those tagged by more than six users. Bookmarks were collected from BibSonomy, CiteULike, and Delicious (a general bookmarking platform). They found that papers receive tags earlier than citations. Although the number of citations these papers received by 2008 was usually higher than the number of users who tagged them, there were some cases where the number of users who tagged the item in 2006 exceeded the total number of citations the item received by 2008. Overall, they found a positive correlation ($r = 0.65$) between Google Scholar citations and the number of bookmarks on papers from the 2006 WWW Conference.

Based on information from its website (Mendeley, 2012a), in August 2012 Mendeley contained 278 million user documents and had more than 1.8 million members. The number of user documents includes duplicates—that is, if an item is bookmarked by *n* users, it may be counted up to *n* times, since Mendeley does not comprehensively identify duplicate entries from users. In contrast, CiteULike contains only 6.27 million documents, and we could not find information about the sizes of Connotea and BibSonomy. These numbers can be compared with the reported 47 million records in Scopus (Elsevier, 2012), and 46 million records in the Web of Science (WoS) (Thomson Reuters, 2011). In spite of these seemingly huge differences in size, according to Alexa (Alexa, 2012), Mendeley attracted much less traffic than the other three services. During the period between July 19, 2012, and August 18, 2012, Mendeley was ranked 28,250 compared with BibSonomy (7,952), Connotea (9,418), and CiteULike (12,628). There are considerable fluctuations in the relative rankings, however. For example, between September 2011 and mid-July 2012, Mendeley had the best traffic ranking among the four services. The search facilities of the services are not optimal (Bar-Ilan et al., 2012; Li & Thelwall, 2012; Haustein & Siebenlist, 2011), the results are not always consistent, and duplicates are a big problem (Bar-Ilan, 2011; Hull, 2010). Mendeley's coverage was correlated with SCIMago[8] data derived from Scopus (Kraker, Körner, Jack, & Granitzer, 2012). The overall correlation was 0.70, indicating that the documents bookmarked in Mendeley are a fair approximation of the Scopus database, at least for recent years. Coverage was found to be especially good in biology and computer science. For some publishers, Mendeley automatically populates its database (Gunn, personal communication, June 21, 2012), hence its high document count, but currently this is not the case for all journals (specifically, not for the *Journal of the American Society for Information Science & Technology* [*JASIST*], mentioned later).

Mendeley seems to be largest in terms of coverage, and a number of studies have calculated correlations between citation counts and readership counts. Li, Thelwall, and Giustini (2012) considered 1,613 papers published in *Science* and *Nature* about four years after their publication. Citation data were collected from WoS and Google Scholar, and readership counts were collected from Mendeley and CiteULike. *Science* and *Nature* are the two "top publication outlets" listed on the Mendeley site (Mendeley, 2012b); thus it is not surprising that Mendeley had excellent coverage of these journals. Its coverage of *Nature* and *Science* articles was 94% and 93%, respectively, compared with 62% and 60% for CiteULike. Although there are a large number of users bookmarking *Science* and *Nature* articles on Mendeley, the average citation count on WoS was 78 and 69 for *Nature* and *Science* articles, respectively, compared with 2.37 and 2.50 readers on average on CiteULike and 10.71 and 8.89 readers on average on Mendeley. Still, Spearman correlations between WoS citations and Mendeley reader counts were significant and quite high at 0.559 and 0.540 for *Science* and *Nature*, respectively. The correlations were also significant for CiteULike but were much lower (0.366 and 0.304, respectively). Li et al. (2012) conclude that "online reference managers may be useful for the research impact measurement from the point of view of general readers" (p.469). In a more recent study, Li and Thelwall (2012) considered articles in genomics and genetics that were published in 2008 and reviewed on F1000.[9] F1000 is a postpublication peer review site where over 5,000 experts review and recommend top articles in biology and medicine. Li and Thelwall's aim was to compare F1000 scores with Mendeley and CiteULike readership counts and citation data from WoS, Scopus, and Google Scholar. Almost all selected articles appeared in all the databases, except for CiteULike, for which coverage was only 68% (the coverage of the other databases was over 99%). The mean citation count was about twice the mean readership count on Mendeley for WoS and Scopus, and almost 18 times the mean readership count on CiteULike. On average, Google Scholar citation counts were about 30% higher than on WoS and Scopus. In this sample, the Spearman correlations between Mendeley and the citation databases were higher than in the previous study (0.687), while the correlations with CiteULike remained about the same (0.356). Correlations between the Journal Impact Factors (JIFs) of the journals in which the articles were published and readership counts and citations were also calculated. Rather interestingly, the correlation between the JIF and Mendeley (0.521) was quite similar to the correlations between the JIF and the citation databases (around 0.568), while the correlation between the JIF and CiteULike was only 0.121. All correlations were highly significant. This shows that there is a tendency to bookmark items published in journals with high impact factors in Mendeley, while this tendency is much less pronounced in CiteULike, perhaps due to its smaller coverage of the literature.

These two studies considered publications in high-impact journals and articles that received high scores from expert evaluators, but are the correlations this high for an average set of papers? This question was investigated in two studies: the first (Bar-Ilan et al., 2012) searched the publications of 57 presenters at the 2010 Science and Technology Indicators Conference. The presenters were of different ages and academic ranks; some were highly prominent while others were students at the beginning of their careers. All the publications (1,136 in total) retrieved from WoS and Scopus were searched in Mendeley and CiteULike. Some of the articles were relatively highly cited and some were not cited at all; some were relatively old and some were published recently. Even for this mixed set of articles, Mendeley's coverage was 82%, while CiteULike's coverage was only 28%. The mean number of citations for cited articles was 19.5 for Scopus and 21.1 for WoS, while the mean readership counts for Mendeley and CiteULike were 9.5 and 2.4, respectively. The Spearman correlation between Mendeley and Scopus was 0.448, and between CiteULike and Scopus 0.232. These numbers are only slightly lower than the correlations when the dataset consisted of *Science* and *Nature* articles (Li et al., 2012). In another study, Bar-Ilan (2012a, 2012b) used *JASIST* articles published between 2001 and 2010 as a dataset. Even though *JASIST* is one of the most prominent journals in LIS, its JIF (2.081) cannot be compared with those of *Science* (31.201) or *Nature* (36.280). Thus, this set of articles can be considered a "general" set, with about 10% (depending on the database) of the items not cited at all. Still, Mendeley had excellent coverage (97.5%), which is an impressive result of crowdsourcing since *JASIST* is not one of the journals Mendeley automatically indexes. The Spearman correlations between Mendeley reader counts and WoS, Scopus, and Google Scholar citations (0.458, 0.502, and 0.519 respectively) were similar to the correlations found in the previously mentioned studies. Thus, it seems that Mendeley, if it continues to be popular, will be a useful data source for altmetrics, not only for highly visible articles but for average articles as well.

Conclusions

This chapter has provided evidence of the value of science blogs and reference managers for scholarly communication and for use in altmetrics. One of the major problems in social media websites is shifting trends in popularity—for example, MySpace used to be the "hottest" social media website, but it lost this title to Facebook. An example from the scholarly social web is Elsevier's 2collab that was discontinued in April 2011. There is no guarantee that the tools reviewed here will exist in the future, or that new tools for managing references will not overtake existing ones in popularity. However, the altmetric directions mentioned in this chapter seem likely to remain valid, even if the specific tools or their features change over time. Readership counting belongs to the family of usage metrics (Kurtz & Bollen, 2010) and has its roots in circulation and reshelving studies carried out in physical libraries (e.g., Peritz, 1995; Rice, 1979; Tsay, 1998; see also Haustein, chapter 17, this volume). Thus, although the digital environment offers new possibilities, some of these can be measured with techniques analogous to traditional bibliometric ones. On the other hand, the amount of available information and the speed with which information is processed, disseminated, reviewed, and discussed offer new bibliometric/altmetric challenges.

Acknowledgments

This chapter was partially funded by the EU FP7 ACUMEN project (Grant agreement: 266632). We wish to thank Bora Zivkovic, the Blog Editor at *Scientific American*, for his help in identifying the current state of science blogs listed by *Nature News* in 2006.

Notes

1. http://rinarchive.jisc-collections.ac.uk
2. For other science blogging aggregators, see http://scienceblogging.org.
3. http://www.BibSonomy.org
4. http://www.CiteULike.org
5. http://www.connotea.org
6. http://www.mendeley.com/
7. http://www.cmu.edu/joss/content/articles/volume11/7_Borner_Ma_files/09-BibSonomy.JPG
8. http://www.scimagojr.com/index.php
9. http://f1000.com

References

Alexa. (2012). Homepage. Retrieved from www.alexa.com.

Bar-Ilan, J. (2011). *Articles tagged by "bibliometrics" on Mendeley and CiteULike.* Presented at the 2011 ASIS&T SIG/MET Symposium on Informetric and Scientometric Research (Metrics 2011), New Orleans.

Bar-Ilan, J. (2012a). *JASIST@mendeley*. Presented at the ACM Web Science Conference Workshop on Altmetrics, Evanston, IL. Retrieved from http://altmetrics.org/altmetrics12/bar-ilan.

Bar-Ilan, J. (2012b). JASIST 2001–2010. *Bulletin of the American Society for Information Science and Technology*, *38*(6), 24–28.

Bar-Ilan, J., Haustein, S., Peters, I., Priem, J., Shema, H., & Terliesner, J. (2012). Beyond citations: Scholars' visibility on the social Web. In *Proceedings of the 17th International Conference on Science and Technology Indicators*, vol. 1, 98-109 Montreal, Canada: Science-Metrix and Université du Québec à Montréal Retrieved from http://arxiv.org/abs/1205.5611.

Bonetta, L. (2007). Scientists enter the blogosphere. *Cell, 129*, 443–445.

Borrego, A., & Fry, J. (2012). Measuring researchers' use of scholarly information through social bookmarking data: A case study of BibSonomy. *Journal of Information Science, 38*(3), 297–308.

Bukvova, H. (2011). Scientists online: A framework for the analysis of Internet profiles. *First Monday, 16*(10). Retrieved from http://firstmonday.org/htbin/cgiwrap/bin/ojs/index.php/fm/article/viewArticle/3584/3065.

Butler, D. (2006). Top five science blogs. *Nature, 442*, 9. doi:10.1038/442009a.

Costas, R., Leeuwen, T. N., & Bordons, M. (2010). Self-citations at the meso and individual levels: Effects of different calculation methods. *Scientometrics, 82*(3), 517–537.

Cronin, B. (1981). The need for a theory of citing. *Journal of Documentation, 37*(1), 16–24.

Davis, P. (2012, April 10). The emergence of a citation cartel [Blog post]. Retrieved from http://scholarlykitchen.sspnet.org/2012/04/10/emergence-of-a-citation-cartel.

De Vrieze, J. (2012, February 1). Thousands of scientists vow to boycott Elsevier to protest journal prices. *Science Insider*. Retrieved from http://news.sciencemag.org/scienceinsider/2012/02/thousands-of-scientists-vow-to-b.html.

Elsevier (2012). What does it cover? Retrieved from http://www.info.sciverse.com/scopus/scopus-in-detail/facts.

Fowler, J., & Aksnes, D. W. (2007). Does self-citation pay? *Scientometrics, 72*(3), 427–437.

Gowers, T. (2012, January 21). Elsevier—my part in its downfall [Blog post]. Retrieved from http://gowers.wordpress.com/2012/01/21/elsevier-my-part-in-its-downfall.

Grossman, L. (2010, December 7). Doubts brew about NASA's new arsenic life. *Wired*. Retrieved from http://www.wired.com/wiredscience/2010/12/arsenic-life-under-fire.

Groth, P., & Gurney, T. (2010). *Studying scientific discourse on the Web using bibliometrics: A chemistry blogging case study*. Presented at WebSci10: Extending the Frontiers of Society On-Line, Raleigh, NC.

Gruzd, A., Goertzen, M., & Mai, P. (2012). Survey results highlights: Trends in scholarly communication and knowledge dissemination in the age of social media. *Social Media Lab Report*. Retrieved from http://socialmedialab.ca/?p=4308.

Gruzd, A., Staves, K., & Wilk, A. (2011). Tenure and promotion in the age of online social media. In *Proceedings of the American Society for Information Science and Technology Conference*. Retrieved from http://www.asis.org/asist2011/proceedings/submissions/154_FINAL_SUBMISSION.doc.

Harley, D., Acord, S. K., Earl-Novell, S., Lawrence, S., & King, C. J. (2010). Assessing the future landscape of scholarly communication: An exploration of faculty values and needs in seven disciplines. Berkeley: Center for Studies in Higher Education, UC Berkeley. Retrieved from http://escholarship.org/uc/cshe_fsc.

Haustein, S., & Siebenlist, T. (2011). Applying social bookmarking data to evaluate journal usage. *Journal of Informetrics, 5*(3), 446–457.

Hull, D. (2010, September). How many unique papers are there in Mendeley? [Blog post]. Retrieved from http://duncan.hull.name/2010/09/01/mendeley.

Impact. (2012). *Oxford English Dictionary*. Retrieved from http://www.oed.com.

Jha, A. (2012, April 9). Academic spring: How an angry maths blog sparked a scientific revolution. *The Guardian*. Retrieved from http://www.guardian.co.uk/science/2012/apr/09/frustrated-blogpost-boycott-scientific-journals?commentpage=2#start-of-comments.

Katsnelson, A. (2010). Microbe gets toxic response. *Nature, 468*, 741. doi:10.1038/468741a.

Ke, H.-R., & Chen, Y.-N. (2012). Structure and pattern of social tags for keyword selection behaviors. *Scientometrics, 92*(1), 43–62.

Kipp, M. E. I. (2011). User, author and professional indexing in context: An exploration of tagging practices on CiteULike. *Canadian Journal of Information and Library Science, 35*(1), 17–48.

Kousha, K., Thelwall, M., & Rezaie, S. (2010). Using the web for research evaluation: The Integrated Online Impact Indicator. *Journal of Informetrics, 4*(1), 124–135.

Kraker, P., Körner, C., Jack, K., & Granitzer, M. (2012). Harnessing user library statistics for research evaluation and knowledge domain visualization. In *Proceedings of the 21st International Conference Companion on the World Wide Web* (pp. 1017–1024). New York: ACM.

Kurtz, M., & Bollen, J. (2010). Usage bibliometrics. *Annual Review of Information Science & Technology, 44*, 1–64.

Lee, D. H., & Schleyer, T. (2010). A comparison of MESH terms and CiteULike social tags as metadata for the same items. In *Proceedings of the 1st ACM International Health Informatics Symposium* (pp. 445–448). New York: ACM.

Li, X., & Thelwall, M. (2012). F1000, Mendeley and traditional bibliometric indicators. In *Proceedings of the 17th International Conference on Science and Technology Indicators* (vol 2, pp. 451–551). Montreal, Canada, Science Metrix and Université du Québec à Montréal.

Li, X., Thelwall, M., & Giustini, D. (2012). Validating online reference managers for scholarly impact measurement. *Scientometrics, 91*(1), 461–471.

Lin, T. (2012, February 14). Mathematicians organize boycott of a publisher. *New York Times*, p. D7.

Ma, N., Duhon, R. J., Hardy, E. F., & Börner, K. (2009). BibSonomy anatomy. Viszards Session at Sunbelt 2009. Retrieved from http://www.kde.cs.uni-kassel.de/ws/Viszards09/presentations/6_boerner.

Mendeley. (2012a). Homepage. Retrieved from http://www.mendeley.com.

Mendeley. (2012b). Research papers. Retrieved from http://www.mendeley.com/research-papers.

National Science Foundation, Division of Science Resources Statistics. (2010a). *Doctorate recipients from U.S. universities: 2009* (Special Report NSF 11–306). Arlington, VA. Retrieved from http://www.nsf.gov/statistics/doctorates.

National Science Foundation, Division of Science Resources Statistics. (2010b). *Science & engineering indicators: 2010, Chapter 2: Higher education in science and engineering.* Retrieved from http://www.nsf.gov/statistics/seind10/c2/c2s4.htm.

Nicholas, D., & Rowlands, I. (2011). Social media use in the research workflow. *Information Services & Use, 31*, 61–83. Retrieved from http://iospress.metapress.com/content/23032g726121kqw4/fulltext.pdf.

Oransky, I. (2012, July 5). A first? Papers retracted for citation manipulation [Blog post]. Retrieved from http://retractionwatch.wordpress.com/2012/07/05/a-first-papers-retracted-for-citation-manipulation.

Overbye, D. (2010, December 14). Poisoned debate encircles a microbe study's result. *New York Times*, p. D4.

Perfecting peer review? [Editorial]. (2011, January 7). *Nature Medicine, 17*, 1–2. doi:10.1038/nm0111-1

Peritz, B. C. (1995). On the association between journal circulation and impact. *Journal of Information Science, 21*(1), 63–67.

Ponte, D., & Simon, J. (2011). Scholarly communication 2.0: Exploring researchers' opinions on Web 2.0 for scientific knowledge creation, evaluation and dissemination. *Serials Review, 37*(3), 149–156. doi:10.1016/j.serrev.2011.06.002.

Priem, J., Taraborelli, D., Groth, P., & Neylon, C. (2010). Altmetrics: A manifesto. Retrieved from http://altmetrics.org/manifesto.

Rice, B. A. (1979). Science periodicals use study. *Serials Librarian, 4*(1), 35–47.

RIN. (2009). *Patterns of information use and exchange: Case studies of researchers in the life sciences.* A Research Information Network Report. Retrieved from http://rinarchive.jisc-collections.ac.uk/our-work/using-and-accessing-information-resources/patterns-information-use-and-exchange-case-studie.

RIN. (2011a). *Collaborative yet independent: Information practices in the physical sciences.* A Research Information Network Report. Retrieved from http://rinarchive.jisc-collections.ac.uk/our-work/using-and-accessing-information-resources/physical-sciences-case-studies-use-and-discovery-.

RIN. (2011b). *Reinventing research? Information practices in the humanities.* A Research Information Network Report. Retrieved from http://rinarchive.jisc-collections.ac.uk/our-work/using-and-accessing-information-resources/information-use-case-studies-humanities.

Saeed, A. U., Afzal, M. T., Latif, A., Stocker, A., & Tochtermann, K. (2008). Does tagging indicate knowledge diffusion? An exploratory case study. In *Proceedings of the 2008 Third International Conference on Convergence and Hybrid Information Technology* (pp. 605–610). Washington, DC: IEEE.

Schiermeier, Q. (2012). Arsenic-loving bacterium needs phosphorus after all. *Nature*. doi:10.1038/nature.2012.10971

Shema, H., Bar-Ilan, J., & Thelwall, M. (2012a). Research blogs and the discussion of scholarly information. *PLoS ONE, 7*(5), e35869. doi:10.1371/journal.pone.0035869

Shema, H., Bar-Ilan, J., & Thelwall, M. (2012b). Self-citation of bloggers in the science blogosphere. In A. Tokar, M. Beurskens, S. Keuneke, M. Mahrt, I. Peters, C. Puschmann, et al. (Eds.), *Proceedings of the 1st International Conference on Science and the Internet (CoSci12)* (pp. 183-192). Düsseldorf, Germany: Düsseldorf University Press.

Thomson Reuters. (2011). *The definitive resource for global research: Web of Science*. Retrieved from http://thomsonreuters.com/content/science/pdf/Web_of_Science_factsheet.pdf.

Tsay, M.-Y. (1998). The relationship between journal use in a medical library and citation use. *Bulletin of the Medical Library Association, 86*(1), 31–39.

Whitfield, J. (2012). Elsevier boycott gathers pace. *Nature*. doi:10.1038/nature.2012.10010.

Wolfe-Simon, F., Blum, J. S., Kulp, T. R., Gordon, G. W., Hoeft, S. E., Pett-Ridge, J., et al. (2010). A bacterium that can grow by using arsenic instead of phosphorus. *Science* [Online early access]. Retrieved from http://www.sciencemag.org/content/332/6034/1163.short.

Zivkovic, B. (2011, September 30). #Arseniclife link collection [Blog post]. Retrieved from http://blogs.scientificamerican.com/a-blog-around-the-clock/2011/09/30/arseniclife-link-collection.

17 Readership Metrics

Stefanie Haustein

Journal Perception and Use

The value of a journal is often primarily, if not solely, measured by the degree to which the journal garners citations. However, the audience for scientific research is not confined to those who cite; many readers are not producers of research. It is therefore necessary to incorporate metrics of "pure" readership—that is, those who read a journal article, but do not cite it—into journal evaluation. The notion of incorporating the readers' perspective into journal evaluation has been repeatedly promoted and plays an important role for a number of actors in the scholarly communication system (e.g., Butkovich, 1996; Rousseau, 2002; Langlois & Von Schulz, 1973; Bustion, Eltinge, & Harer, 1992; Schlögl & Gorraiz, 2006). Authors use notions of audience in selecting journals to which they would like to submit their work, librarians rely on perceived readership for collection management, and editors and publishers apply readership metrics to monitor the performance of periodicals in the scholarly marketplace (Rousseau, 2002).

Traditionally, librarians gathered information about the reading behavior of users either through surveys or by monitoring circulation data. Statistics on the amount of usage per title were collected via reshelving exercises, patron observations, or interlibrary loan and document delivery data; statistics derived from these data were used to evaluate and optimize local collections (Butkovich, 1996). Although compiling reading statistics was time consuming and the results often inaccurate, the data helped librarians determine how much various journal subscriptions were used, and, as a result, allowed them to cancel underutilized titles (Langlois & Von Schulz, 1973; Bustion et al., 1992; Schlögl & Gorraiz, 2006).

Qualitative reader surveys are even more costly but deliver detailed insights into reading behavior (in terms of who reads what, how often, for how long, and for what purpose). Surveys may also ask participants to rank periodicals according to subjectively perceived importance (e.g., Kohl & Davis, 1985). Tenopir has conducted several reader surveys since 1977 and has published important results about the reading behavior of scientists (e.g., Tenopir & King, 2000).

Quantitative library data analysis and qualitative reader surveys are time consuming to conduct, but provide a useful snapshot of the reading behavior of the local patron base. While reader statistics are influenced by particular institutional biases (reflecting the thematic scope of the institution or whether it is research-oriented or education-oriented), surveys may be based on a broader audience and reflect the subjective perspective of the participants (Butkovich, 1996).

Measuring Usage with Citations

A different and indirect method of estimating journal usage is citation analysis. Citations reflect formal scholarly communication globally and thus allow the drawing of more general conclusions about a journal's impact on the international scientific community. Gross and Gross (1927) were the first to apply citation-based journal evaluation to the management of library holdings, when analyzing 3,663 references listed in the 1926 volume of the *Journal of the American Chemical Society*.

With the development of the *Science Citation Index*, the first large-scale citation database, citation analysis became more feasible (Garfield, 1955, 1972). Since then, citation analysis has become a popular tool in research evaluation and has, to a large extent, replaced reading statistics. In journal evaluation today, citation-based rankings and indicators and (above all) the Journal Impact Factor (JIF) play an important role in the selection process of journals by librarians, authors, and readers. With the availability of citation databases such as Web of Science (WoS) and Scopus, journal impact has been equated with journal citations, and the JIF has become the gold standard for evaluating scholarly journals, as well as one of the most popular bibliometric indicators (Glänzel & Moed, 2002). The problem with using citations as a substitute for usage is that we do not capture the influence of journal articles on "pure" readers. Even though pure readers do not publish (and hence do not cite), they may apply journal content in their daily work and in the development of new technologies (Schlögl & Stock, 2004; Rowlands & Nicholas, 2007; Stock, 2009). In universities, pure readers are represented largely by undergraduate and graduate students (Duy & Vaughan, 2006).

Pure readers are not the only people who may read but not cite a work, as shown by studies on citation behavior and motivation (e.g., MacRoberts & MacRoberts, 1989). When using citations to measure journal impact one should remember that one is not analyzing the total readership population, which consists of authors and readers who "have different needs and different requirements of the journals literature" (Rowlands & Nicholas, 2007, p. 223). As Tenopir and King (2000) put it, "[The] scientific community represents the universe of potential authors of scientific and technical journal articles, but not all scientists publish regularly or even one time. The characteristics of the author community are therefore somewhat different from the characteristics of the scientific community as a whole" (Tenopir & King, 2000, p. 142).

Price and Gürsey (1976) distinguish between seven different types of researchers, based on publication frequency. Nonpublishing researchers are estimated to constitute one-third of the scientific community (Price & Gürsey, 1976; Tenopir & King, 2000). Pure (i.e., nonpublishing) readers represent an element of journal usage not covered by citation analysis. Hence, citation-based indicators cannot reflect the entire influence of scholarly journals and documents in general, as noted by Cronin and colleagues:

> While traditional citation analysis can tell us a lot about the formal bases of intellectual influence, it, quite naturally, tells us nothing about the many other modalities of influence, which comprise the total impact of an individual's ideas, thinking, and general professional presence. (Cronin, Snyder, Rosenbaum, Martinson, & Callahan, 1998, p. 1326)

This was reinforced by the Research Councils UK, which acknowledge that in addition to "academic" impact, research can have "economic and societal" effects (Research Councils UK, 2011, para. 1).

The ineffectiveness of citation analysis, and particularly the JIF, for industry-oriented journals was mentioned in the 1980s by Scanlan (1987), the senior editor at Pergamon Press. While the commercial impact of research can, to a certain extent, be covered by patent analysis, Thelwall (2012) lists other examples of instances in which citation-based journal metrics are insufficient. He points out a systematic bias of the JIF against profession-, education-, and policy-oriented research journals, and in favor of theoretical periodicals. A more accurate way of analyzing scholarly journals is to consider being read and being cited as two different aspects of influence, and thus journal usage and journal citations as two separate dimensions of journal evaluation. Such a multidimensional approach has the advantage of highlighting journals' strengths in one area and weaknesses in others so that users (i.e., readers), authors, and librarians can select periodicals that best suit their specific needs (Haustein, 2011, 2012).

Electronic Download Data

With the switch to electronic publishing, data relating to the frequency with which journal articles were accessed became available through logfiles collected on publishers' servers. Electronic reading statistics could be computed on the basis of article download and click rates. Reader-based usage studies were thus rediscovered by libraries for the purposes of collection management (Gorraiz & Gumpenberger, 2010). In a survey by Rowlands and Nicholas (2005), researchers agreed that downloads are a better indicator of the usefulness of research than citations. Furthermore, readership is among the most important factors influencing authors in their selection of journals to which to submit their manuscripts (Tenopir & King, 2000; Rowlands & Nicholas, 2005).

Logfiles collected on publishers' servers provide detailed information about what content is accessed where, when, how, and by whom. They permit direct and continuous acquisition of usage data representing the entire readership. Compared to usage statistics based on reader surveys, patron observations, and re-shelving, which are difficult to conduct or monitor and reflect usage by only a small group of readers, the collection of electronic readership data is automated and expeditious (Nicholas et al., 2008; Schlögl & Gorraiz, 2011). High correlations between local print and online usage support the assumption that reading behavior can be measured by download statistics (Duy & Vaughan, 2006; Emrani, Moradi-Salari, & Jamali, 2010). Reading statistics now represent a common way of optimizing library subscriptions. Many librarians use statistics reflecting local full-text access to calculate actual cost–benefit ratios, thus helping them identify and justify cancellations (Hahn & Faulkner, 2002; Baker & Read, 2008; Emrani et al., 2010).

Validity as Usage Indicators

The primary question regarding the availability of full-text access counts has to do with how far they actually reflect use (i.e., readership of publications). Although initiatives such as COUNTER[1] (Counting Online Usage of NeTworked Electronic Resources) have largely standardized usage counts, full-text access can range from taking a quick look at an article's title or abstract to reading it carefully; in some cases, full-text access can even involve distributing an article to colleagues or putting it on the reading list of a university course (Nicholas et al., 2008).

When counting full-text access, no metric can tell whether the user actually read the article or how many people gained access as the result of a single download. In addition, access to documents via self-archiving (i.e., green open access) and print are ignored. Download and click rates *estimate* readership; they do not *measure* it (Thelwall, 2012). The same is true for the monitoring of print usage through re-shelving; journal usage can range from looking up the table of contents to photocopying several articles in an issue (Duy & Vaughan, 2006). Davis (2004) reports that 70% to 80% of actual usage is not captured by re-shelving statistics.

Citations have similar limitations. Documents may be cited but not read, or read but not cited. Even if read and cited, one citation does not necessarily count as much as another. There is a difference between a reference that is repeatedly mentioned in the methods section and a document listed once in the introduction together with a dozen others, not to mention negative citations (MacRoberts & MacRoberts, 1989). It is true that download statistics can be influenced by authors and publishers, but citations are not free from manipulation either. Although a large portion of self-citation may be warranted (an author writing on a particular topic has likely written on the topic before and would appropriately cite their own work), authors are known to intentionally and inappropriately cite their own documents to increase their citation count and *h*-index. There are also reported cases of editors coercing authors to cite their journal in order to increase its JIF (Yu & Wang, 2007; Wilhite & Fong, 2012). Author and journal self-citations can be excluded for evaluation purposes, but gaming of the system by "citation cartels" is not easy to identify and counteract (Franck, 1999).

Correlation with Citations

Several studies correlating the number of downloads with the number of citations have found positive relationships, but the correlations are too weak to conclude that downloads and citations measure the same thing (Li, Thelwall, & Giustini, 2012). In an early study, Perneger (2004) analyzed download rates of 153 papers one week after publication and compared these with WoS citation rates after 5 years. A correlation of Pearson's $r = 0.50$ was found. As part of the MESUR project, which collected 1 billion usage events to "define, validate and cross-validate a range of usage-based metrics of scholarly impact" (Bollen, Van de Sompel, & Rodriguez, 2008, p. 231), Bollen and Van de Sompel (2008) computed usage impact factors for 3,146 journals. The analysis was based on full-text downloads of nine major institutions from the California State University (CSU) system in 2004 with respect to articles published in 2003 and 2002. A modest negative Spearman correlation was found between the regular JIF and the version calculated with downloads, indicating that the journals used most at CSU have, in general, low JIFs.

This confirms the assumption that local download statistics can differ strongly from citation impact (Bollen & Van de Sompel, 2008). Brody, Harnad, and Carr (2006) found a correlation of $r = 0.46$ between downloads from almost 15,000 physics preprints on arXiv and Citebase citations, and showed that the number of downloads during the first six months after publication has the highest predictive power (Pearson's $r = 0.83$) of citations received after two years. The Spearman correlation between citations and PDF downloads of 4,000 *Public Library of Science (PLoS)* papers published in 2008 was 0.48 (Yan & Gerstein, 2011). Liu, Fang, and Wang (2011) found a Spearman correlation of 0.49 between download and citation counts for 1,622 documents published in Chinese ophthalmology journals in 2005, and found that 3% of the papers had large differences between download and citation counts. Those publications with high usage but low citation rates had an application orientation, or else contained news and summaries about important conferences.

Yan and Gerstein (2011) defined two phases in the aging patterns of usage rates. Unlike citations, usage counts peak shortly after publication, usually within the first two months. After a rapid decrease from the first to the second month, they decline slowly—fitting a power law. These two phases mirror the typical dissemination process where researchers become aware of the documents shortly after publication and pass them on to colleagues at the same time. The older a paper gets, the less attention it receives, and thus the smaller its chances of being disseminated (Yan & Gerstein, 2011). Wan, Hua, Rousseau, and Sun (2010) show that correlations are strongest when they are computed through normalized cross-covariance between citation and download curves (i.e., when the citation curve is shifted backward by two or three years). The annual distributions of downloads and citations received by *PLoS Biology* papers can be seen in figure 17.1. The number of citations usually peaks some years

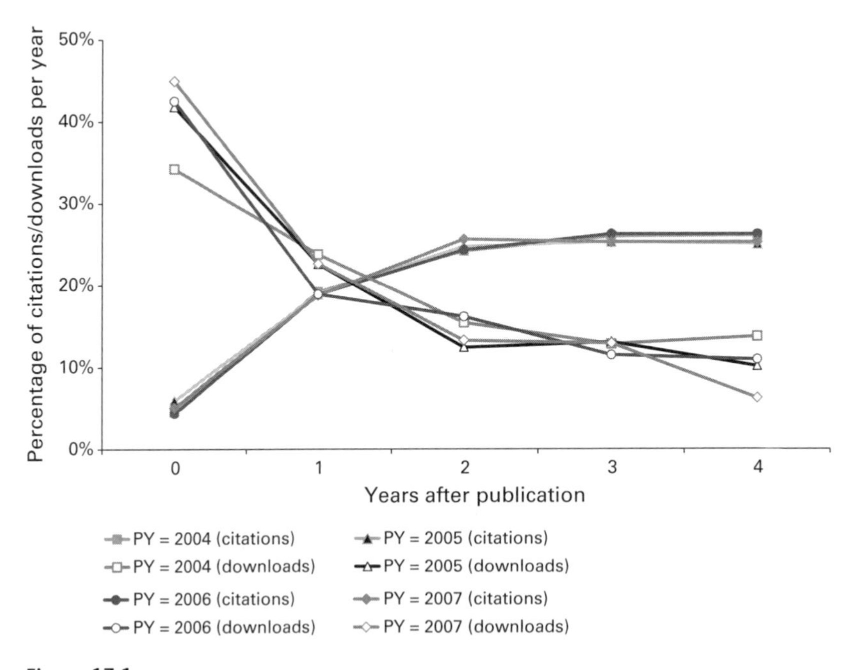

Figure 17.1
Annual distributions of citations and downloads received by *PLoS Biology* documents published between 2004 and 2007. While the majority of downloads are generated during the year of publication, citations peak two to four years later.

postpublication (depending on the discipline), after which it steadily decreases, indicating the obsolescence of the published content (Moed, Van Leeuwen, & Reedijk, 1998). However, the scholarly communication process (i.e., the read-cite-read cycle) has accelerated due to electronic publishing and the availability of preprints (Brody et al., 2006). Differences between cited and usage half-lives were examined by Moed (2005) and Schlögl and Gorraiz (2010, 2011): medium correlations confirm that downloads measure a different impact than citations. Nonetheless, these should be seen as complementary indicators of influence because a fuller picture of impact is provided if both are used.

Practical Limitations

Although quite a number of indicators have been introduced, all of which are based on usage statistics and calculated analogously to citation measures (Rowlands & Nicholas, 2007; Bollen et al., 2008; Schlögl & Gorraiz, 2010; Wan et al., 2010), data aggregation is still problematic. The problem with global usage statistics is that, except for a limited number of case studies exploring the potential of such data, they are not available. In general, commercial publishers are reluctant to make usage transparent and share usage data with the public. Because low usage may affect negatively both manuscript submissions and library subscriptions, usage data are highly sensitive from a commercial perspective (Borrego & Fry, 2012), leading publishers to restrict global download statistics to a few selective lists for promotional purposes, such as the "Top 25 Hottest Articles"[2] on ScienceDirect.

What publishers do provide to subscribing institutions are local download statistics. Even though standards such as COUNTER exist and have improved the quality of data to a certain extent, local statistics are often still incomparable, lack consistency, and above all are not detailed enough to provide meaningful usage information (Baker & Read, 2008; Haustein, 2012). Even if accurate article-based usage statistics were provided, local data would be biased and would not necessarily reflect a journal's impact on the wider research community.

The case is different with open-access journals, which, unlike traditional subscription journals, do not have to worry about low usage reflecting badly on profits. *PLoS* journals are exemplary when it comes to usage statistics. Since 2009 they have provided detailed article-based usage data that, in contrast to COUNTER journal reports, can be used to calculate meaningful relative indicators of journal usage (e.g., figure 17.2). The idea behind *PLoS'* article-level metrics (ALM) is that readers can easily identify popular documents through different criteria by tracking their "overall performance and reach" (*PLoS*, 2012, para. 1) without delay. Yan and Gerstein (2011) and Jamali and Nikzad

(2011) have used these data to analyze the global reading behavior associated with the *PLoS* journals.

As long as global download statistics are not available for subscription journals, an alternative, publisher-independent data source is needed to access data relating to the use of scholarly journals and documents. Online reference managers and social bookmarking sites[3] with an academic focus may be considered a possible alternative.

Altmetrics: Scholarly Influence Reflected on the Social Web

With the development of Web 2.0, online platforms could be based on crowdsourced (i.e., user-generated) content. Peters (2009) differentiates between sharing services (i.e., YouTube and Flickr) and social bookmarking services (Delicious) as the two basic types of Web 2.0 tools. Along with CiteULike, the latter found their way into academia in late 2004, allowing researchers to manage, search, and share scholarly literature online (Hammond, Hannay, Lund, & Scott, 2005; Reher & Haustein, 2010). Many similar reference managers (e.g., Connotea,[4] BibSonomy, Zotero, and Mendeley) have followed. Mendeley claims to be the largest research catalog with 281 million bookmarks to 68 million unique documents. As of August 2012, these included 32 million unique PDFs (Ganegan, 2012) crowdsourced by 1.8 million users.[5] By comparison, the citation databases of WoS currently contain about 52 million documents.[6]

With these new sources comes the possibility of analyzing online usage of scholarly resources independently of publishers. Tracking the usage of scholarly content in social media means that researchers are able to analyze impact more broadly (Li et al., 2012). The altmetrics movement promotes the investigation of alternative impact metrics and retrieval methods based on tweets, blogs, user comments, and social networking platforms (Taraborelli, 2008; Neylon & Wu, 2009; Priem & Hemminger, 2010; Priem, Taraborelli, Groth, & Neylon, 2010; see also Priem, chapter 14, this volume).

Representativeness of Web 2.0 Data in Scholarly Communication

As with the introduction of electronic download statistics, the availability of Web 2.0 usage data makes us ask what these data actually measure. How far does activity on the social Web reflect actual use and usefulness of scholarly output? Who uses these new online tools, and how much? Are the user communities of Mendeley and CiteULike representative of the scholarly community, or are they biased toward young, Web 2.0-savvy researchers and students? Even if there is no demographic bias, is the user community large enough to be representative of global readership?

Surveys show that Web 2.0 tools are not very popular among scholars. In 2007, less than 10% of surveyed high-energy and condensed matter physicists had tried social bookmarking services, and only 1% thought that they were useful (Ginsparg, 2007). In a survey of 3,040 academics by Ware and Monkman (2007), overall 7% used social bookmarking. Usage was higher among young academics (10% of respondents younger than 36 used social bookmarking), physicists and engineers (9%), and participants from Asia (14%). These and other studies (see Bar-Ilan and colleagues, chapter 16, this volume) show that academics are not necessarily at the leading edge of social media and that more time is needed for researchers to adopt these systems. In due course, social bookmarking and reference managing systems may become a reliable source of global usage data of the scholarly literature (Li et al., 2012).

Although social bookmarking is still in its infancy, developments at Mendeley show that this can change quickly. Similar to the overnight success of arXiv in the physics community, the number of user documents in Mendeley (i.e., number of entries created by users) increased sixfold from 44 to 281 million in less than two years (from October 2010 to August 2012),[7] and case studies report extremely high coverage of scholarly journals (see Bar-Ilan and colleagues, chapter 16, this volume). Hence, Mendeley may be the most promising new source for evaluation purposes because it has the largest user population, greatest coverage, highest number of readers per document, and strongest correlations between usage and citation counts (Li et al., 2012). Although the coverage on CiteULike is lower, it can be considered a useful source for social tagging (Haustein, Peters, & Terliesner, 2011; Bar-Ilan et al., 2012). The possibilities of bookmarks and tags as alternative sources for monitoring scholarly impact are described next. However, more qualitative user research is needed in order to determine the true relevance of Web 2.0 reference management in academia and confirm the value of user counts for research evaluation.

Social Bookmarking as Global Readership Data

Using social bookmarks or bibliographic entries stored in online reference managers as an alternative data source to downloads raises the question as to whether these data actually represent the readership of scholarly documents. Bookmarking data may be a better indicator of usage than download statistics, due to the intentionality and social signalling of bookmarking. This is especially true if users also assign tags. The advantage of bookmarking data is that, as with citations and downloads, usage data are generated as a byproduct of existing workflows. Unlike tweeting, for example, searching for and managing literature is an established part of the scholarly communication process (Taraborelli, 2008; Neylon & Wu, 2009; Priem & Hemminger, 2010; Borrego & Fry, 2012; Li et al., 2012). Taraborelli argues:

> Social bookmarking data are likely to provide more robust indicators than usage factors insofar as they result from the intentional behavior of users interested in marking an item for future use

rather than from pure navigation patterns. Bookmarking an item is a much more relevant (and virtually more spam-resistant) kind of action to estimate user interest than merely following a link. In this sense, social bookmarking systems are likely to provide accurate figures on papers that are frequently read and cited in a given area of science. (Taraborelli, 2008, p. 104)

An extensive deep log analysis by Nicholas et al. (2008) revealed that two-thirds of all article views actually last less than three minutes, indicating that a considerable amount of full-text access is cursory. In contrast, bookmarking a document may be more intentional, especially if users make the effort to annotate documents with tags. Given a representative community of users, the number of bookmarks can thus be used as an indicator of a document's popularity, one that might be more reliable than download statistics. One of the major advantages is that bookmarking data are available on a global level and are independent of publishers; thus, in contrast to citations, the influence on the total readership is measured (Taraborelli, 2008; Haustein & Siebenlist, 2011; Borrego & Fry, 2012).

Bookmarking data provide information about which articles were used, how often, and by how many users. Article-based information allows for adequately normalized global usage statistics that can be computed in similar fashion to their citation-based equivalents. In contrast to downloading, the usage by one user is only counted once. Hence, in addition to the amount of usage, the data also reveal the number of unique users (i.e., impact across the members of the community). Journal impact can thus be determined in terms of diffusion. The information relating to whether the periodical is viewed by a broad or narrow audience is important for authors selecting a publication venue and for editors and publishers interested in monitoring journal perception (Haustein & Siebenlist, 2011).

The possibility of manipulating bookmarking counts exists, but since users need a login and a document can only be bookmarked once per user account, it is less susceptible to gaming than are download statistics (Taraborelli, 2008). Borrego and Fry (2012) found that the majority (78%) of bookmarks in BibSonomy were created by only 14 users; it was suspected that these users were managers of digital libraries keen to enhance usage, given that the entries were created within a few days. Such artificially generated usage could be easily removed.

A more serious problem is the incompleteness and errors found in the metadata of bibliographic entries in online reference managers. This often causes an article bookmarked by more than one user not to be recognized as one and the same publication. Thus, not only is the social aspect of the bookmarking service lost (i.e., similar content or users cannot be identified), but it becomes difficult to retrieve bookmarks and match usage to documents. Haustein and Siebenlist (2011) showed that it was best to apply a search strategy based on different metadata fields to retrieve bookmarks on CiteULike, Connotea, and BibSonomy. Similarly, Bar-Ilan and colleagues (2012) and Bar-Ilan (2012) showed that the Mendeley Application Program Interface (API) returned only one record, which could result in the loss of a significant number of users in the case of multiple entries. Thirty-three percent of the records retrieved from Mendeley in the case study by Bar-Ilan and colleagues (2012) did not contain a document object identifier (DOI) and would be missed by the API altogether. The retrieval of usage data from these platforms currently involves a certain amount of data cleaning.

Social Tagging as Crowdsourced Subject Indexing

Most academic social bookmarking systems allow users to annotate the documents in their collections with keywords (i.e., tags). Tags can either be taken from other users or chosen on the fly by the user himself or herself without adhering to any indexing rules. The concept of crowdsourced annotations of resources is referred to as social tagging. The constellation of tags assigned to all resources by all users of a platform is called a folksonomy (Vander Wal, 2005; Peters, 2009). All the tags assigned by one user are called a personomy, and all of the tags assigned to a document a docsonomy. The docsonomy constitutes a new user perspective on a publication—that is, social tagging generates an additional layer of metadata, one that reflects the readers' points of view (Peters, Haustein, & Terliesner, 2011; Haustein & Siebenlist, 2011; Haustein et al., 2011). Tags and tag frequencies can offer direct channels to readers' opinions. Due to the uncontrolled nature of tags, meaningful frequency distributions require a certain amount of term unification. Taraborelli (2008) recommends tag density (i.e., the number of tags per document) as a possible measure of semantic relevance.

Several studies have compared social tags with traditional index terms, including Medical Subject Headings (MeSH) (Lin, Beaudoin, Bul, & Desai, 2006; Kipp, 2011; Lee & Schleyer, 2012), Library of Congress Subject Headings (LCSH) (Yi & Chan, 2009; Lu, Park, & Hu, 2010), and terms from document titles and abstracts. Peters et al. (2011) and Haustein et al. (2011) compared tags assigned to physics publications on CiteULike, Connotea, and BibSonomy with title and abstract terms, author keywords, INSPEC subject headings, and automatic index terms from WoS. To improve accuracy and term matching, terms were compared at the level of single documents instead of the whole folksonomy, as proposed by Noll and Meinel (2007); terms were also extensively preprocessed (spelling unifications and stemming) in order to obtain a more homogeneous collection. Similarity values between tags and terms were low on average. The greatest overlap was found between tags and title terms, while tags and professionally assigned vocabulary differed the most. Lin et al. (2006) suspected that this was due to

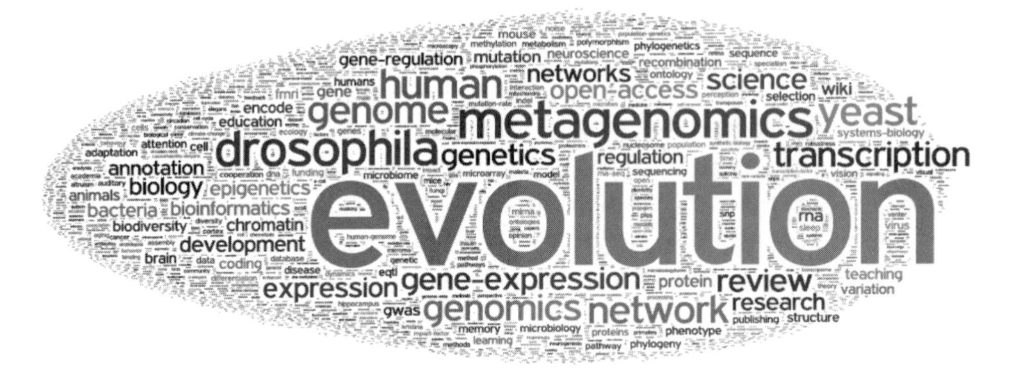

Figure 17.2

Tag cloud representing 9,575 tag applications by users of CiteULike assigned to *PLoS Biology* documents published between 2004 and 2007. Overall, 842 unique publications were bookmarked 3,634 times, which equals a coverage of 67.0% and an average number of 4.3 users per document.

the different aims of traditional indexing and social tagging. While professional indexers try to represent the content of the whole document with controlled terms, users seem to "highlight specific content or facts most interesting to them" (Lin et al., 2006, p. 6). Thus, social tags provide the users' perspectives on article content. Aggregated at the journal level, tag frequencies (e.g., visualized in tag clouds) can be used to represent a reader-specific view on journal content. Figure 17.2 shows the tag cloud for 842 documents—67% of 1,256 documents had at least one user in CiteULike—published in *PLoS Biology* from 2007 to 2011. If tags are analyzed according to the date of the tag application or publication of the document, they can reveal thematic trends or help discover trending topics (Haustein, 2012).

Conclusions

Including the readers' perspective in journal evaluation is crucial when scholarly influence on the entire community is to be measured. Citation analysis, which has become a substitute for time-consuming, costly, and biased local-usage studies, reflects only the impact of a document on those readers who are authors themselves and have correctly indicated its influence by referencing it. With the shift from print to electronic publishing, large-scale usage studies have become technically feasible. Stored in log-files on publishers' servers, usage data at the article level can be monitored, providing information about which content is used when, where, and how often. Such article-based download data make it possible to compute journal usage indicators similar to citation-based indicators, reflecting impact in a broader sense, such as influence on technical applications, education, health policies, and other social issues. However, in practice there are several problems associated with download data, the greatest being that of availability. With the exception of a few open-access journals, global statistics are not available, because publishers fear that they are too commercially sensitive to share with the public. Local data, which are provided to subscribing institutions in the form of COUNTER reports, are biased and, since provided at the journal and not the article level, insufficiently detailed to allow for the necessary normalization. With the introduction of online social reference managers and bookmarking services for scholarly literature, alternative sources of journal usage are now available. The advantage of bookmarking data is that they are created as a byproduct of existing workflows, namely, reading and managing scholarly literature. Article bookmarks capture influence more broadly than citations and, in contrast to downloads, are assumed to be more intentional and harder to manipulate.

The main problem with altmetrics is the representativeness of data in terms of who is using the resource and how much usage is covered. More qualitative research is needed to explore to what extent the user population represents the wider scholarly community, as well as to find out for which purposes bookmarking systems are used. If, in the future, researchers move the reference management function from their local hard drives (or even from folders on their shelves) to the social Web, bookmarking data can be used to monitor article-based journal usage, and tags can provide the readers' perspectives on content.

The rapid growth of the social Web will inevitably lead to a greater availability of data for representing the readership base of the scholarly literature. It is important that these data are captured and reflected in journal and research assessment exercises to provide a holistic view of the actual impact of scholarship beyond the confines of the citing readership. This will also allow scholars—and those who fund, promote, and assess them—to examine the wider impact of their work on scholars and the public alike.

Notes

1. http://www.projectcounter.org

2. http://top25.sciencedirect.com

3. In terms of usage of scholarly documents, no distinction is made between storing bibliographic metadata in an online reference manager or bookmarking it to a social bookmarking platform. Hence, scholarly bookmarking services and Web 2.0 reference managers are regarded as the same.

4. Connotea discontinued its service in March 2013.

5. Number of bookmarks ("user documents") and users were retrieved from http://www.mendeley.com on August 28, 2012.

6. Number of documents in SCI-E, SSCI, A&HCI, CPCI-S were retrieved from http://apps.webofknowledge.com on August 28, 2012.

7. Number of bookmarks (user documents) and users were retrieved from http://www.mendeley.com on August 28, 2012.

References

Baker, G., & Read, E. J. (2008). Vendor supplied usage data for electronic resources: A survey of academic libraries. *Learned Publishing, 21*(1), 48–57.

Bar-Ilan, J. (2012). *JASIST@mendeley*. Presented at ACM Web Science Conference 2012 Workshop. Retrieved from http://altmetrics.org/altmetrics12/bar-ilan.

Bar-Ilan, J., Haustein, S., Peters, I., Priem, J., Shema, H., & Terliesner, J. (2012). Beyond citations: Scholars' visibility on the social Web. In *Proceedings of the 17th International Conference on Science and Technology Indicators* (Vol. 1, pp. 98–109). Montreal, Canada.

Bollen, J., & Van de Sompel, H. (2008). Usage impact factor: The effects of sample characteristics on usage-based impact metrics. *Journal of the American Society for Information Science and Technology, 59*(1), 136–149.

Bollen, J., Van de Sompel, H., & Rodriguez, M. A. (2008). Towards usage-based impact metrics: First results from the MESUR project. In *Proceedings of the 8th ACM/IEEE-CS Joint Conference on Digital Libraries* (pp. 231–240). New York: ACM.

Borrego, A., & Fry, J. (2012). Measuring researchers' use of scholarly information through social bookmarking data: A case study of BibSonomy. *Journal of Information Science, 38*(3), 297–308.

Brody, T., Harnad, S., & Carr, L. (2006). Earlier web usage statistics as predictors of later citation impact. *Journal of the American Society for Information Science and Technology, 57*(8), 1060–1072.

Bustion, M., Eltinge, J., & Harer, J. (1992). On the merits of direct observation of periodical usage—an empirical study. *College & Research Libraries, 53*(6), 537–550.

Butkovich, N. J. (1996). Use studies: A selective review. *Library Resources & Technical Services, 40*(4), 359–368.

Cronin, B., Snyder, H. W., Rosenbaum, H., Martinson, A., & Callahan, E. (1998). Invoked on the Web. *Journal of the American Society for Information Science, 49*(14), 1319–1328.

Davis, P. M. (2004). For electronic journals, total downloads can predict number of users. *Portal-Libraries and the Academy, 4*(3), 379–392.

Duy, J., & Vaughan, L. (2006). Can electronic journal usage data replace citation data as a measure of journal use? An empirical examination. *Journal of Academic Librarianship, 32*(5), 512–517.

Emrani, E., Moradi-Salari, A., & Jamali, H. R. (2010). Usage data, e-journal selection, and negotiations: An Iranian consortium experience. *Serials Review, 36*(2), 86–92.

Franck, G. (1999). Scientific communication—a vanity fair? *Science, 286*(5437), 53–55.

Ganegan, F. (2012, August). Filtering the research record and farming big data. Retrieved from http://www.swets.com/blog/filtering-the-research-record-and-farming-big-data#.

Garfield, E. (1955). Citation indexes for science: A new dimension in documentation through association of ideas. *Science, 122*(3159), 108–111.

Garfield, E. (1972). Citations as a tool in journal evaluation: Journals can be ranked by frequency and impact of citations for science policy studies. *Science, 178*(4060), 471–479.

Ginsparg, P. (2007). Next-generation implications of Open Access. *CTWatch Quarterly, 2*(3). Retrieved from http://www.ctwatch.org/quarterly/articles/2007/08/next-generation-implications-of-open-access.

Glänzel, W., & Moed, H. F. (2002). Journal impact measures in bibliometric research. *Scientometrics, 53*(2), 171–193.

Gorraiz, J., & Gumpenberger, C. (2010). Going beyond citations: SERUM—a new tool provided by a network of libraries. *Library Quarterly, 20*(1), 80–93.

Gross, P. L. K., & Gross, E. M. (1927). College libraries and chemical education. *Science, 66*(1713), 385–389.

Hahn, K., & Faulkner, L. (2002). Evaluative usage-based metrics for the selection of e-journals. *College & Research Libraries, 63*(3), 215–227.

Hammond, T., Hannay, T., Lund, B., & Scott, J. (2005). Social bookmarking tools (I). *D-Lib Magazine, 11*(4). Retrieved from http://www.dlib.org/dlib/april05/hammond/04hammond.html.

Haustein, S. (2011). Taking a multidimensional approach toward journal evaluation. In *Proceedings of the 13th International Conference of the International Society for Scientometrics and Informetrics* (Vol. 1, pp. 280–291). Durban, South Africa.

Haustein, S. (2012). *Multidimensional journal evaluation: Analyzing scientific periodicals beyond the impact factor*. Berlin: De Gruyter Saur.

Haustein, S., Peters, I., & Terliesner, J. (2011). Evaluation of reader perception by using tags from social bookmarking systems. In *Proceedings of the 13th International Conference of the International Society for Scientometrics and Informetrics* (Vol. 2, pp. 999–1001). Durban, South Africa.

Haustein, S., & Siebenlist, T. (2011). Applying social bookmarking data to evaluate journal usage. *Journal of Informetrics, 5*(3), 446–457.

Jamali, H. R., & Nikzad, M. (2011). Article title type and its relation with the number of downloads and citations. *Scientometrics, 88*(2), 653–661.

Kipp, M. E. I. (2011). Tagging of biomedical articles on CiteULike: A comparison of user, author and professional indexing. *Knowledge Organization, 38*(3), 245–261.

Kohl, D. F., & Davis, C. H. (1985). Ratings of journals by ARL library directors and deans of library and information science schools. *College & Research Libraries, 46*(1), 40–47.

Langlois, D. C., & Von Schulz, J. V. (1973). Journal usage survey—method and applications. *Special Libraries, 64*(5–6), 239–244.

Lee, D. H., & Schleyer, T. (2012). Social tagging is no substitute for controlled indexing: A comparison of Medical Subject Headings and CiteULike tags assigned to 231,388 papers. *Journal of the American Society for Information Science and Technology, 63*(9), 1747–1758.

Li, X., Thelwall, M., & Giustini, D. (2012). Validating online reference managers for scholarly impact measurement. *Scientometrics, 91*(2), 461–471.

Lin, X., Beaudoin, J., Bul, Y., & Desai, K. (2006). Exploring characteristics of social classification. In *Proceedings of the 17th Annual ASIS&T SIG/CR Classification Research Workshop*. Austin, TX.

Liu, X., Fang, H., & Wang, M. (2011). Correlation between download and citation and download-citation deviation phenomenon for some papers in Chinese medical journals. *Serials Review, 37*(3), 157–161.

Lu, C., Park, J.-R., & Hu, X. (2010). User tags versus expert-assigned subject terms: A comparison of LibraryThing tags and Library of Congress Subject Headings. *Journal of Information Science, 36*(6), 763–779.

MacRoberts, M. H., & MacRoberts, B. R. (1989). Problems of citation analysis—a critical review. *Journal of the American Society for Information Science, 40*(5), 342–349.

Moed, H. F. (2005). *Citation analysis in research evaluation*. Dordrecht: Springer.

Moed, H. F., Van Leeuwen, T. N., & Reedijk, J. (1998). A new classification system to describe the ageing of scientific journals and their impact factors. *Journal of Documentation, 54*(4), 387–419.

Neylon, C., & Wu, S. (2009). Article-level metrics and the evolution of scientific impact. *PLoS Biology, 7*(11), e1000242.

Nicholas, D., Huntington, P., Jamali, H. R., Rowlands, I., Dobrowolski, T., & Tenopir, C. (2008). Viewing and reading behaviour in a virtual environment—the full-text download and what can be read into it. *Aslib Proceedings, 60*(3), 185–198.

Noll, M. G., & Meinel, C. (2007). Authors vs. readers: A comparative study of document metadata and content in the WWW. In *Proceedings of the 2007 ACM Symposium on Document Engineering* (pp. 177–186). Winnipeg, Canada.

Perneger, T. V. (2004). Relation between online "hit counts" and subsequent citations: Prospective study of research papers in the BMJ. *British Medical Journal, 329*(7465), 546–547.

Peters, I. (2009). *Folksonomies: Indexing and retrieval in Web 2.0*. Berlin: De Gruyter Saur.

Peters, I., Haustein, S., & Terliesner, J. (2011). Crowdsourcing in article evaluation. In *Proceedings of the 3rd International Conference on Web Science—ACM Web Science Conference*. New York: ACM.

Price, D. J. S., & Gürsey, S. (1976). Studies in Scientometrics I. Transience and continuance in scientific authorship. *International Forum on Information and Documentation, 1*(2), 17–24.

Priem, J., & Hemminger, B. (2010). Scientometrics 2.0: Toward new metrics of scholarly impact on the social Web. *First Monday, 15*(7). Retrieved from http://firstmonday.org/htbin/cgiwrap/bin/ojs/index.php/fm/article/viewArticle/2874/2570.

Priem, J., Taraborelli, D., Groth, P., & Neylon, C. (2010, October). Altmetrics: A manifesto. Retrieved from http://altmetrics.org/manifesto.

Public Library of Science (PLoS). (2012). Article level metrics. Retrieved from http://article-level-metrics.plos.org/alm-info.

Reher, S., & Haustein, S. (2010). Social bookmarking in STM: Putting services to the acid test. *ONLINE: Exploring Technology & Resources for Information Professionals, 34*(6), 34–42.

Research Councils UK. (2011, March). Types of impact. Retrieved from http://www.rcuk.ac.uk/documents/impacts/TypologyofResearchImpacts.pdf.

Rousseau, R. (2002). Journal evaluation: Technical and practical issues. *Library Trends, 50*(3), 418–439.

Rowlands, I., & Nicholas, D. (2005, September). *New journal publishing models: An international survey of senior researchers*. Newbury, UK: Publishers Association and International Association of STM Publishers.

Rowlands, I., & Nicholas, D. (2007). The missing link: Journal usage metrics. *Aslib Proceedings, 59*(3), 222–228.

Scanlan, B. D. (1987). Coverage by Current Contents and the validity of impact factors—ISI from a journal publisher's perspectives. *Serials Librarian, 13*(2–3), 57–66.

Schlögl, C., & Gorraiz, J. (2006). Document delivery as a source for bibliometric analyses: The case of Subito. *Journal of Information Science, 32*(3), 223–237.

Schlögl, C., & Gorraiz, J. (2010). Comparison of citation and usage indicators: The case of oncology journals. *Scientometrics, 82*, 567–580.

Schlögl, C., & Gorraiz, J. (2011). Global usage versus global citation metrics: The case of pharmacology journals. *Journal of the American Society for Information Science and Technology, 62*(1), 161–170.

Schlögl, C., & Stock, W. G. (2004). Impact and relevance of LIS journals: A scientometric analysis of international and German-language LIS journals—citation analysis versus reader survey. *Journal of the American Society for Information Science and Technology, 55*(13), 1155–1168.

Stock, W. G. (2009). The inflation of impact factors of scientific journals. *ChemPhysChem, 10*(13), 2193–2196.

Taraborelli, D. (2008). Soft peer review: Social software and distributed scientific evaluation. In *Proceedings of the 8th International Conference on the Design of Cooperative Systems* (pp. 99–110). Carry-le-Rouet, France.

Tenopir, C., & King, D. W. (2000). *Towards electronic journals: Realities for scientists, librarians, and publishers*. Washington, DC: Special Libraries Association.

Thelwall, M. (2012). Journal impact evaluation: A webometric perspective. *Scientometrics, 92*(2), 429–441.

Vander Wal, T. (2005, January). Folksonomy explanations. Retrieved from http://www.vanderwal.net/random/entrysel.php?blog=1622.

Wan, J. K., Hua, P. H., Rousseau, R., & Sun, X. K. (2010). The journal download immediacy index (DII): Experiences using a Chinese full-text database. *Scientometrics, 82*(3), 555–566.

Ware, M., & Monkman, M. (2007). *Peer review in scholarly journals: Perspective of the scholarly community—an international study*. Publishing Research Consortium. Retrieved from http://www.publishingresearch.net/documents/PeerReviewFullPRCReport-final.pdf.

Wilhite, A. W., & Fong, E. A. (2012). Coercive citation in academic publishing. *Science, 335*(6068), 542–543.

Yan, K.-K., & Gerstein, M. (2011). The spread of scientific information: Insights from the web usage statistics in PLoS article-level metrics. *PLoS ONE, 6*(5), e19917.

Yi, K., & Chan, L. M. (2009). Linking folksonomy to Library of Congress Subject Headings: An exploratory study. *Journal of Documentation, 65*(6), 872–900.

Yu, G., & Wang, L. (2007). The self-cited rate of scientific journals and the manipulation of their impact factors. *Scientometrics, 73*(3), 321–330.

18 Evaluating the Work of Judges

Peter A. Hook

Introduction

Interest in metrics is by no means restricted to the world of academic research and scholarship. Indeed, modern bibliometrics owes much to early developments in legal citation indexing. Courts as institutions and the judges that comprise courts have a tremendous impact on the lives of their respective citizens. This may be seen in rulings that prescribe individual, corporate, and government behavior. Courts influence one another at either the state, federal, or international level. Additionally, individual judges,[1] litigants, and scholars have a measureable impact on other judges and multimember courts as a whole. This chapter reviews the metrics, empirical observations, and computational techniques used to measure the impact of one court on another and one jurist on another, as well as the general level of agreement of jurists within multimember courts from year to year, from term to term, or when the court experiences turnover in membership. It is intended to build on Shapiro's (1992)[2] initial effort to expose the scientometric community to metric work, both conventional and alternative, being done by legal academics and political scientists. The studies surveyed, as well as the original contributions made, are built on a long history of explicit, systematized citation and indicia of opinion agreement in legal systems.

Precedent-Based Legal Systems and the Rise of Citation and Topical Infrastructures

The common-law legal system, developed in the British Isles and exported to its colonies, relies heavily on case precedent (Von Nessen, 1992).[3] Opinions issued by courts have binding effects on lower courts that are subject to their authority ("vertical precedent"; Tiersma, 2010, p. 185). Those same opinions have persuasive value in other courts of the same level as well as courts in other jurisdictions. Furthermore, courts are generally constrained by their own prior decisions and afford them great deference prior to overruling them ("stare decisis or horizontal precedent"; Tiersma, 2010, p. 185 [emphasis removed]).

Accordingly, it is imperative that lawyers, judges, and citizens are able to find and use cases that address their particular legal issues and factual scenarios. Prior to the standardization of case reporting and publishing (see Tiersma, 2007), the first tools to assist attorneys in finding relevant cases were "abridgments" (Fitzherbert, 1565), "institutes" (Coke, 1628), and "commentaries" (Blackstone, 2001). These are works that collocate, describe, and organize cases based on either legal or factual themes. Cooper (1982), Shapiro (1992), and Ogden (1993) survey other early tools from which lawyers learned of the existence of relevant case law.

Debt of Gratitude to Frank Shepard's 1875 Legal Case Citator

Scientometrics owes a debt of gratitude to the citation infrastructure of the United States (U.S.) legal system. While it is disputed as to which year Frank Shepard created his citation index for Illinois cases (*Illinois Annotations*)—1873 (Dabney, 2008) or 1875 (Ogden, 1993; Surrency, 1990)—it is not disputed that the company Shepard founded subsequently offered case and statute citators for all U.S. jurisdictions. While not the first legal citators, Shepard's citators became a staple of legal practice and the term *Shepardize* became commonplace in the field. Initially, citation references to subsequent cases were published on gummed labels and affixed in the margins of the case reporter adjacent to the beginning of the case the citations updated. In this way, the citation references worked as manual hyperlinks. Scientometric pioneer Eugene Garfield (1961) is explicit in his acknowledgment that the idea for the *Science Citation Index* and subsequent indexes for the other areas of knowledge was derived from Shepard's legal citators (Garfield, 1955; 1979, pp. 6–18). Garfield's works greatly accelerated the development of scientometrics and other areas of information science—for example, through the intermediary scholarship of Kleinberg (1998, 1999), Garfield's work influenced Page and Brin's *PangeRank* algorithm (Brin & Page, 1998), the successful relevance ranking algorithm at the heart of the Google web search engine (Battelle, 2005; Hopkins, 2005).

Atomistic Specificity of the West Topic and Key Number System

Since 1897, West Publishing has produced an index and abstract service for U.S. case law (*Century Edition of the American Digest*, 1897) that seeks to identify all discreet holdings in cases, excerpt and abstract them, and assign them topics from West's taxonomy (Doyle, 1992; Hanson, 2002; Snyder, 1999; Thomson/West, 2011). The practice continues to this day, and in this manner legal cases are atomistically indexed to a greater extent than most other disciplines or literatures. Furthermore, these same abstracts are published as headnotes and appear as editorial front matter immediately preceding the printed case language in the National Reporter System (Surrency, 1990). Thus, an attorney can easily find specific points of law in the language of a case, because the material covered by each West headnote is indicated in the language of the case. Several scholars have done analysis using the rich subject matter infrastructure that is the West Topic and Key Number System (e.g., Ho & Quinn, 2010; Hook, 2007b, 2007c).

Citation Coding Treatments (Relationship Indicators)

Lipetz (1965) proposed 29 relationship indicators to augment the usefulness of science literature citators. He explicitly acknowledged that Shepard's legal citation indexes had already employed such relationship indicators for generations. As early as 1903, Shepard's citators included "a system of letters at the left of the volume number [that] showed whether the case had been affirmed, criticized, distinguished, explained, followed, harmonized, limited, modified, or overruled" (Ogden, 1993, p. 34). Furthermore, Shepard's citators subsequently included references to the subset of content in the initial case being cited. This was done by referencing the West headnote number. Thus, if a case had multiple holdings, an attorney only had to investigate citations to cases that discussed the holding covered by his or her specific headnote(s) of interest. These two features, (1) relationship indicators, and (2) specific content indicators, greatly enhanced the utility and efficiency of legal citators and predated innovations called for by Lipetz (1965), Frost (1979), and Duncan, Anderson, and McAleese (1981).[4]

In 1997, after Shepard's partnered with (and was subsequently bought out by) LexisNexis (a large legal database company and publisher), West Publishing (the other large legal database purveyor in the United States) introduced a competing legal citator known as KeyCite (Dabney, 2008). One of the innovations of the new citator was depth-of-treatment stars. Employing a scale of one to four stars, KeyCite informed its users how extensively their original case was discussed in subsequent cases. At least one researcher has used the depth-of-treatment indictors when producing citation maps of a particular topic (Hook, 2007b). Furthermore, the depth-of-treatment indicators were in addition to the other legal relationship indicators, long employed by Shepard's, and were subsequently simplified on both online platforms. Shepard's (LexisNexis) and KeyCite (West Publishing) both started using red, yellow, and green warning symbols (a traffic-light metaphor) to inform attorneys of subsequent case and statute treatment. In 2005, West also introduced Graphical KeyCite (Gordon, 2005, p. 6), which visually portrayed how a case worked its way through a multilevel appellate system and the resultant documents (with citations) along each step of the process. Lexis recently introduced similar functionality in its *LexisAdvance* online search platform.

Studies Evaluating the Influence of Courts on One Another

While on a structural level, some courts are constitutionally obligated to follow the precedents of their superior courts, what is more interesting is the comparative impact of one court on another in a voluntary context. The common-law system of justice, the interpretation of statutes or codes, and the establishment of international norms necessarily require an evolving treatment of past legal issues as well as new responses to novel legal issues and fact patterns. Accordingly, legal memes percolate among different courts in the same political subdivision, across a particular country, as well as between countries and the various international courts. Empirical studies are made possible by the fact that judicial opinions are almost always published, widely disseminated, and frequently available electronically in full-text databases. Citation conventions that standardize the way a particular court's output is referenced facilitate counts of the use of judicial opinions of one court by another.

Influence of Courts—International

Numerous studies have tracked the influence of courts in one country on those in other countries. Aft (2011) surveyed empirical works that study the influence of the U.S. Supreme Court on foreign and international courts, and responded to criticism that the Court's influence is waning. Courts that have similar historical backgrounds (i.e., those in the same jurisprudential family) or are in close geographic proximity are more likely to cite one another. There have been several empirical studies (with metrics) of the use of American case law in Canadian courts (Bushnell, 1986; Liptak, 2008; MacIntyre, 1966; Manfredi, 1990; McCormick, 1997, 2009). Other studies exploring the use of U.S. cases in foreign jurisdictions include Australia (Smyth, 2008b; Von Nessen, 1992, 2006), Israel (Gorney, 1955), and the European Court of Human Rights (De Wolf & Wallace, 2009).

There also are studies that analyze the use of foreign precedent in the United States. Shapiro (1992) identified what are perhaps the first two metric works that analyze the frequency of U.S. courts citing other courts, both U.S. and British (Committee on Law Reporting, 1895; Committee on Library and Legal Literature, 1895). In light of recent debate about the appropriateness of the U.S. Supreme Court citing foreign and international precedent, Calabresi and Zimdahl (2005) surveyed the practice over 200 years of the Court's history and enumerated instances in which the Court was more or less likely to cite international precedents. Zaring (2006) expanded the analysis of the use of foreign cases to include all levels of the U.S. federal judiciary. There are also studies that examine the influence of international courts (International Court of Justice, European Court of Human Rights, etc.) on other international courts and the courts of individual countries (Lupu & Voeten, 2010; Miller, 2002; Voeten, 2010). Additionally, there have been empirical studies and essays examining the international use of cases that do not focus on the United States: Australia (Smyth, 2002; Topperwien, 2002), Canada (Roy, 2004; Smithey, 2001), New Zealand (Mathieson, 1963), and South Africa (Smithey, 2001).

Influence of Courts—Intranational

There are also empirical studies that evaluate the influence of courts within the same country. These studies are particularly interesting in large federal systems such as the United States, in which there are numerous geographic subentities functioning as jurisprudential "laboratories" (see Brandeis's dissent in "New State Ice Co. v. Liebmann," 1932). Friedman, Kagan, Cartwright, and Wheeler (1981) conducted a large empirical study of the influence of U.S. state courts on other state courts; the study also included a longitudinal analysis of the average length of state Supreme Court opinions. Hanson (2002) conducted a 25-year analysis (1975–2000) of the opinions of four state Supreme Courts and their rates of citations to out-of-state cases to test the hypothesis that keyword searching (as opposed to the West Topic and Key Number System) has allowed for more discovery of relevant cases. His results were inconclusive. Black and Spriggs conducted a study of the depreciation of the frequency with which U.S. Supreme Court opinions are cited by subsequent U.S. Supreme Court opinions and opinions of the federal courts of appeal (2013). Other U.S. state intercourt citation studies include Blumberg (1998), Caldeira (1983, 1985, 1988), Friedman et al. (1981), Harris (1982, 1985), Mott (1936), and Nagal (1962). Additionally, there is at least one study that identified the diffusion of specific ideas (tort concepts) across state courts (Canon & Baum, 1981). Scholars in Australia have also conducted intranational court citation analyses (Fausten, Nielsen, & Smyth, 2007; Smyth, 1999a, 1999b, 2007, 2008a, 2009b; Smyth & Fausten, 2008).

Influence of Specific Jurists

There have been studies that evaluated and ranked the number of times specific jurists are cited. These seek to empirically identify jurists who have had a significant impact on the development of the law and on their judicial colleagues. Cross and Spriggs

(2010) identified the most cited U.S. Supreme Court Justices (as well as opinions) at all levels of the federal judiciary. Kosma (1998) undertook a similar analysis. Metric studies have also been conducted to establish the most cited U.S. federal courts of appeal judges (Choi & Gulati, 2004; Klein & Morrisroe, 1999; Landes, Lessig, & Solimine, 1998). Similar work has been done to evaluate the justices of the Australian High Court (Smyth, 2000b).

What Do Courts Cite?

Practitioners appearing before courts, as well as academics, want to know the types of works particular courts cite in their opinions. Consequently, there have been studies that break down and quantify the different types of resources that courts cite. These include primary resources (cases, statutes, constitutions, administrative regulations) as well as secondary resources (treatises, law review articles, dictionaries, etc.).

Merryman's (1954, 1978) studies of the California Supreme Court are two of the first that analyzed the primary and secondary materials cited by a particular court. Newton (2012) examined the frequency with which law review articles were cited by the U.S. Supreme Court from 2001 to 2011, broken down by individual justices, scholars cited, and law reviews cited. A study by Marangola (1998) surveyed the frequency with which the Supreme Judicial Court of Massachusetts cited the Massachusetts State Constitution in the context of fundamental rights (as opposed to the Constitution of the United States). This concept, judicial federalism, was also explored by Bierman (1995) for the highest court in the state of New York. Posner and Sunstein (2006) wrote an essay noting the correlation between the relative age of the state and its frequency of out-of-state citations. Manz (2002) examined the relationship between authorities cited in advocacy materials (the briefs of the parties) and what the U.S. Supreme Court ultimately cited in its opinions. This work also contains an extensive bibliography of the secondary material cited by courts. Subsequent studies include analyses of the U.S. Supreme Court (Acker, 1990; Berring, 2000; Petherbridge and Schwartz, 2012) and courts in Australia (Smyth, 2000a, 2009a).

Case Citation Network Studies

A growing number of studies analyze the output of courts from the perspective of network analysis. These studies analyze the citation linkages (links) between court opinions (nodes) and make inferences based on network measuring techniques. One of the first network studies of courts was performed by Caldeira (1988), who used clustering techniques and multidimensional scaling to analyze the court case citation connectivity of all 50 U.S. state court systems and the District of Columbia for one calendar year (i.e., 1975). Fowler, Johnson, Spriggs, Jeon, and Wahlbeck (2007) used a large dataset of U.S. Supreme Court cases to distinguish between inwardly and outwardly directed important cases. They also demonstrated the correlation between cases of high network centrality and future citation frequency.

Cross, Smith, and Tomarchio (2006, 2008) analyzed a dataset of 48,000 substantive (nonprocedural) U.S. Supreme Court cases from 1937 to 2005. They studied network cohesion measures (average degree, density, diameter, and clustering) and how they varied over time. The authors hypothesized that greater network cohesion indicated greater reliance on precedent, and identified timespans in which cohesion measures increased and decreased. Cross and a different group of collaborators (Cross, Spriggs, Johnson, & Wahlbeck, 2010) conducted a study of citations in U.S. Supreme Court opinions in which they identified the network centrality of the case being cited at the time it was cited. Network centrality was seen as "how deeply embedded a Supreme Court opinion [was] within the network of all citations among Supreme Court opinions" (Cross et al., 2010, p. 524). Other law related citation network studies include Bommarito, Katz, and Zelner (2009), Boulet, Mazzega and Bourcier (2011), Chandler (2005, 2007), Katz et al. (2011), Katz and Stafford (2010), Lupu and Voeten (2010), Smith (2007), and Whalen (2013).

Measures of Agreement within Multimember Courts

Co-occurrence analysis has been applied to courts. Higher-level courts frequently consist of multiple jurists who each vote for or against a specific outcome. In many legal systems, multimember courts issue opinions that indicate the agreement of the individual jurists. The resultant covoting statistics allow for empirical studies of a court and its members in order to identify the ideological affinities of its members and possible swing justices. This is also known as bloc analysis—the quest "to identify subgroups of justices of a collegial court who consistently vote together" (Bradley & Ulmer, 1980, p. 257). Swing justices, also known as median justices, are justices who do not consistently side with either of two predominant voting blocs; swing justices frequently determine the outcome of cases (Martin, Quinn, & Epstein, 2005; Schultz & Howard, 1975). Judicial covoting statistics that allow for this type of analysis have appeared in the popular press (Greenhouse, 2005b; Hossain & Cox, 2006).

Judicial covoting analysis has been conducted for decades and has been surveyed by Hook (2007a). Beginning with the 1956 term, the *Harvard Law Review* has annually published covoting matrices for the nine justices of the U.S. Supreme Court ("Supreme Court, 1956 Term," 1957; "Supreme Court 2010 Term: The Statistics," 2011). The *Harvard Law Review*'s method for reporting the various levels of agreement among Supreme Court Justices (different permutations of joins, concurrences, dissents, and total number of cases heard together) has also been utilized to report the covoting of state supreme courts (Betz, 1992; Crandley, Stephenson, Kerridge, & Peabody, 2011). Additional studies not utilizing the *Harvard Law Review* method also provide co-voting data of state courts (Blackwell, 2009; Martin, 1996). Other sources break down Supreme Court co-voting by specific issue areas (Epstein, Segal, Spaeth, & Walker, 2012; Riggs, 1988; Schultz & Howard, 1975; Wilkins, Worthington, Reynolds, & Nielsen, 2005). Particularly helpful for scholars is the fact that 64 years of U.S. Supreme Court co-voting data (1946 to 2010 terms) is freely available in *The Supreme Court Database* (Spaeth et al., 2012). However, at least one scholar has criticized the topical coding of cases in the database (Shapiro, 2009). Other scholars have published an analysis of Supreme Court co-voting data from 1838 to 2009 (Chabot & Chabot, 2011).

Traditional statistical analysis of the U.S. Supreme Court focused on nonunanimous cases (Pritchett, 1948; Schultz & Howard, 1975), because only these reveal discord in the reasoning and ideology of the Court. However, missing from the analysis was a measure that consistently reported the overall agreement (or lack thereof) among the Justices of the Court. Hook (2007a) created the Aggregate Harmony Metric to capture the overall consensus on the U.S. Supreme Court for a particular term, including unanimous cases. This measure, a percentage, is the matrix sum of all the main opinions joined by any two justices on the Court for a particular term (*Harvard Law Review* O Method), divided by the matrix sum of the overall number of cases those same justices heard together for a particular year (*Harvard Law Review* N Method). Updates as to the most recent five years of the Aggregate Harmony Metric are included in tables 18.1 and 18.2. Schultz and Howard (1975) reported two similar measures of agreement—the percentage of unanimous decisions and average dissenting votes per decision, in five-year intervals from 1930 to 1965 and yearly from 1968 to 1973.

Visual Representations of the Work of Courts

Early efforts to spatially visualize the relationship of U.S. Supreme Court justices to one another based on their covoting behavior were discussed in White (2005). Perhaps the first of these efforts was Pritchett's (1941) linear distribution of the justices. Thurstone and Degan (1951) used factorial analysis to produce three-dimensional vector-space representations of the justices' co-voting patterns in the 1943 and 1944 terms. Schubert

Table 18.1
Aggregate Harmony Metric (ΣO/ΣN) (Hook, 2007a, Table 19.1A)

Term	Natural Court	Aggregate Percentage Agreement (O Method)	Cumulative O Count	Cumulative N Count
2006	Roberts 2	**56**	1443	2573
2007	Roberts 2	**58**	1428	2473
2008	Roberts 2	**54**	1514	2808
2009	Roberts 3	**56**	1709	3052
2010	Roberts 4	**61**	1661	2712

Table 18.2
Superlatives (highest and lowest percentage agreement per term) (O Method) (Hook, 2007a, Table 19.1C)

Term	Highest percentage agreement (O Method)	Highest percentage agreement Justice 1	Highest percentage agreement Justice 2	Lowest percentage agreement (O Method)	Lowest percentage agreement Justice 1	Lowest percentage agreement Justice 2
2006	**79**	Roberts	Alito	**35**	Stevens	Thomas
2007	**75**	Kennedy Roberts	Roberts Alito	**36**	Stevens	Thomas
2008	**74**	Kennedy	Roberts	**35**	Stevens	Alito
2009	**77**	Kennedy	Roberts	**34**	Stevens Stevens	Scalia Thomas
2010	**84**	Kennedy	Roberts	**44**	Thomas	Ginsburg

also used factorial analysis to produce spatial distributions of the justices (Schubert, 1962, 1963). Subsequently, many other scholars have used covoting data to produce spatial representations of the voting relationships among the justices. Spaeth and Altfeld (1985) produced nonautomated diagrams of the influence relationships among the justices. Other scholars have used Markov chain Monte Carlo methods with a Bayesian measurement model to produce visual representations of the Court (Epstein, Knight, & Martin, 2003; Epstein, Martin, Quinn, & Segal, 2007; Martin & Quinn, 2002; Martin et al., 2005). Still others used statistical scaling techniques to produce visualizations of the voting patterns of courts of appeal (Epstein, Martin, Segal, & Westerland, 2007). Johnson, Borgatti, and Romney (2005) used network visualization techniques, and Sirovich (2003) used vector models and singular value decomposition. Hook (2007a) employed multidimensional scaling and network layout techniques.

The utility of covoting visualizations stems from their effective summarizing of large datasets and capability of succinctly conveying information to the viewer. Greenhouse (2005a) described how two childhood friends ended up on the U.S. Supreme Court together—Burger and Blackmun. In their first term together (1970), the two justices agreed 78% of the time. This was the highest percentage for that particular term. In 1985, the last year the two served on the Court together, Blackmun had moved far to the liberal end of the spectrum and he and his childhood friend agreed only 48% of the time. Multidimensional scaling representations of these two terms clearly illustrate the change (figures 18.1 and 18.2).

Conclusion

There is a large literature employing metrics, empirical observations, and computational techniques that evaluates the work of judges. The scientometric community

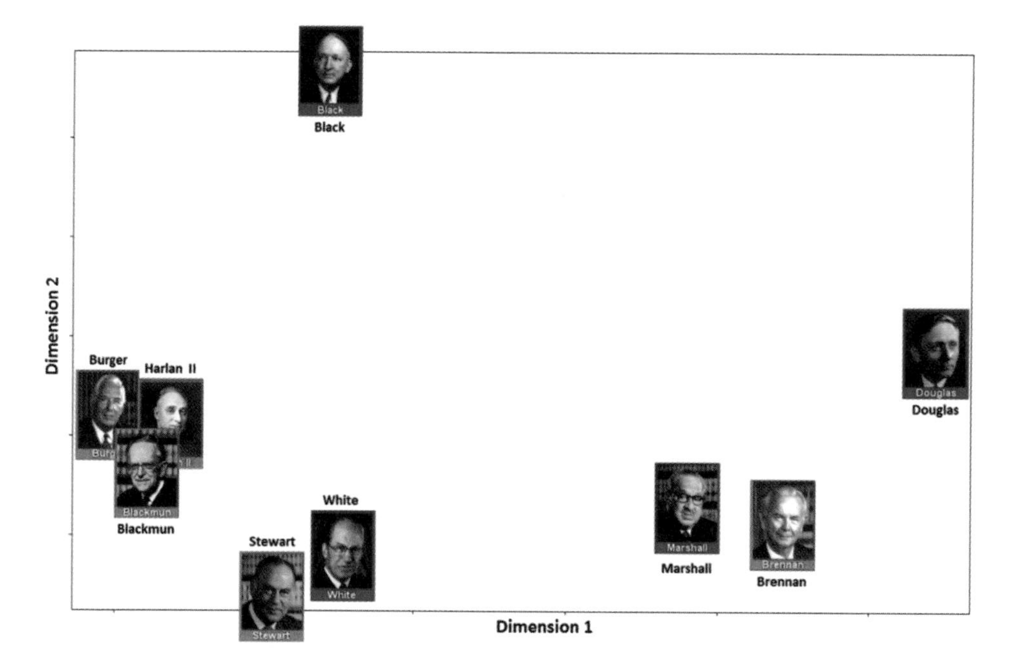

Figure 18.1
1970 U.S. Supreme Court term. Justices Blackmun and Burger are ideologically adjacent in this MDS representation of their voting agreement during the first year both were on the Court. (Photos used by permission of the Office of the Curator, Supreme Court of the United States.)

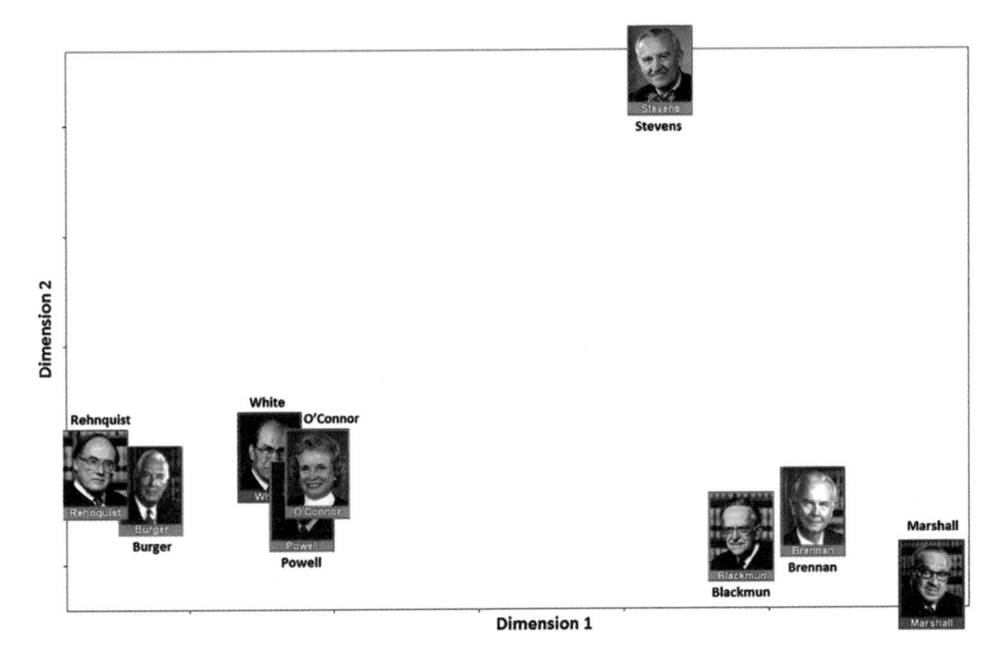

Figure 18.2
1985 U.S. Supreme Court term. Justice Blackmun has abandoned the intellectual tutelage of Burger, his childhood friend, and has sided with the liberals on the Court. (Photos used by permission of the Office of the Curator, Supreme Court of the United States.)

could benefit by being more aware of the metric work conducted by legal scholars and political scientists, and vice versa. In particular, the scientometric community should know and understand the innovations to the citator that originated in the legal community. After all, "whereas, in science, publications and their interconnections are by-products of the research enterprise, in law, publications and their interconnections are at the very heart of the discipline" (Shapiro, 1992, p. 337).

Notes

1. *Jurist* and *judge* are used herein interchangeably.

2. Fred Shapiro is perhaps the most active and best-known legal bibliometrician (Shapiro, 1992, 2001; Shapiro & Pearse, 2012). He has analyzed the citation rankings of law review articles, legal books (Shapiro, 2000a), and legal academics (Shapiro, 2000b) in the United States.

3. In contrast, civil law jurisdictions (e.g., France and Germany) have historically relied on statutory law (codes) as opposed to judge-made law. Presently however, most Western legal traditions are a hybrid of both (Merryman & Pérez-Perdomo, 2007).

4. The reader is referred to Cross (2012) and Yung (2012) for additional citation relationship indicators that would be useful to legal academics.

References

Acker, J. R. (1990). Thirty years of social science in Supreme Court criminal cases. *Law & Policy, 12*(1), 1–23.

Aft, A. B. (2011). Respect my authority: Analyzing claims of diminished U.S. Supreme Court influence abroad. *Indiana Journal of Global Legal Studies, 18*(1), 421–454.

Battelle, J. (2005). *The search: How Google and its rivals rewrote the rules of business and transformed our culture*. New York: Portfolio.

Berring, R. C. (2000). Legal information and the search for cognitive authority. *California Law Review, 88*(6), 1673–1708.

Betz, K. W. (1992). Examination of the Indiana Supreme Court docket, dispositions, and voting in 1991. *Indiana Law Review, 25*(4), 1469–1483.

Bierman, L. (1995). Dynamics of state constitutional decision-making: Judicial behavior at the New York Court of Appeals. *Temple Law Review, 68*(3), 1403–1456.

Black, R. C., & Spriggs, J. F. (2013). Citation and depreciation of U.S. Supreme Court precedent. *Journal of Empirical Legal Studies, 10*(2), 325–358.

Blackstone, W. (2001). *Blackstone's commentaries on the laws of England in four volumes*. Edited with an introduction by Wayne Morrison (9th ed. in modernized English). London: Cavendish.

Blackwell, K. (2009). Shipping up to Boston: Voting of the Massachusetts Supreme Judicial Court in non-unanimous criminal cases from 2001–2008. *Albany Law Review, 72*(3), 673–700.

Blumberg, D. (1998). Influence of the Massachusetts Supreme Judicial Court on state high court decisionmaking 1982–1997: A study in horizontal federalism. *Albany Law Review, 61*(5), 1583–1624.

Bommarito, M. J., Katz, D. M., & Zelner, J. (2009, June 08–12). *Law as a seamless web? Comparison of various network representations of the United States Supreme Court corpus (1791–2005)*. Paper presented at the 12th International Conference on Artificial Intelligence and Law—ICAIL '09, Barcelona, Spain.

Boulet, R., Mazzega, P., & Bourcier, D. (2011). Network approach to the French system of legal codes—part I: Analysis of a dense network. *Artificial Intelligence and Law, 19*(4), 333–355.

Bradley, R., & Ulmer, S. S. (1980). Examination of voting behavior in the Supreme Court of Illinois: 1971–1975. *Southern Illinois University Law Journal, 5*(3), 245–262.

Brin, S., & Page, L. (1998). The anatomy of a large-scale hypertextual Web search engine. *Computer Networks and ISDN Systems, 30*(1–7), 107–117.

Bushnell, S. I. (1986). The use of American cases. *University of New Brunswick Law Journal, 35*, 157–181.

Calabresi, S. G., & Zimdahl, S. D. (2005). The Supreme Court and foreign sources of law: Two hundred years of practice and the juvenile death penalty decision. *William and Mary Law Review, 47*(3), 743–909.

Caldeira, G. A. (1983). On the reputation of state supreme courts. *Political Behavior, 5*(1), 83–108.

Caldeira, G. A. (1985). The transmission of legal precedent: A study of state supreme courts. *American Political Science Review, 79*(1), 178–194.

Caldeira, G. A. (1988). Legal precedent: Structures of communication between state supreme courts. *Social Networks, 10*(1), 29–55.

Canon, B. C., & Baum, L. (1981). Patterns of adoption of tort law innovations: An application of diffusion theory to judicial doctrines. *American Political Science Review, 75*(4), 975–987.

Century edition of the American Digest. (1897). St. Paul, MN: West Publishing Co.

Chabot, C. K., & Chabot, B. R. (2011). Mavericks, moderates, or drifters—Supreme Court voting alignments, 1838–2009. *Missouri Law Review, 76*(4), 999–1044.

Chandler, S. J. (2005). *The network structure of the Uniform Commercial Code: It's a small world after all.* Paper presented at the 2005 Wolfram Technology Conference, Champaign, IL. Retrieved from http://library.wolfram.com/infocenter/Conferences/5800.

Chandler, S. J. (2007). The network structure of supreme court jurisprudence. *Mathematica Journal, 10*(3), 501–526.

Choi, S. J., & Gulati, G. M. (2004). Choosing the next Supreme Court justice: An empirical ranking of judge performance. *Southern California Law Review, 78*(1), 23–117.

Coke, E. (1628). *The first part of the institutes of the laws of England: Or, commentary upon Littleton.* London: Printed for the Societie of Stationers.

Committee on Law Reporting. (1895, August 27–30). *Report of the Committee on Law Reporting.* Paper presented at the Eighteenth Annual Meeting of the American Bar Association, Detroit.

Committee on Library and Legal Literature. (1895, August 6–8). *Report of the Committee on Library and Legal Literature.* Paper presented at the Seventh Annual Meeting of the Virginia State Bar Association, White Sulphur Springs, WV.

Cooper, B. D. (1982). Anglo-American legal citation: Historical development and library implications. *Law Library Journal, 75*(1), 3–33.

Crandley, M. J., Stephenson, P. J., Kerridge, J., & Peabody, J. (2011). Examination of the Indiana Supreme Court docket, dispositions, and voting in 2010. *Indiana Law Review, 44*(4), 993–1007.

Cross, F. B. (2012). The ideology of Supreme Court opinions and citations. *Iowa Law Review, 97*(3), 693–751.

Cross, F. B., Smith, T. A., & Tomarchio, A. (2006). *Determinants of cohesion in the Supreme Court's network of precedents*. San Diego Legal Studies Paper No. 07–67. University of Texas at Austin. Retrieved from http://papers.ssrn.com/sol3/papers.cfm?abstract_id=924110.

Cross, F. B., Smith, T. A., & Tomarchio, A. (2008). The Reagan revolution in the network of law. *Emory Law Journal, 57*(5), 1227–1258.

Cross, F. B., & Spriggs, J. F. (2010). The most important (and best) Supreme Court opinions and justices. *Emory Law Journal, 60*(2), 407–502.

Cross, F. B., Spriggs, J. F. I., Johnson, T. R., & Wahlbeck, P. J. (2010). Citations in the U.S. Supreme Court: An empirical study of their use and significance. *University of Illinois Law Review*, (2): 489–576.

Dabney, L. C. (2008). Citators: Past, present, and future. *Legal Reference Services Quarterly, 27*(2–3), 165–190.

De Wolf, A. H., & Wallace, D. H. (2009). The overseas exchange of human rights jurisprudence: The U.S. Supreme Court in the European Court of Human Rights. *International Criminal Justice Review, 19*(3), 287–307.

Doyle, J. (1992). Westlaw and the American digest classification scheme. *Law Library Journal, 84*, 229–257.

Duncan, E. B., Anderson, F. D., & McAleese, R. (1981). *Qualified citation indexing: Its relevance to educational technology*. Paper presented at the First Symposium on Information Retrieval in Educational Technology, Aberdeen, Scotland.

Epstein, L., Knight, J., & Martin, A. D. (2003). The political (science) context of judging. *Saint Louis University Law Journal, 47*(3), 783–817.

Epstein, L., Martin, A. D., Quinn, K. M., & Segal, J. A. (2007). Ideological drift among Supreme Court justices: Who, when, and how important. *Northwestern University Law Review, 101*(4), 1483–1541.

Epstein, L., Martin, A. D., Segal, J. A., & Westerland, C. (2007). The judicial common space. *Journal of Law Economics and Organization, 23*(2), 303–325.

Epstein, L., Segal, J. A., Spaeth, H. J., & Walker, T. G. (2012). *The Supreme Court compendium: Data, decisions, and developments* (5th ed.). Thousand Oaks, CA: SAGE/CQ Press.

Fausten, D., Nielsen, I., & Smyth, R. (2007). A century of citation practice on the Supreme Court of Victoria. *Melbourne University Law Review, 31*(3), 733–804.

Fitzherbert, A. (1565). *La graunde abridgement: Collect par le judge tresreuerend monsieur Anthony Fitzherbert, dernierment conferre auesq[ue] la copy escript, et per ceo correct, aueques le nombre del fueil, per quel facilement poies trouer les cases cy abrydges en les lyuers dans, nouelment annote, iammais deuaunt imprimee: Auxi vous troues les residuums de lauter liuer places icy in ceo liuer en le fyne de lour apte titles*. London: In aedibus Ricardi Tottell, duodecimo Nouembris.

Fowler, J. H., Johnson, T. R., Spriggs, J. F., Jeon, S., & Wahlbeck, P. J. (2007). Network analysis and the law: Measuring the legal importance of Supreme Court precedents. *Political Analysis, 15*(3), 324–346.

Friedman, L. M., Kagan, R. A., Cartwright, B., & Wheeler, S. (1981). State supreme courts: A century of style and citation. *Stanford Law Review, 33*(5), 773–818.

Frost, C. O. (1979). The use of citations in literary research: A preliminary classification of citation functions. *Library Quarterly, 49*(4), 399–414.

Garfield, E. (1955). Citation indexes for science. *Science, 122*(3159), 108–111.

Garfield, E. (1961). *Science citation index.* Philadelphia: Institute for Scientific Information.

Garfield, E. (1979). *Citation indexing: Its theory and application in science, technology, and humanities.* New York: Wiley.

Gordon, S. L. (2005). Update XXII: What's new on LexisNexis, Westlaw, Loislaw, and VersusLaw. *Legal Information ALERT, 24*(10), 1–14.

Gorney, U. (1955). American precedent in the Supreme Court of Israel. *Harvard Law Review, 68*(7), 1194–1210.

Greenhouse, L. (2005a). *Becoming Justice Blackmun: Harry Blackmun's Supreme Court journey.* New York: Times Books / Henry Holt.

Greenhouse, L. (2005b, July 2). Court in transition: News analysis: Consistently, a pivotal role: Groundbreaking justice held balance of power. *New York Times,* p. A1.

Hanson, F. A. (2002). From key words to key numbers: How automation has transformed the law. *Law Library Journal, 94*(4), 563–600.

Harris, P. (1982). Structural change in the communication of precedent among state supreme courts, 1870–1970. *Social Networks, 4*(3), 201–212.

Harris, P. (1985). Ecology and culture in the communication of precedent among state supreme courts, 1870–1970. *Law & Society Review, 19*(3), 449–486.

Ho, D. E., & Quinn, K. M. (2010). How not to lie with judicial votes: Misconceptions, measurement, and models. *California Law Review, 98*(3), 813–876.

Hook, P. A. (2007a). The aggregate harmony metric and a statistical and visual contextualization of the Rehnquist Court: 50 years of data. *Constitutional Commentary, 24*(1), 221–264.

Hook, P. A. (2007b). *Network derived domain maps of the United States Supreme Court: 50 years of co-voting data and a case study on abortion.* Paper presented at the International Workshop and Conference on Network Science 2007, Queens, NY.

Hook, P. A. (2007c, June 25–27). *Visualizing the topic space of the United States Supreme Court.* Paper presented at the 11th International Conference of the International Society for Scientometrics and Informetrics, CSIC, Madrid, Spain.

Hopkins, K. (2005). Most highly cited. *Scientist (Philadelphia), 19*(20), 22–27.

Hossain, F., & Cox, A. (2006, Sunday, July 2). Percentage of times that pairs of justices agreed in nonunanimous decision in the 2005–6 term. *New York Times,* p. 22.

Johnson, J. C., Borgatti, S. P., & Romney, K. (2005). *Analysis of voting patterns in U.S. Supreme Court decisions*. Paper presented at the Sunbelt XXV, International Sunbelt Social Network Conference, Redondo Beach, CA. Abstract available at http://www.socsci.uci.edu/~ssnconf/conf/SunbeltXXVProgram.pdf.

Katz, D. M., Gubler, J. R., Zelner, J., Bommarito, M. J., Provins, E., & Ingall, E. (2011). Reproduction of hierarchy?: A social network analysis of the American law professoriate. *Journal of Legal Education, 61*(1), 76–103.

Katz, D. M., & Stafford, D. K. (2010). Hustle and flow: A social network analysis of the American Federal Judiciary. *Ohio State Law Journal, 71*(3), 457–509.

Klein, D., & Morrisroe, D. (1999). The prestige and influence of individual judges on the U.S. courts of appeals. *Journal of Legal Studies, 28*(2), 371–391.

Kleinberg, J. (1998, January 25–27). *Authoritative sources in a hyperlinked environment.* Paper presented at the Ninth Annual ACM-SIAM Symposium on Discrete Algorithms, San Francisco.

Kleinberg, J. (1999). Authoritative sources in a hyperlinked environment. *Journal of the ACM, 46*(5), 604–632.

Kosma, M. N. (1998). Measuring the influence of Supreme Court justices. *Journal of Legal Studies, 27*(2), 333–372.

Landes, W. M., Lessig, L., & Solimine, M. E. (1998). Judicial influence: A citation analysis of federal courts of appeals judges. *Journal of Legal Studies, 27*(2), 271–332.

Lipetz, B.-A. (1965). Improvement of the selectivity of citation indexes to science literature through inclusion of citation relationship indicators. *American Documentation, 16*(2), 81–90.

Liptak, A. (2008, September 18). U.S. court is now guiding fewer nations: American exception: A loss of influence. *New York Times*, p. A1.

Lupu, Y., & Voeten, E. (2010). Precedent on international courts: A network analysis of case citations by the European Court of Human Rights. *British Journal of Political Science, 42*(2), 413–439.

MacIntyre, J. M. (1966). Use of American cases in Canadian courts. *University of British Columbia Law Review, 2*(3), 478–490.

Manfredi, C. P. (1990). The use of United States decisions by the Supreme Court of Canada under the Charter of Rights and Freedoms. *Canadian Journal of Political Science / Revue canadienne de science politique, 23*(3), 499–518.

Manz, W. H. (2002). Citations in Supreme Court opinions and briefs: A comparative study. *Law Library Journal, 94*(2), 267–300.

Marangola, R. A. (1998). Independent state constitutional adjudication in Massachusetts: 1988–1998. *Albany Law Review, 61*(5), 1625–1679.

Martin, A. D., & Quinn, K. M. (2002). Dynamic ideal point estimation via Markov chain Monte Carlo for the U.S. Supreme Court, 1953–1999. *Political Analysis, 10*(2), 134–153.

Martin, A. D., Quinn, K. M., & Epstein, L. (2005). The median justice on the United States Supreme Court. *North Carolina Law Review, 83*(5), 1275–1322.

Martin, H. C. (1996). Statistical compilation of the opinions of the Supreme Court of North Carolina terms 1993–1994 through 1994–1995. *North Carolina Law Review, 74*(6), 1851–1862.

Mathieson, D. L. (1963). Australian precedents in New Zealand courts. *New Zealand Universities Law Review, 1*, 77–112.

McCormick, P. (1997). The Supreme Court of Canada and American citations 1945–1994: A statistical overview. *Supreme Court Law Review, 8*, 527.

McCormick, P. (2009). American citations and the McLachlin Court: An empirical study. *Osgoode Hall Law Journal, 47*(1), 83–129.

Merryman, J. H. (1954). Authority of authority: What the California Supreme Court cited in 1950. *Stanford Law Review, 6*(4), 613–673.

Merryman, J. H. (1978). Toward a theory of citations: An empirical study of the citation practice of the California Supreme Court in 1950, 1960, and 1970. *Southern California Law Review, 50*(3), 381–428.

Merryman, J. H., & Pérez-Perdomo, R. (2007). *The civil law tradition: An introduction to the legal systems of Europe and Latin America* (3rd ed.). Stanford, CA: Stanford University Press.

Miller, N. (2002). An international jurisprudence? The operation of "precedent" across international tribunals. *Leiden Journal of International Law, 15*(3), 483–526.

Mott, R. (1936). Judicial influence. *American Political Science Review, 30*, 295–315.

Nagal, S. S. (1962). Sociometric relations among American courts. *Southwestern Social Science Quarterly, 43*(2), 136–142.

New State Ice Co. v. Liebmann, 285 U.S. 262 (1932).

Newton, B. E. (2012). Law review scholarship in the eyes of the twenty-first century Supreme Court justices: An empirical analysis. *Drexel Law Review, 4*(2), 399–416.

Ogden, P. (1993). "Mastering the lawless science of our law": A story of legal citation indexes. *Law Library Journal, 85*(1), 1–48.

Petherbridge, L., & Schwartz, D. L. (2012). Empirical assessment of the Supreme Court's use of legal scholarship. *Northwestern University Law Review, 106*(3), 995–1032.

Posner, E. A., & Sunstein, C. R. (2006). The law of other states. *Stanford Law Review, 59*(1), 131–179.

Pritchett, C. H. (1941). Divisions of opinion among justices of the U.S. Supreme Court, 1939–1941. *American Political Science Review, 35*(5), 890–898.

Pritchett, C. H. (1948). *The Roosevelt Court: A study in judicial politics and values 1937–1947*. New York: Macmillan.

Riggs, R. E. (1988). Supreme Court voting behavior: 1986 term. *BYU Journal of Public Law, 2*(1), 15–34.

Roy, B. (2004). An empirical survey of foreign jurisprudence and international instruments in charter litigation. *University of Toronto Faculty of Law Review, 62*(2), 99–148.

Schubert, G. (1962). The 1960 term of the Supreme Court: A psychological analysis. *American Political Science Review, 56*(1), 90–107.

Schubert, G. (1963). Judicial attitudes and voting behavior: The 1961 term of the United States Supreme Court. *Law and Contemporary Problems, 28*(1), 100–142.

Schultz, W. B., & Howard, P. K. (1975). Myth of swing voting: Analysis of voting patterns on the Supreme Court. *New York University Law Review, 50*(4), 798–868.

Shapiro, C. (2009). Coding complexity: Bringing law to the empirical analysis of the Supreme Court. *Hastings Law Journal, 60*(3), 477–543.

Shapiro, F. R. (1992). Origins of bibliometrics, citation indexing, and citation analysis: The neglected legal literature. *Journal of the American Society for Information Science and Technology, 43*(5), 337–339.

Shapiro, F. R. (2000a). The most-cited legal books published since 1978. *Journal of Legal Studies, 29*(1), 397–407.

Shapiro, F. R. (2000b). The most-cited legal scholars. *Journal of Legal Studies, 29*(1), 409–426.

Shapiro, F. R. (Ed.). (2001). *Collected papers on legal citation analysis*. Littleton, CO: Fred B. Rothman.

Shapiro, F. R., & Pearse, M. (2012). The most-cited law review articles of all time. *Michigan Law Review, 110*(8), 1483–1520.

Sirovich, L. (2003). A pattern analysis of the second Rehnquist U.S. Supreme Court. *Proceedings of the National Academy of Sciences of the United States of America, 100*(13), 7432–7437.

Smith, T. A. (2007). Web of law. *San Diego Law Review, 44*(2), 309–354.

Smithey, S. I. (2001). A tool, not a master: The use of foreign case law in Canada and South Africa. *Comparative Political Studies, 34*(10), 1188–1211.

Smyth, R. (1999a). What do intermediate appellate courts cite? A quantitative study of the citation practice of Australian state supreme courts. *Adelaide Law Review, 21*(1), 51–80.

Smyth, R. (1999b). What do judges cite? An empirical study of the authority of authority in the Supreme Court of Victoria. *Monash University Law Review, 25*(1), 29–53.

Smyth, R. (2000a). The authority of secondary authority: A quantitative study of secondary source citations in the Federal Court. *Griffith Law Review, 9*(1), 25–51.

Smyth, R. (2000b). Who gets cited: An empirical study of judicial prestige in the High Court. *University of Queensland Law Journal, 21*(1), 7–22.

Smyth, R. (2002). Citations by court. In M. Coper, G. Williams, & A. Blackshield (Eds.), *The Oxford companion to the High Court of Australia* (p. 98–99). Melbourne: Oxford University Press.

Smyth, R. (2007). The citation practices of the Supreme Court of Tasmania, 1905–2005. *University of Tasmania Law Review, 26*(1), 34–62.

Smyth, R. (2008a). A century of citation: Case-Law and secondary authority in the Supreme Court of Western Australia. *University of Western Australia Law Review, 34*(1), 145–167.

Smyth, R. (2008b). *Citations of foreign decisions in Australian state supreme courts over the course of the twentieth century: An empirical analysis*. Monash University. ExpressO. Retrieved from http://works.bepress.com/russell_smyth/1.

Smyth, R. (2009a). *Citing outside the law reports: Citations of secondary authorities on the Australian state supreme courts over the twentieth century*. Monash University. ExpressO. Retrieved from http://works.bepress.com/russell_smyth/2.

Smyth, R. (2009b). Trends in the citation practice of the Supreme Court of Queensland over the course of the twentieth century. *University of Queensland Law Journal, 28*(1), 39–80.

Smyth, R., & Fausten, D. (2008). Coordinate citations between Australian state supreme courts over the 20th century. *Monash University Law Review, 34*(1), 54–74.

Snyder, F. (1999). The West Digest System: The Ninth Circuit and the Montana Supreme Court. *Montana Law Review, 60*, 541–597.

Spaeth, H. J., & Altfeld, M. F. (1985). Influence relationships within the Supreme Court: A comparison of the Warren and Burger Courts. *Western Political Quarterly, 38*(1), 70–83.

Spaeth, H. J., Epstein, L., Ruger, T., Whittington, K., Segal, J. A., & Martin, A. D. (2012). The Supreme Court Database. Retrieved from http://scdb.wustl.edu.

Supreme Court, 1956 term. (1957). *Harvard Law Review, 71*(1), 94–106.

Supreme Court 2010 term: The statistics. (2011). *Harvard Law Review, 125*(1), 362–377.

Surrency, E. C. (1990). *A history of American law publishing*. New York: Oceana Publications.

Thomson/West. (2011). *West's analysis of American law: With key number classifications* (2011th ed.). St. Paul, MN: West.

Thurstone, L. L., & Degan, J. W. (1951). A factorial study of the Supreme Court. *Proceedings of the National Academy of Sciences of the United States of America, 37*(9), 628–635.

Tiersma, P. M. (2007). The textualization of precedent. *Notre Dame Law Review, 82*(3), 1187–1278.

Tiersma, P. M. (2010). *Parchment, paper, pixels: Law and the technologies of communication*. Chicago: University of Chicago Press.

Topperwien, B. (2002). Foreign precedents. In M. Coper, G. Williams, & A. Blackshield (Eds.), *The Oxford companion to the High Court of Australia* (p. 280). Melbourne: Oxford University Press.

Voeten, E. (2010). Borrowing and nonborrowing among international courts. *Journal of Legal Studies, 39*(2), 547–576.

Von Nessen, P. E. (1992). Use of American precedents by the High Court of Australia, 1901–1987. *Adelaide Law Review, 14*(2), 181–218.

Von Nessen, P. E. (2006). Is there anything to fear in transnationalist development of law? The Australian experience. *Pepperdine Law Review, 33*(4), 883–924.

Whalen, R. (2013). Modeling annual Supreme Court influence: the role of citation practices and judicial tenure in determining precedent network growth. In R. Menezes, A. Evsukoff, & M. C. González (Eds.), *Complex Networks* (pp. 169–176). Heidelberg: Springer.

White, G. E. (2005). Unpacking the idea of the judicial center. *North Carolina Law Review, 83*(5), 1089–1186.

Wilkins, R. G., Worthington, S., Reynolds, J., & Nielsen, J. J. (2005). Supreme Court voting behavior: 2004 term. *Hastings Constitutional Law Quarterly, 32*(4), 909–986.

Yung, C. R. (2012). Supreme Court opinions and the justices who cite them: A response to Cross. *Iowa Law Review Bulletin, 97*, 41–50.

Zaring, D. (2006). Use of foreign decisions by federal courts: An empirical analysis. *Journal of Empirical Legal Studies, 3*(2), 297–331.

19 Academic Genealogy

Cassidy R. Sugimoto

Introduction

Academic genealogy is the quantitative study of intellectual heritage operationalized through chains of students and their advisors. In most cases, this is the doctoral student and advisor, although the concept of doctoral advisor has been liberally interpreted in some projects. For example, the Mathematics Genealogy Project traces genealogy back as far as 1380—centuries before the establishment of the research dissertation or the doctor of philosophy degree (Clark, 2006). Although academic genealogy has occasionally been termed "intellectual genealogy" or "scientific genealogy," it should not be confused with the former, which is a term used to describe any kind of intellectual influence (not necessarily a formal or institutional mentor) (Lubek et al., 1995), or the latter, which is a term used to elevate the studies of family histories to the level of a science (Davenport, 1915).

Academic genealogy has been criticized for having low levels of generalizability and rigor, used only to honor eminent scholars or to satisfy a scholar's curiosity about their academic ancestors. Similar criticisms were made regarding early citation analyses, which, due to the painstaking manual data collection, were often no more than local case studies with limited generalizability or use. However, as with the rise of citation indexes, there are now sources and means for collecting large-scale data on academic genealogy that make possible rigorous studies in this area. Academic genealogy is no longer the exclusive purview of academics attempting to trace their own roots, but is also of interest to those who study science—from historical, philosophical, sociological, and scientific perspectives.

The underlying assumption of academic genealogy is that disciplines are propagated through knowledge transfer activities (Abbott, 1999; Turner, 2000)—in this case, through doctoral mentoring. As noted by Kuhn (1996) and others, disciplines are governed by norms or paradigms—cultural rules that inform members of appropriate practices in the discipline. It is assumed that many of these practices are transferred both tacitly and explicitly during interactions with trusted mentors (Girves & Wemmerus, 1988). Academic genealogy provides a means to measure and analyze these interactions and to study the relationship between mentoring and "disciplining" (Abbott, 2001; Foucault, 1975/1995). As a proxy for studying the transfer of disciplinary knowledge, academic genealogy can also be used to examine interactions between and among disciplines as faculty members "disciplined" in one area move to another (Sugimoto, Ni, Russell, & Bychowski, 2011). The migration of doctoral students in one discipline to faculty members in another has a direct impact on the knowledge landscape, potentially changing the topical trajectory of the next generation of acolytes.

As an evaluative metric, academic genealogy can serve to counteract the "neglect of silent evidence" (Taleb, 2010, p. 103), by making visible the contributions of mentors, particularly those with high mentorship fecundity—that is, those who produce a large number of protégés (Malmgren, Ottino, & Amaral, 2010). Metrics of academic genealogy (Russell & Sugimoto, 2009) serve to demonstrate and contextualize these contributions within the larger academic and disciplinary spheres. The rationale underlying the use of academic genealogy as an evaluative metric is the belief that the "most effective way for a scientist's work to live beyond their time is for them to populate the next generation of academics with people that they have mentored . . . [in order that] their ideas, contributions, and views will continue to influence scientific thought" (Andraos, 2005, p. 1405). A scholar's lifetime is finite, but his or her contribution is amplified, enhanced, and extended through successive generations of mentees.

The objectives of this chapter are to introduce the concept of academic genealogy, review previous studies of academic genealogy, provide a typology of academic genealogies with corresponding motivations and outcomes, outline appropriate methods for the conduct of academic genealogy, and discuss implications and future research directions.

State of Existing Research

Published studies of academic genealogy have examined neuroscience (David & Hayden, 2012), organic chemistry (Andraos, 2005), mathematics (Chang, 2010; Malmgren et al., 2010), physiology (Bennett & Lowe, 2005; Jackson, 2011), energy expenditure (Durnin, 1991), primatology (Kelley & Sussman, 2007), exercise and sports sciences (Mitchell, 1992; Montoye & Washburn, 1980), pharmaceutical sciences (Stella, 2001; Tyler & Tyler, 1992), library and information science (LIS) (Marchionini, Solomon, Davis, & Russell, 2007; Russell & Sugimoto, 2009; Sugimoto, Ni, et al., 2011), and psychology (Boring & Boring, 1948; Lubek et al., 1995; Newton, 1995; Robertson, 1994; Williams,

1993). Online projects range in size and coverage. Bennett and Lowe (2005) include a link from their article to a companion website recording the academic descendants of a single individual. The University of Illinois at Urbana-Champaign hosts a database for chemists,[1] and the Department of Computer Science at the University of Texas at Austin collects academic genealogy information for the artificial intelligence community.[2] The largest single-discipline database is the Mathematics Genealogy Project,[3] hosted by the Department of Mathematics at North Dakota State University. As of September 1, 2012, the project contained data on 163,855 mathematicians.

Two projects that began as single-discipline databases have expanded to include other disciplines. The MPACT Project,[4] hosted jointly by Indiana University Bloomington and the University of North Carolina at Chapel Hill, began as a study of LIS, but contains, at the time of writing, data on 9,037 individuals across 217 disciplines, 319 schools, and 156 countries. The Academic Family Tree,[5] an extension of the neuroscience-focused Neurotree.org (for an analysis of which, see David and Hayden, 2012), is an interdisciplinary project currently listing 28 fields of study. The number of people in the database per area ranges from 37,387 for neuroscience to 51 for ingestive behavior. The content is entirely user-supplied and, although the trees are networked between fields, each field has its own portal. The growth rate for neuroscience is estimated at 150 new additions per week, and the website maintains a page demonstrating these growth rates.[6] However, this source, as with the previously mentioned ones, is heavily biased toward a single discipline.

Types of Academic Genealogy

Academic genealogy has experienced some of the same criticisms that have plagued other scientometric methods—that these are mere navel-gazing exercises done for personal pleasure or to celebrate an individual. Ego and honor do play a part; however, there are a number of motivations for conducting this type of research and, more importantly, a variety of research questions that can be answered by the construction of an academic genealogy. Therefore, I propose a five-part typology for academic genealogies: honorific, egotistical, historical, paradigmatic, and analytic. Each is briefly described below. It should be noted, however, that these are not exclusive categories; most academic genealogies function in at least two of these ways.

Honorific

Many academic genealogies focus on the descendants of a single individual, in order to honor this individual and demonstrate the scholar's impact on generations of academics (e.g., Bennett & Lowe, 2005). At the conclusion of one such study, readers were "encouraged to explore their own academic legacies as a way of honoring those who prepared the way for us" (Jackson, 2011, p. 120). Honorific genealogies are typically prepared as part of a Festschrift, workshop, or conference aimed at honoring a canonical, eminent, or recently deceased member of a field. Those honoring canonical figures are likely to be historical (e.g., Zittoun, Perret-Clermont, & Barrelet, 2008), while those honoring eminent or recently deceased figures tend to focus on one generation of descendants and contemporary accomplishments.

Egotistical

Egotistical academic genealogies take as their point of departure a single individual or small set of scholars and trace antecedents. Typically, the author of the study conducts a personalized search to understand his or her academic ancestors (e.g., Durnin, 1991; Jackson, 2011; Robertson; 1994; Williams, 1993), although it could also involve a department, school, or other group of individuals tracing their collective past. Curiosity and community building (Newton, 1995) can motivate egotistical genealogy. There is also a sense of establishing oneself or one's unit by delineating connections with prestigious predecessors. As Lubek et al. (1995) noted regarding academic genealogies of psychologists:

> Amateur genealogists, in pursuing their family trees, often hope to uncover some illustrious ancestor who might add a measure of glamour to their lineage. Collective genealogies, like those of departments of psychology, may serve similar functions. The recognition that we have "descended," in a manner of speaking, from some celebrated figure in the history of psychology may infuse our circumscribed daily activities with some transcendental value, with the recognition that although we are ordinary workers in the fields of Kuhnian normal science, we are nonetheless part of an historical endeavour led by persons of some renown. (p. 52)

Demonstrating that one descended from Wilhelm Wundt (Lubek et al., 1995; Robertson, 1994; Williams, 1993) is a form of credentialing. Demonstrating that one of your forebears won the Nobel Prize (Tyler & Tyler, 1992) is a way to flaunt your academic genes.

Historical

The historical nature of an academic genealogy is perhaps unavoidable, regardless of the type of genealogy constructed. Explicit historical genealogies are those that use academic genealogy as a platform for a historical analysis. In practice, these often involve a historical narrative around academic influences in an honorific genealogy and potted histories of each advisor in an egotistical genealogy (e.g., Durnin, 1991; Jackson, 2011; Tyler & Tyler, 1992). The underlying motivation of a historical genealogy is not to honor or confer prestige on any individual, but rather to describe the development of a field or area of study through academic lineages.

"Founding Fathers" are often the centerpiece of such studies. In empirical studies, the goal is to identify precursors by tracing current members of a field back to a few key individuals (e.g., Boring & Boring, 1948). For example, Lubek et al. (1995) examined the academic genealogy of faculty in five Canadian psychology departments and found that 75% of the faculty members could be traced to nine pioneer figures. In this way, one can empirically identify the canonical figures in a field. Alternatively, canonical scholars can serve as the starting point of the analysis and the development of the field can be demonstrated by tracing academic descendants (e.g., Tyler & Tyler, 1992).

Other studies revolve around contemporary elders (e.g., Mitchell, 1992)—that is, current eminent scholars in the field. Tracing the ancestry of these individuals provides insights into the formation of the field. These studies can be used to show the growth of an intellectual area (e.g., Kelley & Sussman, 2007), and the networked nature of such studies can throw "into relief several facets of our collective history" (Lubek et al., 1995, p. 66). This collective history can have pedagogical value for neophytes. As Andraos (2005) has commented: "[The] predictability of what is important and worth looking into as a viable research question or area markedly improves when one knows the chronology of ideas and associates themselves with the key players in shaping those ideas or at least their direct scientific descendants" (p. 1405). Presenting new doctoral students with this research or requiring them to search their own pedigrees can be a useful initiation into the discipline.

Paradigmatic

Paradigmatic genealogy can be used to not only define formal relationships between people, but also to study the extent to which knowledge and epistemic practices are transmitted through these relationships. It has been cautioned that institutional genealogy (i.e., tracing genealogy through formal lineage) should not be conflated with intellectual genealogy (i.e., tracing how ideas move through mentors, collaborators, and colleagues over time) (Lubek et al., 1995). However, as De Mey (1992) remarked on Zuckerman's (1977) study of Nobel laureates:

> Though innovation might seem to be coupled to erratic and unpredictable movement, there are also impressive chains of continuity in science through master-apprentice relationships. . . . Apprentices need not remain in the area of their master to exemplify this continuity. . . . The conservation and transmission of a tradition of scientific leadership and innovativeness has again both sociological and cognitive aspects. Through their position in the social organization of science, masters can introduce their protégé more effectively into the leading circles and innovative areas.

> . . . [High quality mentoring] is probably the most effective way of installing cognitive structures which remain intact when the apprentice migrates to other fields or encounters revolutionary turmoil. (p. 146)

Studies have supported the claim of "impressive chains of continuity," finding evidence of shared methods or epistemology across academic generations (e.g., Jackson, 2011; Kelley & Sussman, 2007; Robertson, 1994). However, Lubek et al. (1995, p. 65) warned that genealogical trees can "sometimes give a false picture of continuity" when they fail to take into account the "model of discontinuous discoveries in science" and urged for more "critical approaches . . . sensitized to the tension between continuity and discontinuity" to address this limitation.

Many studies of a paradigmatic nature have lacked rigor, merely exploring the "self-evident" replication of theories, methods, and practices in an academic lineage. However, the introduction of robust topic-modeling techniques and large-scale dissertation data provides a new lens for exploring paradigmatic genealogies. For example, Sugimoto, Li, Russell, Finlay, and Ding (2011) analyzed the topics in LIS dissertations over an 80-year timespan by applying topic-modeling techniques to dissertation titles and abstracts, revealing the main topic areas by decade. Given that each of the titles and abstracts is matched with a student who is matched with an advisor, one can apply network analysis to this to reveal underlying topic communities (Yan, Ding, Milojević, & Sugimoto, 2012) within and across academic family trees. Qualitative analyses might also be used to reveal the transmission of ideas and practices from one academic to another—interviews with a cohort of students advised by a single advisor provides insight on the nature of knowledge, skills, and practices transferred through formal mentoring relationships.

Analytic

The growth of large-scale databases and rigorous statistical analyses has made possible a new type of genealogy. Analytic academic genealogies scientifically address explicit research questions and provide evaluative assessments of scholars and scholarship. Such studies are typically descriptive and evaluative in nature and, in some cases, predictive. Early analytic genealogies sought to establish sets of metrics for quantifying academic genealogy—calculating the number of times an individual served as an advisor and committee member and various weighted approaches to these calculations, as well as metrics for calculating the breadth and depth of academic lineage (Marchionini et al., 2007; Russell & Sugimoto, 2009). Subsequent studies have also sought to examine distances in relationships among individuals in the family tree (e.g., David & Hayden, 2012). Metrics of genealogy have been used to investigate the relationship between quantitative advising metrics and other metrics of academic productivity (Sugimoto, Russell, Meho, & Marchionini, 2008). The evidence suggested that the relationship was not linear, but rather individual and dependent on the time in the academic lifecycle. Malmgren et al. (2010) similarly noted that mentorship fecundity (that is, the number of protégés a mentor trains) for mathematicians had different rates over the course of the academic lifecycle: fecundity was higher than expected for the first two-thirds of a high-performing academic's career and lower than expected in the last third of the academic's mentoring career. That is, a student mentored in a high-performing scholar's early years would be more likely to have a large number of academic progeny than those produced later.

Many indicators of interdisciplinarity are static—describing the distribution of units (e.g., authors or documents) across a discipline at a given point in time. However, interdisciplinarity can also be measured using the backgrounds of advisors represented at various points in time, providing a diachronic indicator of the heterogeneity of a field. The use of academic genealogy to study interdisciplinary was introduced by Sugimoto, Ni, et al. (2011), who investigated the discipline in which advisors and committee members on LIS dissertations had received their degrees. This work demonstrated that the degree of the advisor had a direct impact on the intellectual content of the advisee's dissertation—when an advisor received their degree in one discipline and then moved to another, the subsequent advisees' use of interdisciplinary research was significantly different from those mentored by "within-field" advisors. This provides evidence of the transmission of knowledge through advisee-advisor chains and the possible impact of interdisciplinarity on advisorship networks. The diachronic nature of academic genealogy studies provides the added benefit of being able to study not only the heterogeneity of a discipline, but also its permeability at various points across time (Klein, 1996). Depictions of the dynamic nature of the discipline may provide better descriptions of "life cycle of scientific specialties" (De Mey, 1992, p. 148).

Network science provides additional opportunities for analytic academic genealogy. In some fields, it is common for people to do postdoctoral training with their advisor's advisor (Andraos, 2005). This form of academic inbreeding produces strong networks and can be "particularly effective in securing consistent and credible recommendation letters from people who know one another professionally as well as personally" (Andraos, 2005, p. 1404). Studies have also shown that doctoral origins work to create strong "power cliques" in scholarly communication. Yoels (1971) found a disproportionate number of editors coming from a few doctoral programs and showed that this distribution was not proportional to the rate of production of doctoral graduates across all programs; furthermore, Yoels revealed that the editorial boards' doctoral origins tended to align with the doctoral origin of the editor-in-chief. These studies suggest that the "Matthew effect" (Merton, 1973) not only results in cumulative advantages for the advisor, but also residual benefits for the student; quite simply, "student[s] who have trained with any of the key scientific contributors or their immediate descendants are highly sought after as the next generation of academics" (Andraos, 2005, p. 1405). Many direct correlations have been identified with respect to mentorship fecundity, including publications, citations, and membership in prestigious associations (Malmgren et al., 2010; Sugimoto et al., 2008). Future studies of academic genealogy could identify more ways in which the productivity and success of the doctoral student are directly related to the academic network in which the student was trained.

Method

Academic genealogies appear, on the surface, to be fairly simple: create links between individuals based on formal academic affiliations. However, several approaches can be taken and have various advantages and disadvantages. There is also the possibility of error and misinterpretation if variables are not explicitly operationalized. As Boring and Boring (1948) noted in their construction of an academic genealogy of psychologists: "Our experience with the living shows that we may easily have made some errors with the dead" (p. 527). The next section reviews the methods of academic genealogy, including initializing a search, operationalizing links, identifying data sources, and visualizing results.

Initializing

Academic genealogies typically begin with a purposive sample of one or more individuals. In the case of an honorific genealogy, a single individual is selected and all the descendants of this individual are identified (e.g., Bennett & Lowe, 2005). In the case of an egotistical genealogy, a contemporary scholar (typically the researcher of the study) identifies his or her own lineage (i.e., one's advisors and advisor's advisor, and so on until no more information can be found) (e.g., Jackson, 2011; Williams, 1993). The goal of a single-individual study is either to demonstrate the importance of a key scholar or to trace a personal genealogy.

Multiple individuals are usually selected when the motivation is to study the growth and development of a field or discipline. In similar fashion to single-individual studies, these can be done either from the perspective of descendants or antecedents. Certain canonical authors may be chosen and their collective genealogical descendants traced to outline the parameters of the discipline. Conversely, canonical authors could be identified by studying the genealogies of all current members of the discipline (e.g., Lubek et al., 1995). Contemporary scholars may be identified using prolific authors in a high-impact venue (e.g., Mitchell, 1992; Montoye & Washburn, 1980), doctoral graduates from representative schools in the discipline (e.g., Sugimoto, Ni, et al., 2011), or current faculty members at representative schools or in representative associations.

Great care should be exercised in selecting the individual(s) used to "seed" the genealogy. Given that the majority of academic genealogy studies employ purposive sampling, caution should be taken when generalizing the results (unless the goal is explicitly honorific or egotistical). Studies of a historical, paradigmatic, or analytic nature should always justify the sample and, when a census is not possible, explore possible sampling options.

Operationalizing Links

It is critical that any academic genealogy study begin with a clear operationalization of the criteria necessary to establish a link between student and mentor. For the strictest studies, this implies a degree of formality in the relationship, typically as academic advisor, dissertation chair, major professor, or supervisor—that is, the person formally and primarily responsible for guiding the student through the process of completing a doctoral degree (e.g., Kelley & Sussman, 2007; Lubek et al., 1995; Mitchell, 1992; Robertson, 1994; Williams, 1993). However, this operationalization is often not followed in strict fashion, particularly when academic genealogies reach beyond the modern doctoral education system and the modern dissertation.

There is some criticism of equating an academic advisor with a mentor and with the exclusive use of doctoral advisors to denote links between students and mentors. It is argued that many individuals can serve as the primary mentor for an individual and that individuals are often influenced by a "mentoring constellation" (Sugimoto, 2012) rather than an individual. Despite this, general consensus and empirical evidence (Sugimoto, 2012) suggest that the doctoral advisor is the primary mentor in most cases and is an adequate proxy when examining large-scale genealogies. It may be useful in microlevel analyses, however, to look beyond formal doctoral student-advisor relationships to other informal intellectual connections (e.g., postdoctoral advisors (e.g., Andraos, 2005; Bennett & Lowe, 2005; Jackson, 2011; David & Hayden, 2012), collaborators, or other informal mentors and students (e.g., Zittoun et al., 2008; Bennett & Lowe, 2005; Durnin, 1991).

The links in the genealogy end when no additional verified information can be located on an advisor/advisee. Other potential discontinuities in the family tree are the rare advisors who have nonacademic origins ("self-starters," as Lubek et al. [1995] label them), and students who bear no academic children themselves (largely as a result of going into a nonacademic position). There is also a practice of maintaining an intradisciplinary perspective and taking as a "point of disciplinary discontinuity" any branch in the tree where someone was trained by someone from outside the discipline (Lubek et al., 1995, p. 53). The goal of these (typically historical or paradigmatic) genealogies is to create a disciplinarily homogeneous genealogy. This can be useful in describing the extent and breadth of disciplinary histories. However, it has been suggested that academic genealogies should not avoid these discontinuities, but rather use them to inform our understanding of the birth, maturation, and interaction of disciplines (Sugimoto, Ni, et al., 2011). Such interdisciplinary genealogies integrate homogeneous academic family trees into a "canopy of trees" ("The Academic Family Tree," n.d.) that display the interconnectivity of doctoral education and, by extension, of knowledge.

Identifying Data Sources

The most reliable source of information on doctoral advisors and committee members is the actual dissertation. Ideally, names and roles will be typed underneath or next to a signature line on the cover page. If neither a cover page nor a typed name can be found, and if the signatures (assuming they are available) are indecipherable, the next best source is the acknowledgments section, where the authors will often explicitly thank their advisors and committee members. However, acknowledgments can be ambiguous, as when the author thanks a number of faculty members without specifying their roles. Acknowledgments can also shed light on transitions: the author may thank multiple advisors who guided them during one point in their doctoral career, but who did not sign the final dissertation for various reasons (e.g., leaving the university, death, etc.). In some cases, the final advisor may be more of an administrative signatory, rather than a true intellectual counsel. However, operationalization must remain strict because the researcher cannot be responsible for identifying "true mentors" in each situation (see the earlier section on operationalization).

The most comprehensive data source for dissertations is ProQuest's Dissertations and Theses database, "covering 40% of all dissertations from major universities" (Andersen & Hammarfelt, 2011, p. 374). For the years 1848 to 2009, this includes about 2.3 million dissertations conferred at 1,490 research institutions in 66 countries (Ni & Sugimoto, 2012). However, ProQuest is heavily biased toward English-language degrees conferred at North American universities and should thus be used with caution in global studies. An additional limitation for historical, paradigmatic, and analytic studies is the lack of information on the school or department in which the dissertation was conferred. Although recent dissertations contain this information, most do not. Therefore, to gain a "disciplinary sample," a researcher must rely on ProQuest Subject Categories. While these may be useful for identifying some disciplines, they are poorly suited to identifying highly interdisciplinary areas. Sugimoto, Russell, and Grant (2009) detail the substantial number of false positives generated when searching by Subject Categories and then validating by examining the cover page of the dissertation. There are also false negatives that can only be validated if one begins with an accurate sampling frame, generated either from the graduating school or a directory of recent graduates. In short, ProQuest Subject Categories are mere proxies for disciplinarity, and inferences based on them should acknowledge the limitations.

The ProQuest database can be useful in identifying a sample, but it has certain limitations for conducting academic genealogies: only half of the dissertations contain information on advisors, and even fewer contain information about committee members. In addition, the majority of dissertations containing advisorship information are from the last 20 years, making the study of multiple generations difficult. Nonetheless, ProQuest does have some advantages. For many dissertations, it offers a "Preview" that provides the first 24 pages of the dissertation—including the cover page and acknowledgments section where information for an academic genealogy can be found. Unfortunately, this means manual data collection and entry—a tedious and time-consuming activity for a large genealogy project. In addition, "Previews" tend to be available only for recent dissertations. Therefore, conducting an academic genealogy before 1950 requires using the physical resources of a library—either from a local collection or via interlibrary loan. Researchers should keep in mind that many of these items will only be available on microfiche, so access to a microfiche reader will be necessary.

Scholars may also be interested in exploring the data provided by open-access sources. For dissertations, the largest of these is the Networked Digital Library of Theses and Dissertations (NDLTD),[7] which contains more than one million theses. This site also lists a number of other electronic thesis and dissertation (ETD) sources, many of which are country specific. Country-specific sources are often government-mandated and therefore fairly reliable and current.

In many cases, the dissertation itself does not provide the information needed for the genealogy; other sources must be consulted. Several studies have used direct communication with scholars listed in the genealogy (Kelley & Sussman, 2007; Lubek et al., 1995; Montoye & Washburn, 1980), standard bibliographic reference material (Andraos, 2005; Robertson, 1994; Williams, 1993), obituaries (Williams, 1993), and solicitations for people to contact the authors with corrections or additional information (Montoye & Washburn, 1980). Recent projects have utilized the Web to solicit and crowdsource this information. For example, two large academic genealogies on the

Web, the Mathematics Genealogy Project and Academic Family Tree, use crowdsourced content as their primary data source.

User-generated content, whether via the Web or personal contact with scholars, has large potential for error. There is a tendency to "obscure or to disown inadvertently a non-prestigious institutional supervisor in favour of a highly visible intellectual mentor," leading to "mythical creation of a genealogical link" (Lubek et al., 1995, pp. 59–60). In some cases, these are unintentional. Previous studies (Boring & Boring, 1948) have based their analyses on vague questions, such as "Who was it who influenced you most in psychology up to the time you got the PhD?" or even the more explicitly phrased "Whose student were you?" Both questions tend to yield answers that are not precisely in keeping with the notion of doctoral advisors as understood in an academic genealogy. Biographies can often provide erroneous information based on historical myths, or provide ambiguous statements (e.g., "x was a student of y") that do not identify precisely the relationship between the individuals. Therefore, validation studies (e.g., David & Hayden, 2012) are necessary to ensure the accuracy of crowdsourced academic genealogy data. Lastly, the often highly emotional relationships between students and their advisors can cause problems—in cases where the relationship ended on poor terms, either the student or advisor may be unwilling to acknowledge the relationship. In such instances, the student often declares that another member of the dissertation committee was more of a "mentor" than the official chair. This brings up issues of ethics pertaining to genealogies, since anonymity is impossible. However, given the public nature of the data, studies typically do not require permission from institutional review boards (IRBs) or those named in the genealogy.

Visualizing Results

Published visualizations of academic genealogies mimic the structure of genetic family trees: names are provided, beneath these names a place and date are given, vertical lines connect "parents" to "children," and academic "siblings" are connected via horizontal lines. There is software available for the construction of academic genealogies (e.g., Kelley and Sussman [2007] used Microsoft Visio), although many scholars simply visualize by hand or use word processing software. Some academic genealogies contain birth and death dates (e.g., Stella, 2001; Tyler & Tyler, 1992); however, it is more common to list the graduation date and institution at which a doctoral degree was conferred (e.g., Andraos, 2005; Mitchell, 1992). There are alternative visualization approaches—for example, listing a series of advisors in a table, without any vertical lines (e.g., Robertson, 1994), providing photographs of each advisor (e.g., Jackson, 2011), crafting a descending matrix in tabular form (e.g., Bennett & Lowe, 2005), or extending the tree metaphor by generating a tree with the listing of descendants in branches (e.g., Bennett & Lowe, 2005). However, there are limits to what can be done in print (or traditionally print-based) publications.

Online projects provide more flexibility in generating visualizations of academic genealogies. The MPACT project uses the DOT language (rendered by the open-source software Graphviz) to provide the directed graphs that represent the academic genealogies. This allows automatic rendering of the representations whenever new information is added to the database. Furthermore, users can interact with the visualization by selecting any node in the graph and generating a new graph based on this node (where all descendants are visualized and each ancestor is provided). Neurotree visualizes the family tree through a set of PHP scripts and also includes a brief biographic information page for each individual (David & Hayden, 2012). However, there are limitations, because visualizations only provide links between advisors and not between committee members or additional intellectual influences. Other online projects have used variants on family genealogy software (e.g., Bennett & Lowe, 2005) or simple web programming language, using hyperlinks to navigate the genealogy (e.g., the Mathematics Genealogy Project). Many of these online projects offer personalized genealogy posters (e.g., personal genealogy posters cost $60 from the Mathematics Genealogy Project; departmental posters cost $165).

Conclusion

Dissertations are an underutilized source of information on the growth and evolution of science (Andersen & Hammarfelt, 2011). What we know about the emergence and development of disciplines, the diffusion of knowledge, and the evolution of science has been predominantly charted through analysis of journal articles. Symbolic capital is calculated by ascertaining the number of publications, the impact factors of the venues in which one is published, and the number of citations one receives. However, as noted by many (e.g., Cronin & La Barre, 2004; Larivière, Archambault, Gingras, & Vignola Gagné, 2006; Sula, 2012), journal articles do not tell the entire story or, for some disciplines, even a significant part of the story. In addition, contributions to scholarship made by those who engage in substantial amounts of mentoring are largely overlooked. Academic genealogy provides a means to recognize and metricize these contributions and to provide a new lens for understanding knowledge diffusion.

This chapter provides the first typology of academic genealogies and the outcomes of each. Although of limited scholarly value, there is certainly a place for honorific and egotistical genealogies: these highlight the importance of mentoring in an academic career, help to contextualize one's work in the larger scholarly sphere, and can be used as pedagogical tools. However, most scientific value is likely to be derived from historical, paradigmatic, and analytic genealogies, which provide opportunities to systematically depict disciplinary histories, interdisciplinarity, and knowledge diffusion processes.

Recent developments in network science (see West and Vilhena, chapter 8, this volume) and big data initiatives should serve to enhance the academic genealogy research front. Many timely and important topics can be addressed by positioning academic genealogy research in the context of network science, such as by examining bias and power cliques in doctoral mentoring networks. For example, a report on National Institutes of Health (NIH) funding demonstrated that white applicants were awarded a disproportionate number of awards (Ginther et al., 2011). A social networking explanation was given by an *Economist* summary suggesting that advisors may unconsciously choose advisees of their own race, providing a perpetuating advantage in scholarship ("Racial Discrimination in Science," 2011). This could be empirically tested by adding race and other social variables to an academic genealogy analysis, which is particularly important given the change in doctoral student demographics in recent decades (Thurgood, Golladay, & Hill, 2006). In addition, anecdotal evidence suggests that pedigree and academic genealogy create power cliques in scholarly communication and the academic labor market, with the ability to constrict academic and disciplinary mobility. Analytic genealogy could be used to examine, explain, and hopefully eradicate these and other violations of the universal norms of science (Merton, 1973).

Mentors and doctoral students leave traces of their scholarly interaction in acknowledgments (Cronin, 1991) and in patterns of coauthorship (Sugimoto & Cronin, 2012; Sugimoto, 2011). However, for many advisor-advisee relationships, there is no trace besides the cover page of the dissertation. Bringing these relationships to light could reveal underlying paths of knowledge diffusion, academic mobility, and disciplinary developments, particularly when academic genealogy is used in conjunction with other scientometric and social variables to provide a multidimensional view of the scholarly landscape.

Acknowledgment

Work on this chapter was enabled by the Faculty Research Support Program at Indiana University Bloomington and NSF grant no. 1158670.

Notes

1. http://www.scs.illinois.edu/~mainzv/Web_Genealogy/index.php. At the time of writing, the database contained data from the faculty of UIUC and nine other chemistry departments. In total, the database included 1,070 advisors, 1,032 students, and 44 scientists (as detailed in a personal email communication from Vera Mainz).

2. http://aigp.eecs.umich.edu/about. Although the site is no longer active, in early 2013 it claimed to have 16,741 researchers, from 1,569 schools and 52 countries.

3. http://genealogy.math.ndsu.nodak.edu

4. http://www.ibiblio.org/mpact/

5. http://academictree.org/about.php

6. http://neurotree.org/neurotree/growth.php

7. http://www.ndltd.org/find

References

Abbott, A. (1999). *Department and discipline: Chicago sociology at one hundred.* Chicago: University of Chicago Press.

Abbott, A. (2001). *Chaos of disciplines.* Chicago: University of Chicago Press.

The academic family tree. (n.d.). Retrieved from http://Academictree.org.

Andersen, J. P., & Hammarfelt, B. (2011). Price revisited: On the growth of dissertations in eight research fields. *Scientometrics, 88*(2), 371–383.

Andraos, J. (2005). Scientific genealogies of physical and mechanistic organic chemists. *Canadian Journal of Chemistry, 83*(9), 1400–1414.

Bennett, A. F., & Lowe, C. (2005). The academic genealogy of George A. Bartholomew. *Integrative and Comparative Biology, 45*(2), 231–233.

Boring, M. D., & Boring, E. G. (1948). Masters and pupils among the American psychologists. *American Journal of Psychology, 61*(4), 527–534.

Chang, S. (2010). *Academic genealogy of mathematics.* Singapore: World Scientific Publishing Co.

Cronin, B. (1991). Let the credits roll: A preliminary examination of the role played by mentors and trusted assessors in disciplinary formation. *Journal of Documentation, 47*(3), 227–239.

Cronin, B., & La Barre, K. (2004). Mickey Mouse and Milton: Book publishing in the humanities. *Learned Publishing, 17*, 85–98.

Davenport, C. B. (1915). The value of scientific genealogy. *Science, 41*(1053), 337–342.

David, S. V., & Hayden, B. Y. (2012). Neurotree: A collaborative, graphical database of the academic genealogy of neuroscience. *PLoS ONE, 7*(10), e46608.

De Mey, M. (1992). *The cognitive paradigm: An integrated understanding of scientific development*. Chicago: University of Chicago Press.

Durnin, J. V. G. A. (1991). Practical estimates of energy requirements. *Journal of Nutrition, 121*(11), 1907–1913.

Foucault, M. (1995). *Discipline and punish: The birth of the prison* (A. Sheridan, Trans.). New York: Vintage Books. (Original work published 1975)

Ginther, D. K., Schaffer, W. T., Schnell, J., Masimore, B., Liu, F., Haak, L. L., et al. (2011). Race, ethnicity, and NIH Research awards. *Science, 333*(6045), 1015–1019.

Girves, J. E., & Wemmerus, V. (1988). Developing models of graduate student degree progress. *Journal of Higher Education, 59*(2), 163–189.

Jackson, D. C. (2011). Academic genealogy and direct calorimetry: A personal account. *Advances in Physiology Education, 35*(2), 120–128.

Kelley, E. A., & Sussman, R. W. (2007). An academic genealogy on the history of American field primatologists. *American Journal of Physical Anthropology, 132*(3), 406–425.

Klein, J. T. (1996). *Crossing boundaries: Knowledge, disciplinarities, and interdisciplinarities*. Charlottesville: University Press of Virginia.

Kuhn, T. S. (1996). *The structure of scientific revolutions*. Chicago: University of Chicago Press (3rd ed.). (Original work published 1962).

Larivière, V., Archambault, É., Gingras, Y., & Vignola Gagné, É. (2006). The place of serials in referencing practices: Comparing natural sciences and engineering with social sciences and humanities. *Journal of the American Society for Information Science and Technology, 57*(8), 997–1004.

Lubek, I., Innis, N. K., Kroger, R. O., McGuire, G. R., Stam, H. J., & Herrmann, T. (1995). Faculty genealogies in five Canadian universities: Historiographical and pedagogical concerns. *Journal of the History of the Behavioral Sciences, 31*(1), 52–72.

Malmgren, R. D., Ottino, J. M., & Amaral, L. A. N. (2010). The role of mentorship in protégé performance. *Nature, 465*(3), 622–627.

Marchionini, G., Solomon, P., Davis, C., & Russell, T. (2006). Information and library science MPACT: A preliminary analysis. *Library & Information Science Research, 28*(4), 480–500.

Merton, R. K. (1973). *The sociology of science: Theoretical and empirical investigations*. Chicago: University of Chicago Press.

Mitchell, M. F. (1992). A descriptive analysis and academic genealogy of major contributors to JTPE in the 1980s. *Journal of Teaching in Physical Education, 11*(4), 426–442.

Montoye, H. J., & Washburn, R. (1980). Research quarterly contributors: An academic genealogy. *Research Quarterly for Exercise and Sport, 51*(1), 261–266.

Newton, F. B. (1995). Academic genealogy: A staff development activity. *Journal of College Student Development, 36*(1), 89–90.

Ni, C., & Sugimoto, C. R. (2012, October 25–31). *Using doctoral dissertations for a new understanding of disciplinarity and interdisciplinarity* [Poster]. Presented at the Annual Meeting of the American Society for Information Science & Technology, Baltimore.

Racial discrimination in science. (2011). *The Economist.* Retrieved from http://www.economist.com/node/21526320.

Robertson, J. M. (1994). Tracing ideological perspectives through 100 years of an academic genealogy. *Psychological Reports, 75*(2), 859–879.

Russell, T. G., & Sugimoto, C. R. (2009). MPACT family trees: Quantifying academic genealogy in library and information science. *Journal of Education for Library and Information Science, 50*(4), 248–262.

Stella, V. J. (2001). Invited editorial: My mentors. *Journal of Pharmaceutical Sciences, 90*(8), 969–978.

Sugimoto, C. R. (2011). Collaboration in information and library science doctoral education. *Library & Information Science Research, 33*, 3–11.

Sugimoto, C. R. (2012). Are you my mentor? Identifying mentors and their roles in LIS doctoral education. *Journal of Education for Library and Information Science, 53*(1), 2–19.

Sugimoto, C. R., & Cronin, B. (2012). Biobibliometric profiling: An examination of multifaceted approaches to scholarship. *Journal of the American Society for Information Science and Technology, 63*(3), 450–468.

Sugimoto, C. R., Li, D., Russell, T., Finlay, S., & Ding, Y. (2011). The shifting sands of disciplinary development: Analyzing library and information science (LIS) dissertations. *Journal of the American Society for Information Science and Technology, 62*(1), 185–204.

Sugimoto, C. R., Ni, C., Russell, T. G., & Bychowski, B. (2011). Academic genealogy as an indicator of interdisciplinarity: An examination of dissertation networks in library and information science. *Journal of the American Society for Information Science and Technology, 62*(9), 1808–1828. doi:10.1002/asi.21568.

Sugimoto, C. R., Russell, T. G., & Grant, S. (2009). Library and information science doctoral education: The landscape from 1930–2007. *Journal of Education for Library and Information Science, 50*(3), 190–202.

Sugimoto, C. R., Russell, T. G., Meho, L. I., & Marchionini, G. (2008). MPACT and citation impact: Two sides of the same scholarly coin? *Library & Information Science Research, 30*(4), 273–281.

Sula, C. A. (2012). Visualizing social connections in the humanities: Beyond bibliometrics. *Bulletin of the American Society for Information Science & Technology, 38*(4), 31–35.

Taleb, N. N. (2010). *The black swan: The impact of the highly improbable* (2nd ed.). New York: Random House.

Thurgood, L., Golladay, M. J., & Hill, S. T. (2006). *U.S. Doctorates in the 20th Century: Special Report.* Arlington, VA: National Science Foundation.

Turner, S. (2000). What are disciplines? And how is interdisciplinarity different? In P. Weingart & N. Stehr (Eds.), *Practising interdisicplinarity* (pp. 46–65). Toronto: University of Toronto Press.

Tyler, V. M., & Tyler, V. E. (1992). The academic genealogy of Arthur E. Schwarting, Phamacognosist. *Journal of Natural Products, 55*(7), 833–844.

Williams, R. B. (1993). Contributions to the history of psychology: XCIII. Tracing academic genealogy. *Psychological Reports, 72*(1), 85–86.

Yan, E., Ding, Y., Milojević, S., & Sugimoto, C. R. (2012). Topics in dynamic research communities: An exploratory study for the field of information retrieval. *Journal of Informetrics, 6*(1), 140–153.

Yoels, W. C. (1971). Destiny or dynasty: Doctoral origins and appointment patterns of editors of the *American Sociology Review*, 1948–1968. *American Sociologist, 6*(2), 134–139.

Zittoun, T., Perret-Clermont, A.-N., & Barrelet, J.-M. (2008). The socio-intellectual genealogy of Jean Piaget. In A.-N. Perret-Clermont & J.-M. Barrelet (Eds.), *Jean Piaget and Neuchâtel: The learner and the scholar*, pp. 109–118. New York: Psychology Press.

Part V Perspectives

20 A Publishing Perspective on Bibliometrics

Judith Kamalski, Andrew Plume, and Mayur Amin

Historical Context of Bibliometrics

The birth of scientometrics can be traced to the first half of the 20th century, with J. D. Bernal's *Social Function of Science* (1939) setting forth the blueprint for the nascent study of the "science of science" (Garfield, 2009). Eugene Garfield (1955) published his influential proposal for the creation of citation indexes in science and shortly thereafter Derek de Solla Price, the "father of scientometrics," published his landmark works *Science Since Babylon* (1961) and *Little Science, Big Science* (1963). However, it was Garfield's introduction of the *Science Citation Index* as a five-volume print edition covering 613 journals and 1.4 million citations in 1964 that offered the first widely available resource for scientometric studies in which published documents were linked by their cited references. This provided the catalyst for the invention of journal metrics, including Garfield's Impact Factor. Increasingly, people became interested in the Impact Factor as a proxy for scientific quality.

In the decades that have followed, the corpus of research on scientometrics—and more broadly, bibliometrics—has grown substantially. Over time, the uses to which scientometric indicators have been put have shifted from a focus on journal-level analysis and field-specific studies to applications at author, institute, and country level, reflecting the maturation of scientometrics in research evaluation and research policy circles. As the use of bibliometric indicators has grown, so, too, has the sophistication of those indicators increased (see table 20.1). Despite this, much debate continues to focus on the use, and abuse, of the traditional journal Impact Factor—indeed, an entire special issue of *Scientometrics*, the major journal in the field, was recently devoted to this (see Braun, 2012, for an introduction to the issue).

Table 20.1
Types of bibliometric indicators

Type (generation)	Description	Typical examples
First	Basic indicators; relatively easy to obtain from sources that have been available for decades	Number of publications; number of citations; journal impact metrics
Second	Relative or normalized indicators, correcting for particular biases (e.g., differences in citation practices between subject fields)	Relative or field-normalized citation rates
Third	Based on advanced network analysis using parameters such as network centrality	Influence weights; SCImago Journal Rank; "prestige" indicators

Source: Reproduced with permission from Moed & Plume (2011).

Why Should Publishers Be Involved in This Area of Research?

In this book, different perspectives on bibliometrics are presented. Here, we elaborate on the perspective of publishing organizations. Increasingly, publishers have taken an interest in bibliometrics, for several reasons that will be explained in more detail.

The first reason is that, as publishers, we consider it an important part of our role to help the research communities we serve to better understand the use and value of bibliometrics. Even the relatively simple, first-generation metrics (exemplified by the Impact Factor; see table 20.1) are open to interpretation without appropriate information on what is (and is not) counted and how indicators vary by journal size, subject field, or over time (Amin & Mabe, 2000). Bibliometric indicators can lead the user to important insights, but only when there is knowledge of their methodologies and respect for their limitations. There is no single "perfect" metric; instead, the power of bibliometrics lies in the fact that a variety of indicators can be combined to provide a multidimensional view of research (Moed, Aisati, & Plume, 2012). Bringing other stakeholders in scholarly communication to this realization is an important aspect of our role.

Second, there is a need to evaluate our own offerings as a publisher of research journals, books, and magazines—identifying opportunities, tracking performance, and assessing our position in the market on the basis of both quantity and quality. As such, we turn the lens of research assessment on ourselves using bibliometric tools.

The third reason is perhaps the broadest one, since it encompasses the desire to build an evidence-based understanding of the research communities we serve and the ways they are changing. It is only by having a complete and unbiased picture of the scholarly landscape that we can provide information solutions that satisfy the needs of those who push the frontiers of human knowledge.

The structure of the remainder of the chapter emerges from the three reasons enumerated above. In the following pages, we discuss the information we provide to stakeholders in research and scholarly communication, then the ways we evaluate and improve on our own offerings as a scholarly publisher, before finally turning to a discussion on how we seek to better understand the research communities we serve.

To Inform Stakeholders in Research and Scholarly Communication

Many of the analyses we perform, and the insights that stem from these, are intended to inform or educate. As outlined above, an important role for a scholarly publisher is to help people understand bibliometric indicators and to ensure that they are used in an informed way. When explaining bibliometric principles, our focus is always on how useful a certain indicator, approach, or method is when applied in a specific context. For example, a treatise on the "use and abuse" of the Journal Impact Factor (Amin & Mabe, 2000) was published to illustrate some of the properties of the Impact Factor as a journal performance evaluation tool and to show how misuse and misinterpretation of this indicator may lead to inappropriate conclusions. At Elsevier we have, since 2001, hosted regular editors' conferences and use these as a forum for discussion and debate about bibliometric indicators commonly applied to journals and authors (the latter typified by the ubiquitous *h*-index; Hirsch, 2005). These conferences have reached over 2,500 journal editors directly over the years, and the insights gained by these esteemed senior researchers and thought leaders are likely to have reached many more in their respective fields. Finally, Elsevier seeks to promote best practice within the scholarly community and across the publishing industry by offering timely and considered guidelines on issues such as ethical citation practices and the effect of deviating from community norms on journal reputation (Huggett, 2012).

We have also partnered with other organizations to spread information on the appropriate use of bibliometric indicators to a wider audience. An example of such a partnership is provided by Elsevier's collaboration with the Global Ocean Ecosystem Dynamics (GLOBEC) research organization (Kamalski & Barranguet, 2010). GLOBEC is a research program initiated in 1990 with the stated aim "to advance our understanding of the structure and functioning of the global ocean ecosystem, its major subsystems, and its response to physical forcing so that a capability can be developed to forecast the responses of the marine ecosystem to global change" (GLOBEC, 2012, para. 3). In practice, this means focusing on shorter-term effects, such as overfishing and the changing ways we use the seas, as well as efforts to monitor and understand long-term global change in the broadest sense. To celebrate GLOBEC's 20th anniversary, a bibliometric study was conducted to identify the topics with the greatest impact among GLOBEC publications, as well as the most productive and highly cited authors and institutes. One of the analyses is reproduced in figure 20.1, where the different trends in impact for the different subject areas are shown.

To Evaluate and Improve Our Offerings

Together with journal editors, scholarly publishers aim at constantly improving the journals they publish for the benefit of the communities of authors and readers they

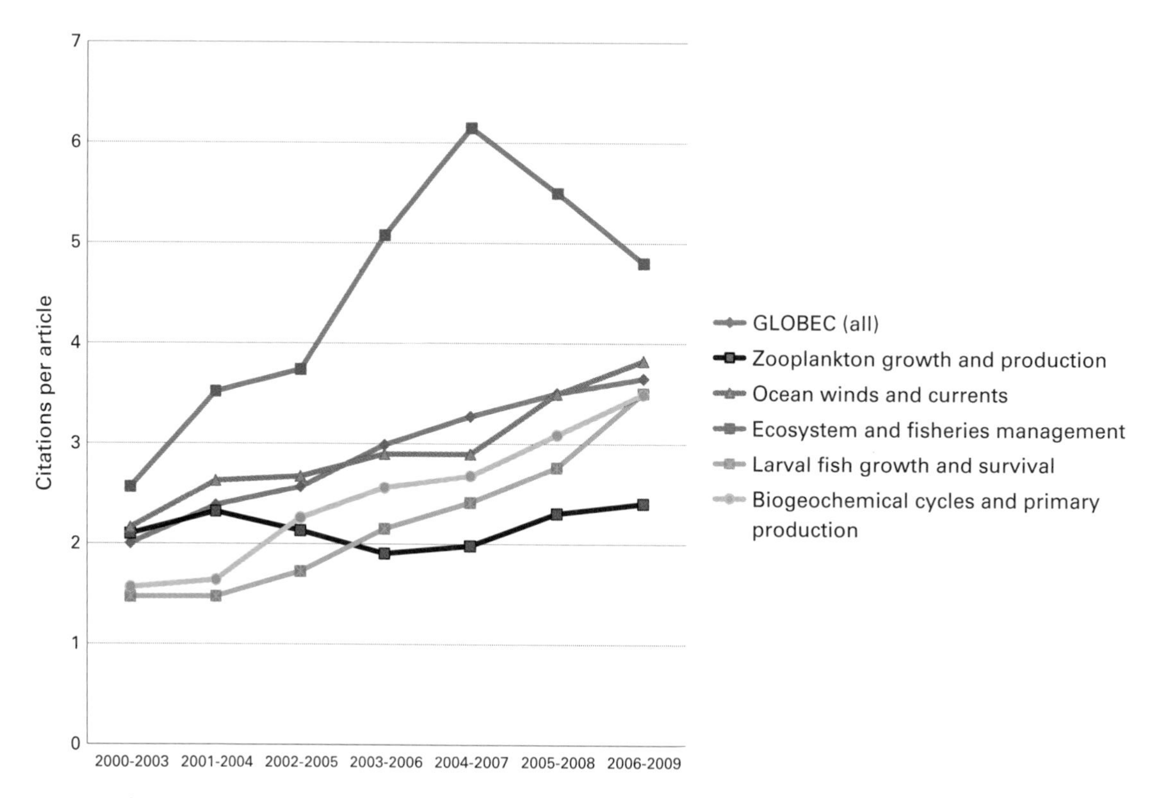

Figure 20.1
Citations per article (on a four-year rolling window, e.g., 2006–2009 data is for cites received in 2006–2009 to articles published in the same period) for each of the five major subject areas identified in GLOBEC output, as well as for GLOBEC as a whole. Reproduced with permission from Kamalski and Barranguet (2010).

serve. But how can one measure whether the implementation of a strategic change, such as a shift in topical scope or the introduction of a new article type, has been a success? Metrics to assess journal performance are important and widely available, but are not without controversy (Corbyn, 2009a, 2009b). There is a growing body of literature suggesting that the Journal Impact Factor, for example, is itself decreasing in impact (Reedijk & Moed, 2008) and becoming even less representative of the distribution of citations across articles within the journal (Lozano, Larivière, & Gingras, 2012). This has led to the development of a slew of new metrics in the last few years, exemplified by Source Normalized Impact per Paper (SNIP) and SCImago Journal Rank (SJR), both using Scopus as the data source (Colledge et al., 2010; González-Pereira, Guerrero-Bote, & Moya-Agenón, 2010; Moed, 2011). Unlike the Impact Factor, which functions as a popularity metric, SJR reflects the prestige transferred to a journal through the citations received from other journals in a weighted network (González-Pereira et al., 2010). SNIP, arguably the most sophisticated of all journal-level indicators, accounts for differences in citation potential between subject fields and so permits the direct comparison of journals across different subjects in an unbiased way (Moed, 2010). In a guest editorial, Jones, Huggett, and Kamalski (2011) presented a more detailed overview of these journal metrics. For journal publishers, SNIP and SJR have the advantage of being calculated for more journals than the Impact Factor, primarily through the more extensive coverage of the global literature in the Scopus database.

In table 20.2, three different metrics are shown for a selection of Elsevier biomaterials and bioengineering journals. While these journals span a range of Impact Factor values of over 7-fold, and SJR values of over 8-fold, note how the SNIP values are contracted around a range of just over 2.5-fold. The rankings reveal how these journals, all with very different scopes of coverage within the broadly defined fields of biomaterials

Table 20.2
2011 metrics for Elsevier biomaterials and bioengineering journals. Values in parentheses show the rank of each journal by each indicator

Journal	Impact Factor (2011)	SJR (2011)	SNIP (2011)
Biomaterials	7.404 (1)	0.633 (1)	3.730 (1)
Acta Biomaterialia	4.865 (2)	0.285 (2)	1.979 (3)
Colloids and Surfaces B: Biointerfaces	3.456 (3)	0.187 (4)	1.222 (6)
Dental Materials	3.135 (4)	0.159 (5)	2.237 (2)
Journal of the Mechanical Behavior of Biomedical Materials	2.814 (5)	0.222 (3)	1.454 (5)
Journal of Bionic Engineering	1.023 (6)	0.073 (6)	1.490 (4)

and bioengineering, fare quite differently when compared using different metrics. For example, *Colloids and Surfaces B: Biointerfaces* ranks 3rd by Impact Factor and 4th by SJR, but ranks 6th of these journals by SNIP. Conversely, *Journal of Bionic Engineering* ranks 6th in both Impact Factor and SJR but rises to 4th under SNIP. Indeed, only *Biomaterials* retains a consistent ranking (1st) across all three metrics when compared against this selection of other journals. These data illustrate the importance of including more than one metric in any comparison in order to offer a more balanced view. However, we do not believe that numbers by themselves can offer a complete picture: bibliometric indicators should always be used in concert with the informed opinions of academic editors and the research community.

Even though the newer metrics such as SNIP and SJR address some of the criticisms that have been expressed about the Impact Factor, we recognize that a journal represents many things to many people, and that citation-based metrics represent only part of the story. For example, journals in fields with a significant professional or applied spectrum (for which peer-reviewed journal publication and citation are not core elements of the job role) may place more emphasis on the usefulness or applicability of an article, irrespective of whether this article is subsequently cited it in the literature. For such communities, exemplified by nursing and allied health and many fields within the social sciences, indicators of article usage such as the forthcoming Usage Factor (COUNTER, 2012), may well serve them better as a measure of a journal's importance to their work.

While bibliometric indicators are used to advise journal publishing staff and editors on how to improve their journal, they can also be applied to the shifting landscape of the wider field to identify emerging areas where new journals or conferences may be appropriate. For example, we are frequently interested in mapping entire research fields, reflecting the output of dozens or even hundreds of journals, identifying the emergence of new research topics and the decline of old ones, and visualizing the citations rates associated with these topics. An example of such an analysis, for a set of journals reflecting the subject of energy research, is shown in figure 20.2.

The map clearly reveals a tripartite structure to the field of energy research, with the chemical aspects of the field, exemplified by solar and fuel cells and dominated by the journals of the American Chemical Society, forming the relatively "hot" corner. In contrast, the engineering perspective on energy generation, conversion, and storage typically published in the journals of the Institute of Electrical and Electronics Engineers (IEEE) forms a "cooler" citation corner at the top of the map. The more applied end of the field, relating to current energy generation from fossil fuels, forms the third corner of the map with the lowest average citations per article. Such maps are used for the

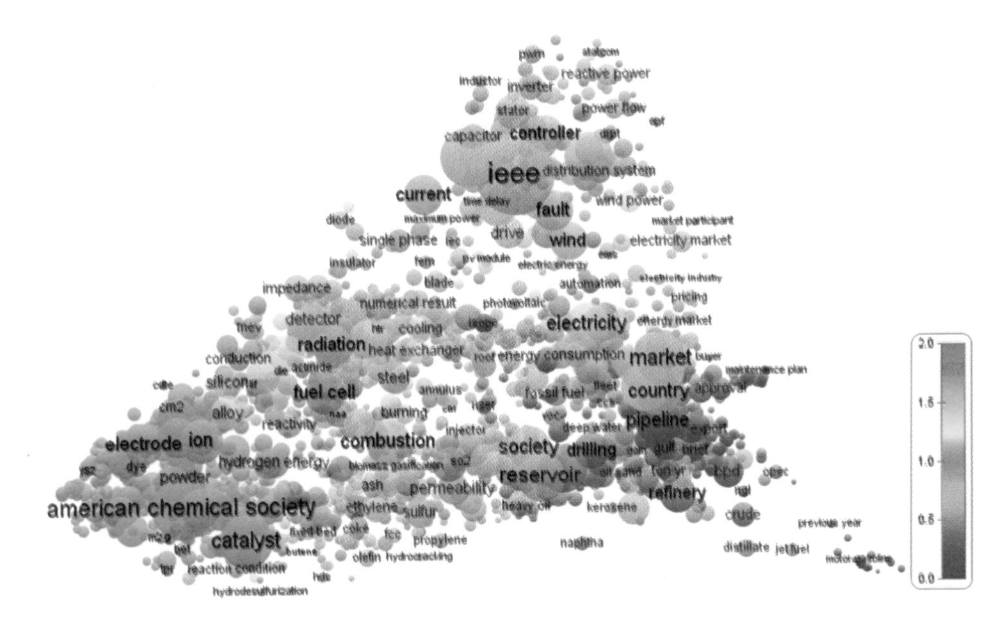

Figure 20.2
A term map for energy research, using VOS Viewer (www.vosviewer.com) to map each noun phrase by their co-occurrence in articles published in 2005–2009 from across dozens of relevant journals. Term proximity reflects co-occurrence frequency, while blob size indicates the number of articles with at least one co-occurrence of the term and blob color indicates citation frequency relative to the aggregate citation frequency for all terms in the map on a heatmap scale.

strategic development of individual journals, and to look for topics suitable for timely invited reviews of the literature, commissioned special issues within existing journals, and opportunities for new journals or conferences.

To Understand the Research Communities We Serve

A clear understanding by scholarly publishers of the fundamental drivers of change in the publication and citation behaviors of authors is one of the keys to the development and delivery of products and services that facilitate research. Such monitoring of underlying trends that have the potential to change the nature of the scholarly publishing landscape is exemplified by much of the work published in the *International Comparative Performance of the UK Research Base* (Elsevier, 2011), a report produced by Elsevier for the UK government's Department for Business, Innovation and Skills (BIS). This report took a holistic view of the United Kingdom's research base from the perspective of inputs (R&D spending and human capital), throughputs (such as research collaboration), and outputs (including articles, usage, citations, patents and more).

One of the key insights from the report was that—despite the United Kingdom's relatively stable researcher population over time (at approximately a quarter-million in recent years)—such figures mask the high level of international mobility exhibited by many of these individuals. Indeed, almost 63% of active researchers publishing under a UK affiliation in the period 1996–2010 also published under an affiliation in one or more other countries in the same period (see figure 20.3). Such mobility, also known as brain circulation, appears to be a key factor in the success of the United Kingdom as

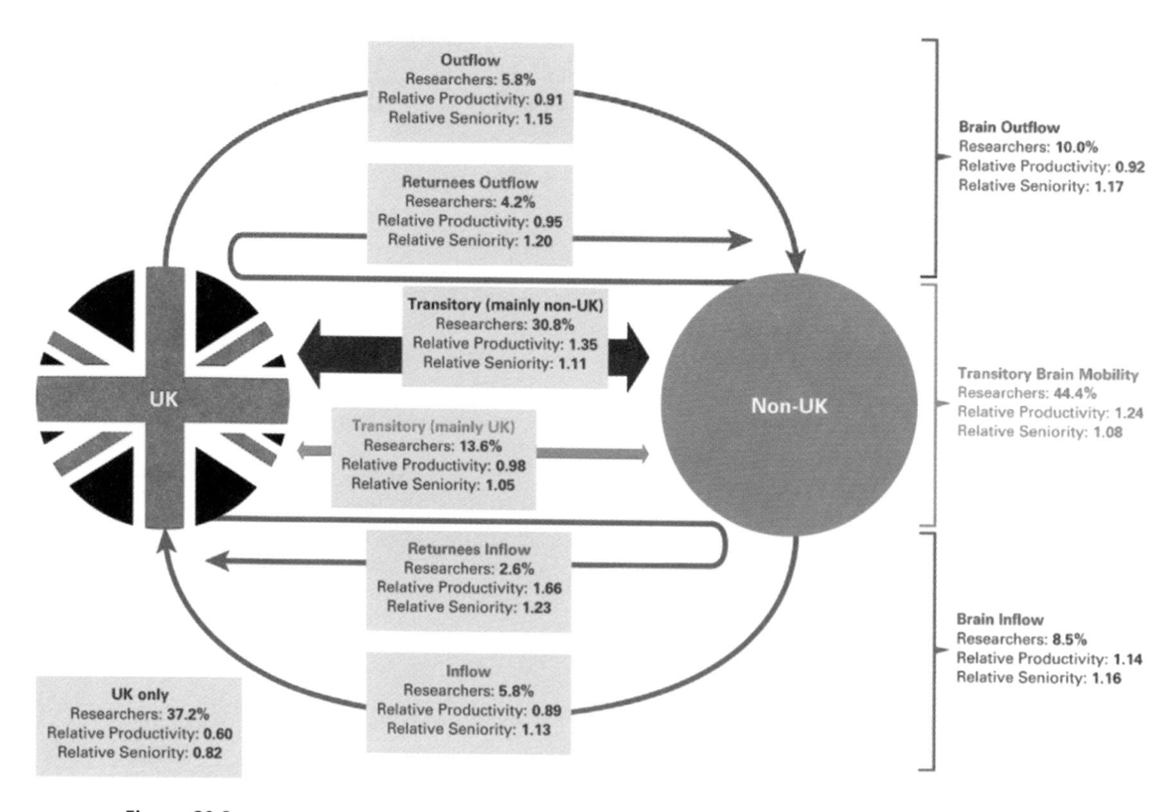

Figure 20.3

International mobility of UK researchers, 1996–2010. Based on affiliations listed in their publications in Scopus, migratory researchers (brain outflow and inflow) are defined as individuals that migrate permanently or who return after two or more years abroad ("returnees"), while transitory researchers (transitory brain mobility) are defined as individuals spending less than two years abroad. Relative Productivity and Relative Seniority are respectively defined as the count of articles published per year, and years of publication activity since first appearance in the data, relative to all UK researchers. Reproduced with permission from Elsevier (2011).

a research nation, since the 37% of active UK researchers who did not appear to have published from another country exhibited lower-than-average productivity and seniority. The largest single mobility group identified in this analysis were the 44% of active UK researchers who showed a pattern referred to as "transitory mobility," with stays of less than two years abroad, but who may primarily publish under a UK affiliation or a non-UK affiliation. The productivity and seniority of these transitory researchers is higher than average. Using a somewhat different approach to mining Scopus author-profile data to track international research mobility, we have extended these observations to other countries beyond the United Kingdom in a recent paper (Moed, Aisati, & Plume, 2013).

Bibliometric analysis also enables deeper understanding of smaller areas of interest. An example of such a study is the analysis of urban studies conducted by Kamalski and Kirby (2012). The study examines how the field of urban studies is constructed by looking at three contexts: first, in the narrowly defined population of journals that constitutes the Thomson Reuters classification of urban studies; second, in the larger population of journals deemed to be within the social and behavioral sciences; and third, in a subset of the applied sciences. The results show that, using keyword analysis, it is possible to identify three distinct spheres of "urban knowledge" that contain some overlap but also significant differences. Such bibliometric analysis may help us understand how a certain field operates, what subgroups exist, and how these subgroups may interact.

Another underlying theme in the science of science on which we often turn our analytic tools is international collaboration. International collaboration rates are rising (He, 2009), and internationally coauthored papers have been shown to have a positive effect on citations (Glänzel, 2001). A letter in *Nature* (Tijssen, Waltman, & Van Eck, 2011) looked at international collaboration distances across different countries and found that between 1980 and 2009, the average collaboration distance (i.e., the linear distance separating two collaborating but not co-located authors) increased more or less linearly from 334 to 1,553 kilometers. Analysis produced by Elsevier for the Royal Society's *Knowledge, Networks & Nations* report (2011) reaffirms the previously established positive association between international collaboration and citation impact; the *International Comparative Performance of the UK Research Base* (Elsevier, 2011) report goes further, demonstrating increased citation impact for internationally collaborative articles over and above that for articles representing domestic collaboration.

Discussion and Conclusion

Bibliometric analysis is a useful tool for scholarly publishers but is not free of potential pitfalls. There is always the risk of oversimplifying a complex system by applying metrics and indicators to it. It is therefore imperative to always complement such numbers with expert opinions. No single metric can provide the whole picture, and a panel of metrics informed by expert opinion is typically the best way to analyze the performance of a journal, the changes in a subject field or country, and indeed the overall position of the portfolio of journals and articles published by a given publishing house. The most insightful bibliometric research typically results from collaboration between bibliometricians and people with expert knowledge about the topic of interest.

In conclusion, the examples of bibliometric analyses in this chapter have shown some of the diversity of points of interest within a commercial publishing environment. Such analyses have a multitude of uses in publishing, from the narrow level of a single journal (e.g., by looking at journal metrics or topics within a journal) to the broadest scope of the research landscape across fields and geographies (e.g., by looking at shifting topics or international researcher mobility). It is for precisely these reasons that we continually look at new ways to gain insights from bibliometric data and make these available as widely as possible.

References

Amin, M., & Mabe, M. (2000). Impact factors: Use and abuse. *Perspectives in Publishing, 1*(1). Retrieved from www.elsevier.com/framework_editors/pdfs/Perspectives1.pdf.

Bernal, J. D. (1939). *The social function of science*. London: Routledge.

Braun, T. (2012). Editorial in "special discussion issue on journal impact factors." *Scientometrics, 92*(2), 207–208.

Colledge, L., Moya-Anegón, F., Guerrero-Bote, V., López-Illescas, C., El Aisati, M., & Moed, H. (2010). SJR and SNIP: Two new journal metrics in Elsevier's Scopus. *Serials, 23*(3), 215–221.

Corbyn, Z. (2009a). Hefce backs off citations in favour of peer review in REF. *Times Higher Education*. Retrieved from http://www.timeshighereducation.co.uk/story.asp?storycode=407041.

Corbyn, Z. (2009b). A threat to scientific communication. *Times Higher Education*. Retrieved from http://www.timeshighereducation.co.uk/story.asp?storycode=407705.

COUNTER. (2012). Usage-based measures of journal impact and quality. Retrieved from http://www.projectcounter.org/usage_factor.html.

Elsevier (2011). *International Comparative Performance of the UK Research Base—2011*. London: Elsevier. Retrieved from http://www.bis.gov.uk/assets/biscore/science/docs/i/11-p123-international-comparative-performance-uk-research-base-2011.pdf.

Garfield, E. (1955). Citation indexes for science. *Science, 122*(3159), 108–111.

Garfield, E. (2009). From the science of science to Scientometrics: Visualizing the history of science with HistCite software. *Journal of Informetrics, 3*(3), 173–179.

Glänzel, W. (2001). National characteristics in international scientific co-authorship relations. *Scientometrics, 51*(1), 69–115.

GLOBEC. (2012). Homepage. Retrieved from http://www.globec.org.

González-Pereira, B., Guerrero-Bote, V., & Moya-Agenón, F. (2010). A new approach to the metric of journals' scientific prestige: The SJR indicator. *Journal of Informetrics, 4*(3), 379–391.

He, T. (2009). International scientific collaboration of China with the G7 countries. *Scientometrics, 80*(3), 571–582.

Hirsch, J. E. (2005). An index to quantify an individual's scientific research output. *Proceedings of the National Academy of Sciences of the United States of America, 102*(46), 16569–16572.

Huggett, S. (2012) Impact factor ethics for editors: How impact factor engineering can damage a journal's reputation. *Editor's Update, 36*. Retrieved from http://editorsupdate.elsevier.com/2012/06/impact-factor-ethics-for-editors.

Jones, T., Huggett, S., & Kamalski, J. (2011). Finding a way through the scientific literature: Indexes and measures. *World Neurosurgery, 769*(1–2), 36–38.

Kamalski, J., & Barranguet, C. (2010). Mapping 20 years of global ocean ecosystem research. *Research Trends, 19*. Retrieved from http://www.researchtrends.com/issue19-september-2010/mapping-20-years-of-global-ocean-ecosystem-research.

Kamalski, J., & Kirby, A. (2012). Bibliometrics and urban knowledge transfer. *Cities (London), 29*(Suppl. 2), S3–S8.

Lozano, G. A., Larivière, V., & Gingras, Y. (2012). The weakening relationship between the Impact Factor and papers' citations in the digital age. Retrieved from http://arxiv.org/abs/1205.4328.

Moed, H. (2010). Measuring contextual impact of scientific journals. *Journal of Informetrics, 4*(3), 265–277.

Moed, H. (2011). The source normalized impact per paper is a valid and sophisticated indicator of journal citation impact. *Journal of the American Society for Information Science and Technology, 62*(1), 211–213.

Moed, H., Aisati, M., & Plume, A. (2013). Studying scientific migration in Scopus. *Scientometrics 94*(3), 929–942.

Moed, H., Colledge, L., Reedijk, J., Moya-Agenón, F., Guerrero-Bote, V., Plume, A., et al. (2012). Citation-based metrics are appropriate tools in journal assessment provided that they are accurate and used in an informed way. *Scientometrics, 92*(2), 367–376.

Moed, H., & Plume, A. (2011). The multi-dimensional research assessment matrix. *Research Assessment,* May 2011 (23). Retrieved from http://www.researchtrends.com/issue23-may-2011/the-multi-dimensional-research-assessment-matrix.

Price, D. J. de Solla. (1961). *Science Since Babylon*. New Haven, CT: Yale University Press.

Price, D. J. de Solla. (1963). *Little science, big science*. New York: Columbia University Press.

Reedijk, J., & Moed, H. F. (2008). Is the impact of journal impact factors decreasing? *Journal of Documentation, 64*(2), 183–192.

Royal Society. (2011). *Knowledge, Networks & Nations: Global Scientific Collaboration in the 21st Century*. London: The Royal Society. Retrieved from http://royalsociety.org/policy/projects/knowledge-networks-nations/report.

Tijssen, R., Waltman, L., & Van Eck, N. J. (2011). Collaborations span 1,533 kilometres. *Nature, 473*, 154.

21 Science Metrics and Science Policy

Julia Lane, Mark Largent, and Rebecca Rosen

Science metrics have many uses. In this chapter we discuss their application in science policy, particularly focusing on how they can be used to inform policymakers' understanding about the way science investments affect the conduct of science.

Metrics are common in all policy domains, and most policy fields have settled on well-known indicators to inform policy. For example, labor policy relies heavily on the unemployment rate as a labor market indicator; economic policy on a plethora of measures, including the growth rate of the gross domestic product (GDP); and education policymakers on international math and science scores as measured by the Programme for International Student Assessment (PISA) test. These measures, while often imperfect, are widely used and the underlying theoretical justifications for their use are well known among policymakers. This, however, is not the case in science policy. There is a proliferation of science metrics, often without a clear rationale for their choice other than that the data exist and metrics can be generated from them.

Generating metrics from existing data can create incentive structures that result in undesirable outcomes. One of the most often cited articles in the management literature, "On the Folly of Rewarding A, while Hoping for B" (Kerr, 1975), describes how attempts by management to tie rewards to invalid outcome measures led to predictable and undesirable results. When the Heinz Company rewarded employees for divisional earnings increases, its managers manipulated the timing of shipments and prepayments; when Dun & Bradstreet rewarded employees for larger subscription increases, its employees deceived customers; and when Sears rewarded mechanics for auto repairs, more auto repairs were "authorized" by customers.

In this chapter we argue that it is important to draw on findings from the National Science Foundation's Science of Science and Innovation Policy (SciSIP) program to guide the construction of a robust theoretical and empirical framework for developing good metrics. The metrics of science need to be grounded within an organizing framework of the scientific enterprise, which means describing the behavior of scientists and

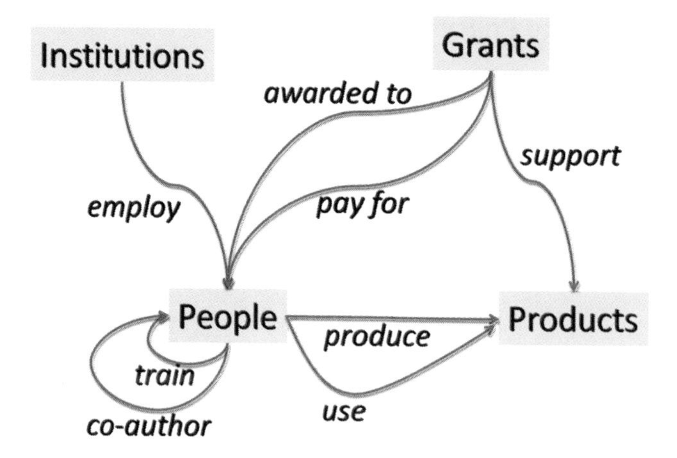

Figure 21.1
Framework for developing metrics. *Source*: Ian Foster, University of Chicago.

scientific networks, not merely the easily quantified scientific publications that have come to dominate much of today's scientific metrics. Such a framework is encapsulated in figure 21.1, which identifies individual scientists (or the scientific community consisting of the networks of scientists) as the "engine" that generates scientific ideas. Here, the theory of change is that there is a link between funding and the way those networks assemble. In turn, there is a link between scientific networks and the way those ideas are created, transmitted, and ultimately generate scientific, social, economic, and workforce "products." This organizing framework should form the basis of any metric. Without such a framework, policymakers are left to rely on scientific publications with the predictable results that incentive structures will be created that may do little more than encourage more scientific publications, rather than more and better science.

This chapter steps back and discusses what might be "good" metrics of scientific performance that might be useful for science agencies. These metrics should be based on multiple units of analysis: individual researchers, research institutions, funding agencies, or even the entire scientific sector. The development of good metrics will require a solid grounding in the theory of the production of science (the creation, transmission, and adoption of knowledge), the development of high-quality data, and examination by a community of practice.

Bibliometrics

Bibliometrics have come to dominate science metrics. They have a great deal of value in many contexts, and certainly contribute substantially to our understanding of the corpus of publications. However, their usefulness for describing the performance of science is limited. Indeed, bibliometrics were not designed to describe the scientific enterprise. Priem and Hemminger summarized a decade of study and asserted:

> Evaluators often rely on numerically-based shortcuts drawn from the closely related fields (Hood & Wilson, 2001) of bibliometrics and scientometrics—in particular, Thompson [sic] Scientific's Journal Impact Factor (JIF). However, despite the popularity of this measure, it is slow (Brody & Harnad, 2005); narrow (Anderson, 2009); secretive and irreproducible (Rosner, et al., 2007); open to gaming (Falagas & Alexiou, 2008); and based on journals, not the articles they contain. (Priem & Hemminger, 2010, para. 1)

There are two major concerns with the focus on bibliometrics. One is that, as noted, they do not do a good job of capturing the creation, transmission, and adoption of knowledge—so they do not capture scientific performance. The other is that they create the wrong incentives and foster bad science. Indeed, Ioannidis has shown that most published findings are false (Ioannidis, 2005); Karr and Young specifically blame the system for their finding that "any claim coming from an observational study is wrong" (Young & Karr, 2011, p. 116). The implications for science policy are profound. Simply put, if the pressure created by funding agencies and universities has created a publication treadmill, and the result is that most research results are wrong, there should be less, not more, funding of science. There should be more thoughtful review of published results, not more publications.

Our view is that bibliometrics were never intended to measure performance; they were created to classify documents via a set of methods that would allow for the quantitative analysis of the scientific and technological literature by library and information science professionals. Scientific performance, on the other hand, is essentially about the creation, transmission, and adoption of knowledge. In the absence of any sensible performance metrics for the creation, transmission, and adoption of knowledge, bibliometrics have been adopted to serve that need.

Useful Science Metrics

The foundations of good scientific metrics should be built on scientific advances in both the social sciences and in fields like graph theory and social network theory, which will enable us to formally describe the way scientists communicate ideas. The production of these metrics should build on new technology that can capture data on the complexities of scientific interaction and the links between scientists and their activities. They should be built on new technological advances that enable us to structure and use data sensibly, such as the development of graph-oriented databases and natural language processing techniques. Finally, the foundations should be created by a community of practice that develops and uses the metrics in making and implementing science policy with a keen eye to the incentive structure these metrics will institute.

The development of an appropriate theory of science metrics is clearly important. Metrics are typically adopted because stakeholders want simple, quantifiable standards to measure and reward both performance and highly visible behavior (Gibbons, 1998). Unfortunately, citation metrics have often been adopted based on their technical characteristics, rather than on their ability to capture the underlying latent variables of interest. To replace weak metrics, alternative measures need to be developed that capture the appropriate features of science. Indeed, as explained in the Federal Science of Science Policy Roadmap ("The Science of Science Policy," 2008, p. 16), "The Federal community lacks a theoretical framework that it can use to assess the impact of science and technology policies on discovery and resultant social welfare outcomes." Other open challenges include developing real-time evaluative and decision-making tools for assessing the contribution of public sector investments in science and technology to economic growth and social well-being.

Theory is urgently needed to link scientific metrics with the study of how science functions. Several fields of scholarly inquiry have a long history of studying the scientific enterprise and the role of science in society. Science and technology studies (STS) scholars, sociologists of knowledge and science, historians of science, and philosophers of science have studied the processes of scientific development and innovation, and they have developed a significant body of literature that advances the theoretical underpinnings of such questions. However, as with economists, science studies scholars' work has rarely been employed by policymakers in the analysis and implementation of science policy. This void was noted by John Marburger, Director of the Office of Science and Technology Policy, and his call for better coordination and development of various disciplines' activities to establish a more rigorous, justifiable, and effective science policy agenda for the federal government led to the development of the Federal Science of Science Policy Roadmap ("The Science of Science Policy," 2008).

Theory

Theories that link scientific metrics and scholarly studies of the scientific enterprise are starting to emerge in a variety of disciplines. Economists have studied the production of scientific information as a commodity, with very peculiar market aspects: it is paid for by public investment, only shared with other scientists under certain conditions (Haeussler, Jiang, Thursby, & Thursby, 2009), offered free to publishers, then resold back to the producers and their underwriters (Young, Ioannidis, & Al-Ubaydli, 2008). Sociologists study the sociology of scientific communication (Leydesdorff, 2010), and statisticians attempt to classify it (Lambe, 2007). Metrics based on these theories are already being developed: to identify "star" researchers and "star" ideas, and to measure the scientific performance of academic communities and countries ("Academic Ranking of World Universities," 2012). Other metrics being developed relate to the adoption and diffusion of academic ideas and the evolution of scientific communities (Paley, 2012). A number of disciplines are examining how the creation of metrics (and the associated changes in incentives) can change the way tasks are performed.

The theoretical foundations of science metrics should be independent of national boundaries, and international cooperation in funding theory could substantially advance collective knowledge. An excellent example of how scientific agencies can fund research into the development of high-quality metrics linked to the appropriate scientific outcomes is provided in table 21.1. This is drawn from a study of a sample of projects that were supported by the National Science Foundation (Kiesler & Cummings, 2007).

Table 21.1
Project outcomes studied in Cummings & Kielser (2007)

Index	Items
Knowledge outcomes ("ideas")	Started new field or area of research; developed new model or approach in field; came up with new grant or spin-off project; submitted patent application; presented at conference or workshop; published article(s), book(s), or proceeding(s); recognized with award(s) for contribution(s) to field(s). Alpha = .63 (7items)
Tools outcomes ("tools")	Developed new methodology; created new software; generated new dataset; generated new materials; created data repository; created website to share data; created collaboratory; created national survey; developed new kind of instrument; created online experiment site. Alpha = .65 (11 items)
Training outcomes ("people")	Grad student finished thesis or dissertation; grad student/post-doc got academic job; grad student/post-doc got industry job; undergrad/grad student(s) received training; undergrad(s) went to grad school. Alpha = .70 (5 items)
Outreach outcomes ("people")	Formed partnership with industry; formed community relationship through research; formed collaboration with researchers; established collaboration with high school or elementary school students; established collaboration with museum or community institution, established collaboration with healthcare institution. Alpha = .45 (6 items)

Data

The development of better *data* is also critical to the development of policy-relevant science metrics. The literature points to four important areas of focus in the development of a science policy data infrastructure. The first is to capture data that describe the universe of ways in which scientists create and transmit knowledge, as well as the ways in which nonscientists assimilate that knowledge. Scientific knowledge is an academic, economic, and social commodity. The exchange of knowledge in each of these policy spheres can be traced, but the data exist in siloed, often proprietary, databases all over the world. There are many important fragmented efforts to create an integrated scientific knowledge data infrastructure, all of which are labor-intensive and require both extensive reporting on the part of scientists and extensive data cleaning on the part of the data developer. Not surprisingly, such efforts have yet to generate the community of practice that is critical for continual validation and currency of the data. One mammoth data creation exercise (COMETS, 2012) links government investment in research and development (NSF, NIH, DoD, and DoE grants) through the path of knowledge creation, its transmission and codification, and then its commercialization by matching all journal articles and citations, high-impact articles, highly cited authors, UMI ProQuest Digital Dissertations, U.S. utility patents, venture capital, IPOs, web-based firm data, and links to major public firms. The STAR METRICS (2012) collaboration between U.S. federal science funding agencies and research institutions is linking research portfolio data with university administrative data as well as U.S. patent data in order to identify who is generating scientific knowledge and who is commercializing that knowledge. Other researchers have expended substantial effort creating and visualizing datasets linking research inputs with outputs (e.g., Börner & Scharnhorst, 2009).

It is essential that such an approach include a thoughtful consideration of newer modes of transmitting scientific knowledge. The bulk of data collection and research on science, technology, and innovation indicators is focused on the article-level and journal-level measures that originated in bibliometrics. Widely collected and maintained only in recent years, online usage statistics enable a multifaceted description of the *digital* reach of knowledge presented in scientific publications. Web usage statistics include page views, article downloads, and follow-on page views for online scientific content. The umbrella term *altmetrics* encompasses the ways one can track social media–driven dissemination of scientific knowledge (as presented in publications, conference proceedings, slide decks, and open datasets, for example) (see Priem, chapter 14, this volume). This includes reposting and discussing scientific items on social media websites and blogs, such as Twitter, Facebook, and ScienceBlogs, and sharing the items on open-exchange platforms like Mendeley. Several early studies demonstrate a positive correlation between web usage metrics and traditional article-level and journal-level metrics (i.e., citation indices and impact factors) (see Bar-Ilan and colleagues, chapter 16, this volume). However, neither traditional nor web-based, article-level measures alone provide sufficient material for science policy analysis and decision making.[1]

A second core requirement is the development of a system that will enable the correct attribution of the creation of knowledge to the right individual. Accurate attribution of knowledge, or author disambiguation, is fundamental to assigning credit. The aforementioned person-centric framework of knowledge transmission in the scientific enterprise suggests that data from disparate databases can be linked together as long as the identity of the researcher or group of researchers associated with the grant, product, collaboration, or organization is captured within the database. Science is networked, but people or groups (networks) of people, do science, and the primary product is knowledge, which is transferred to other people by way of publication, training, and collaboration. Knowledge also is transferred for economic benefit, by way of licensing or adoption into a new process or product, for example. The ability to access researcher-level information from interlinked databases eventually will enable empirical analyses of science funding policies.

Disambiguation is a critical challenge, however, particularly with duplicate names, name changes, and the increasing contribution of Asian scientists, many of whom have identical first and last names. Computer scientists have generated matching algorithms that utilize researcher attribute metadata, such as email address or organization, to provide a reasonably confident level of researcher disambiguation in merged databases. Another approach is to generate persistent research identification numbers or codes that the researcher can use to claim past and future knowledge products, including journal articles, conference proceedings, and patent applications. Several early efforts have demonstrated some technical success (the InCommon single sign-on technology is a good example[2]) but have not generated a broad user community. More generally accepted are persistent identifiers for the knowledge products themselves, including the International Standard Serial Number (ISSN), a numerical code developed in 1970 to identify paper publications, and the digital object identifier (doi) developed in 2000, which also enables the storage of publication metadata. More recently, the Open Researchers and Contributor ID (ORCID[3]) international nonprofit consortium is developing a persistent researcher identifier and metadata repository combined with API technology for integration into worldwide data systems. The deployment of the ORCID system will be critical; if a large and diverse user community develops around the effort, the consequences for science metrics could be profound.

Although the technical means have existed to achieve disambiguation and capture individual contributions, there has been a lack of motivation—and lack of incentives—within the scientific community. The Brazilian Ministry of Science and Technology's approach is a powerful example of an incentive system that supports universal adoption of unique identifiers among national researchers. The Lattes Database[4] was compiled and maintained by the Ministry, with no user costs. After 12 years in use throughout the country, the Lattes platform contains individually curated information describing scientific knowledge creation and communication by more than one million researchers from about 4,000 institutions in the country. There are several critical elements to the success of the Lattes platform. One is that there was a decision by the Brazilian funding agency to invest in the appropriate infrastructure, employing a team of systems and software engineers who extensively iterated with the researcher community to define and respond to their needs. The second was the decision to develop the appropriate incentives. Researchers enter their professional profile data a single time and then use their Lattes profiles to streamline the application and reporting process for federal science funding. All federally funded Brazilian researchers must have a Lattes profile and associated URL. The link to grant applications creates a strong incentive for researchers to regularly log in to Lattes and ensure that their data are complete and correct. The result is one of the cleanest researcher databases in existence today.

A third factor is that the data infrastructure does not need to be developed using a top-down approach. The Lattes system began as a top-down initiative but an online community has since developed and maintains strong social control over data quality and validity. With sufficient incentives, a bottom-up approach could be developed whereby scientists voluntarily contribute content, as has been done by online communities like *Wikipedia* and Facebook. Initial efforts by the Concept Web Alliance have been extremely promising in the life sciences and could be fostered by collective intervention. Companies such as Thomson Reuters and Elsevier have extensive knowledge in this arena, which could be supported and leveraged.

Scientific publishers, funding organizations, research institutions, and the private sector all have invested in the development of research networking software platforms designed to create such communities, although none has developed into the much-hyped vision of a "Facebook for scientists." Researcher networking platforms allow scientists to post and maintain online profiles to describe to their colleagues, organizations, and the public the type of work they perform and the many ways they create and share knowledge.[5] A few examples of proprietary research networking platforms in the United States include Elsevier's SciVal Experts, Thomson Reuters' ResearchInView, and InfoEd's Genius software. The United Kingdom's Symplectic Elements has a substantial number of implementations worldwide. Recently, Elsevier purchased Mendeley, expanding its stake in online research sharing and collaboration tools, and ResearchGate received $35 million in series C capital to further develop its social networking software for researchers. Another notable and rapidly growing online research networking community is the VIVO[6] open-source platform, which was originally funded by NIH grants but is now operating as a nonprofit entity. Uniquely, the VIVO community consists of both researchers and web developers. VIVO is also unique in that it is based entirely on linked open data, a way of encoding data about the scientific enterprise. The U.S. federal science funding agencies also are developing a research networking platform, SciENcv, which intends to use new technology and online collaborative tools to streamline the grant application and reporting processes.

The final task is to ensure that there is open access to the data. Accessible data are central to the replicability of metrics and are central to good science. Open data are particularly generative because application programming interface (API) technology precludes any need to build and maintain a single "science of science" database. Rather, it will encourage analysts to develop new tools and techniques to leverage big data for evidence-based science policy. This is certainly feasible: new computational capacity has emerged that facilitates the analysis of the data in terms of modeling and simulation with an unprecedented breadth, depth, and scale. The incentives to data providers, both researchers and publishing houses, are currently perverse, and could be changed through collective action.

Publishing houses have an incentive to provide data for a fee, but not to provide access to the broader research community so that research could be replicated and generalized. Similarly, researchers have every incentive not to share data. In most fields, there is no credit for collecting, documenting, indexing, and archiving data—and there is a substantial time and monetary cost involved in doing so. International funding agencies could act to change these norms. In the former case, they could pool resources to reimburse publishing houses for the very real costs associated with collecting and documenting citation data, and then provide access to academic researchers. In the latter case, the international funding agencies could act jointly to change incentives. They could require that scientific credit be given by data users to the creator, provider, and documentor of data. This could be instituted as simply as a citation in journal articles, just as references are made to other scientific contributions. Failure to cite the data provider/creator/documentor could be treated as plagiarism. Users could be required to acknowledge this as part of a user agreement when data are accessed. Biographic sketches submitted to funding agencies could include "Data creation/production/documentation" as an element, together with a metric such as a count of how many times the data were used.

Practice

The existence of an academic field of science of science policy is a necessary precondition for such a system. Policies can be formed and carried through rationally only when a sufficient number of men and women follow them in a deep, thoughtful, and open way. Science policy, in its broadest sense, has become so important that it deserves the enduring scrutiny from a profession of its own. This is the promise of the academic discipline of the science of science policy. (Marburger, 2011, p. 21).

Advancing the *practice* of metrics is critical. This is best achieved by developing a scientific community that examines the scientific validity of different types of metrics. A plethora of new metrics have been developed since the *h*-index was first suggested (Hirsch, 2005; Zhou, 2008), and there is a lively debate examining the appropriate theoretical constructs that should underlie metric development (Börner & Scharnhorst, 2009; Leydesdorff, 2008). New metrics are being developed. The MESUR project[7], for example, has created a semantic model of the ways scholars communicate based on creating a set of relational and semantic web databases from over one billion usage events and over ten billion semantic statements. The combination of usage, citation, and bibliographic data can be used to develop metrics of scholarly impact that go well beyond the standard bibliometric approaches used by academics (Bollen, Rodriguez, & Sompel, 2007).

Funding agencies can foster this community of practice by supporting national and international workshops that bring active researchers together. In addition, incentives could be provided to encourage the community of practice to take advantage of the new computational capacity that has emerged. This capacity facilitates the analysis of the data in terms of modeling and simulation with an unprecedented breadth, depth, and scale (Lane, 2009; Lazer et al., 2009). New instrumentation provides opportunities for researchers to advance scientific understanding through collaboration with colleagues around the globe (National Science and Technology Council, 2009). Funding agencies could establish an international cooperative venture through a virtual observatory to provide the scientific community with a chance to develop metrics, test their validity, and build a community of practice.

What Might Scientific Metrics Look Like?

Good metrics have many features, but the primary one is that they should be grounded in the theoretical and empirical microfoundations necessary to understand science and technology investments.

Here we sketch an example of a set of metrics that would draw on the literature of industrial organization and innovation theory, as well as empirical advances made possible by cyberinfrastructure investments.[8] A conceptual approach might be sketched as follows:

1. A scientific project can be seen as analogous to a firm—with the major caveat that the "production" process is open. Individuals working on the project (including postdocs and graduate students) can be viewed as workers in the firm.
2. Each project's aim is to create and transmit scientific ideas and push for their adoption.
3. Ideas are transmitted by scientists in a variety of potentially measurable ways, including publications, presentations, blogs, internal project workspaces, and emails.
4. Collaboration is the main vehicle whereby ideas are transmitted.
5. Adoption can be by other scientists, policymakers, or private businesses.

The empirical approach could be informed by cyberinfrastructure advances; much scientific and economic activity can be captured electronically, as has been shown in the STAR METRICS program (Largent & Lane, 2012). The approach might be outlined follows:

1. Data can be collected on project teams from the human resource records of research organizations (Lane & Bertuzzi, 2011).
2. Data on collaborations within and outside networks can be captured through project documents, scientists' curricula vitae, and commonly used workflow tools like Mendeley.
3. Scientific topics and ideas, or the information on the content of text documents, can be summarized using natural language processing and topic modeling (Blei, 2012).
4. Collaborative networks can largely be measured using existing economic, scientometric, and sociological tools.
5. Adoption can be measured by the structure of the networks, such as size, quality, distribution, and openness.

The link between science funding and science outcomes can be formally expressed as the two following econometric relations (or two groups of such interrelated regression equations):

$$Y_{it}^{(1)} = Y_{it}^{(2)}\alpha + X_{it}^{(1)}\lambda + \varepsilon_{it} \text{ and (2) } Y_{it}^{(2)} = Z_{it}\beta + X_{it}^{(2)}\mu + \eta_{it}, \quad (1)$$

where the subscripts i and t denote project teams and quarters and ε and η stand for unobserved factors and errors of measurement and specification (and can possibly include individual unobserved project teams' characteristics). The output variables are measured by $Y^{(1)}$ and team composition variables by $Y^{(2)}$. Both are determined by a set of control variables $X^{(1)}$ and $X^{(2)}$ that can overlap and be truly exogenous or predetermined variables of key interest Z. One way of thinking about this is to consider a team as analogous to a firm, a principal investigator as analogous to an entrepreneur, and the scientific product as being analogous to a physical product. The analogy, like all analogies, should not be carried too far but is a useful organizing construct in terms of describing one important body of knowledge on which we can draw. (This characterization draws on Lane's joint work with Jacques Mairesse and Paula Stephan.)

This formal model can be used as a basis for developing different sets of metrics. The coefficients can be used to analyze how the composition of teams is specifically driven by the presence of funding, and how this differs by scientific research area. New data, like the STAR METRICS data, now exist to examine these questions. There are several key features of the STAR METRICS data. First, they are project-level data. Quarterly information on all individuals who are directly employed on research funding is available from the human resource records of the institution. Second, the data are quite detailed; they include information about the occupation of each individual as well as the proportion of earnings that are generated by the award funding in each period. Third, the data include information about other inputs into project teams beyond the immediate staff. Additional information is also available from the financial system: the data capture the amount of project funds that are allocated to collaborating institutions. They also capture the amount of project funds expended at vendors, as well as on infrastructure support, including financial, information technology, physical space, and research services.

Analysis of collaborations in this sense has hardly ever been done precisely because researchers are not able to identify the status of researchers, due to lack of data at the project/team level. Put simply, while there is a great deal of work on coauthorship patterns of papers, such work is not really an analysis of teams per se. In this sense, a viable set of metrics based on the appropriate units of analysis can be developed and standardized across disciplines. One might be the speed with which ideas are adopted at various stages of the scientific process. Just as in industry, where basic research moves in various stages through to the adoption of an idea (see figure 21.2), the stages of research could be written down and delineated in each field of research.

Other sets of theoretically and empirically grounded metrics could and should be developed. But the key to making the metrics scientific is that the underlying models should be based on human interactions, rather than on the study of documents. The source data should be open and generally available, rather than "black box" in nature. And all data, models, and tools should be tested by the Science of Science Policy

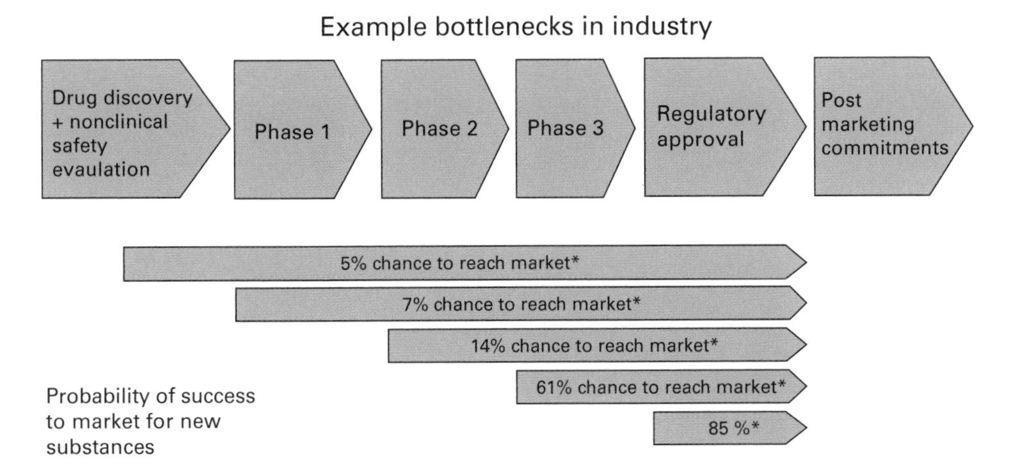

Figure 21.2
Example bottlenecks in industry.

community of practice. It is only in this way that the metrics can be validated and their generalizability and replicability assured, and the only way that the science metrics can be said to be scientific.

Summary

Just as firms get what they pay for, it is likely that science agencies and countries will get what they pay for. It is thus important that scientists and science funding agencies build on the initial foundation to design a science measurement system that is of a quality equal to that of systems designed to study other scientific phenomena. A concerted international endeavor will be essential to the establishment of such a high-quality system.

Notes

1. New work is being done in this area, however. Notably, the Foresight and Understanding from Scientific Exposition (FUSE) project, funded by the Office of the National Director for Intelligence, hopes to integrate technological and subject matter expertise with massive quantities of scientific literature data in order to identify patterns of technical emergence. If successful, this effort would demonstrate that complex algorithmic analysis of very large stores of article-level data could identify emergent properties of the large data systems (Murdick, 2012).

2. www.incommon.org

3. about.orcid.org

4. http://lattes.cnpq.br

5. For a comprehensive comparison of research networking tools, see http://en.wikipedia.org/wiki/Comparison_of_Research_Networking_Tools_and_Research_Profiling_Systems.

6. vivoweb.org

7. MESUR is an acronym for Metrics from Scholarly Usage of Resources; see http://www.mesur.org/MESUR.html.

8. This draws heavily on joint work with Jacques Mairesse, Paula Stephan, and Lee Fleming.

References

Academic Ranking of World Universities. (2012). In *Wikipedia*. Retrieved from http://en.wikipedia.org/wiki/Academic_Ranking_of_World_Universities

Blei, D. (2012). Probabilistic topic models. *Communications of the ACM, 55*(4), 77–84.

Bollen, J., Rodriguez, M., & Sompel, H. V. (2007). MESUR: Usage-based metrics of scholarly impact. In *Proceedings of the 7th ACM/IEEE-CS Joint Conference on Digital Libraries*. Vancouver, CA.

Börner, K., & Scharnhorst, A. (2009). Visual conceptualizations and models of science. *Journal of Informetrics, 3*(3), 161–172.

COMETS. (2012). About COMETS. Retrieved from http://www.kauffman.org/COMETS/About-COMETS.aspx.

Cummings, J. N., & Kiesler, S. (2007). Coordination costs and project outcomes in multi-university collaborations. *Research Policy, 36*(10), 1620–1634. doi: 10.1016/j.respol.2007.09.001.

Gibbons, R. (1998). Incentives in organizations. *Journal of Economic Perspectives, 12*(4), 115–132.

Haeussler, C., Jiang, L., Thursby, J., & Thursby, M. (2009). *General and specific information sharing among academic scientists*. Cambridge, MA: National Bureau of Economic Research.

Hirsch, J. (2005). An index to quantify an individual's scientific research output. *Proceedings of the National Academy of Sciences of the United States of America, 102*(46), 16569–16572.

Ioannidis, J. P. A. (2005). Why most published research findings are false. *PLoS Medicine, 2*(8), e124. doi:10.1371/journal.pmed.0020124.

Kerr, S. (1975). On the folly of rewarding A while hoping for B. *Academy of Management Journal, 18*(4), 769–783.

Kiesler, S., & Cummings, J. (2007). Modeling productive climates for virtual research collaborations [Powerpoint]. Retrieved from http://www.aaas.org/spp/scisip/ppts/SciSIP.3.2009_Cummings.pdf.

Lambe, P. (2007). *Organising knowledge: Taxonomies, knowledge and organisational effectiveness*. Oxford: Neal-Schuman.

Lane, J. (2009). Administrative transactions data. Working Paper No. 52, *Rat für Sozial und Wissenschafts Daten*. Retrieved from http://ideas.repec.org/p/rsw/rswwps/rswwps52.html.

Lane, J. I., & Bertuzzi, S. (2011). Measuring the results of science investments. *Science, 331*(6018), 678–680.

Largent, M. A., & Lane, J. I. (2012). STAR METRICS and the science of science policy. *Review of Policy Research, 29*(3), 431–438.

Lazer, D., Pentland, A., Adamic, L., Aral, S., Barabási, A.-L., Brewer, D., et al. (2009). Computational social science. *Science, 323*(5915), 721–723.

Leydesdorff, L. (2008). Measuring research output with science & technology indicators. *SciVerse Topics*. Retrieved from http://www.scitopics.com/Measuring_Research_Output_with_Science_Technology_Indicators.html.

Leydesdorff, L. (2010). Luhmann reconsidered: Steps towards an empirical research programme in the sociology of communication. In C. Grant (Ed.), *Beyond universal pragmatics: Essays in the philosophy of communication* (pp. 149–173). Oxford: Peter Lang.

Marburger, J. H. (2011). Why policy implementation needs a science of science policy. In K. H. Fealing, J. I. Lane, J. H. Marburger, & S. S. Shipp (Eds.), *The science of science policy: A handbook* (pp. 9–22). Stanford, CA: Stanford University Press.

Murdick, D. (2012). Foresight and Understanding from Scientific Exposition (FUSE). Retrieved from http://dpcpsi.nih.gov/pdf/10_IARPA.pdf.

National Science and Technology Council. (2009). *Harnessing the power of digital data for science and society*. Retrieved from http://www.nitrd.gov/About/Harnessing_Power_Web.pdf.

Paley, W. B. (2012). Map of science in the journal *Nature, SEED* and *Discover* Magazines [website]. Retrieved from http://wbpaley.com/brad/mapOfScience.

Priem, J., & Hemminger, B. (2010). Scientometrics 2.0: Toward new metrics of scholarly impact on the social Web. *First Monday, 15*(7).

Science of Science Policy, The : A Federal Research Roadmap. (2008). Retrieved from http://scienceofsciencepolicy.net/category/tags/roadmap.

STAR METRICS. (2012). What is STAR METRICS? Retrieved from https://www.starmetrics.nih.gov.

Young, N., Ioannidis, J., & Al-Ubaydli, O. (2008). Why current publication practices may distort science. *PLoS Medicine, 5*(10), e201.

Young, S., & Karr, A. (2011). Deming, data and observational studies: A process out of control and needing fixing. *Significance, 8*(3), 116–120. doi:10.1111/j.1740-9713.2011.00506.x.

Zhou, M. (2008). Z factor: A new index for measuring academic research output. *Molecular Pain, 4*, 53.

List of Contributors

Editors

Blaise Cronin is Rudy Professor of Information Science at Indiana University Bloomington and Honorary Visiting Professor at both City University London and Edinburgh Napier University. Previously, he was the Professor of Information Science at the University of Strathclyde, Glasgow. His books include *The Citation Process, The Scholar's Courtesy*, and *The Hand of Science*. Cronin is editor-in-chief of the *Journal of the American Society for Information Science and Technology* and for 10 years was editor of the *Annual Review of Information Science and Technology*. He holds an MA from Trinity College Dublin, a PhD and DSSc from the Queen's University of Belfast, and a DLitt (honoris causa) from Queen Margaret University, Edinburgh. bcronin@indiana.edu

Cassidy R. Sugimoto is an assistant professor at Indiana University Bloomington. She earned her bachelor's, master's, and doctoral degrees from the University of North Carolina at Chapel Hill. She has published more than 60 journal and conference works and serves on the editorial boards of *JASIST* and *Scientometrics*. She is active in the Association for Information Science & Technology (ASIST) and the International Society for Scientometrics and Informetrics (ISSI), and has served on the board of directors of both. Her research has been funded intramurally as well as by professional associations (ALISE, ASIS&T) and national agencies (e.g., NSF). sugimoto@indiana.edu

Authors

Mayur Amin is Senior Vice President of Research in Research & Academic Relations at Elsevier. Amin's work seeks to understand how scientific publishing can better meet the differing expectations of researchers, authors, and academics—issues on which he has published articles in specialist journals and delivered conference presentations. His knowledge of the scholarly publishing industry, acquired over 25 years, has made him an authoritative commentator on this dynamic and fast-changing sector, and a regular contributor to industry debates and policy-formation exercises. By training a scientist (with a degree in biochemistry and toxicology), Amin conducted research in the energy industry (Shell) before moving into scientific publishing, first with Pergamon and then with Elsevier. m.amin@elsevier.com

Judit Bar-Ilan is a professor in the Department of Information Science of Bar-Ilan University in Israel. She received her PhD in computer science from the Hebrew University of Jerusalem and started her research in information science in the mid-1990s at the School of Library, Archive and Information Studies of the Hebrew University of Jerusalem. She moved to the Department of Information Science at Bar-Ilan University in 2002. She is a member of the editorial boards of the *Journal of the American Society for Information Science and Technology*, *Scientometrics*, *Journal of Informetrics*, *Cybermetrics*, and *Online Information Review*. Her areas of interest include informetrics, information retrieval, Internet research, information behavior, the semantic Web, and usability. Judit.Bar-Ilan@biu.ac.il

Johann Bauer received his PhD from the Ludwig-Maximilians University of Munich in 1981. Since 1991 he has been involved in scientific information retrieval and bibliometrics in the Information Retrieval Group at the Max Planck Institute for Biochemistry, serving the Biology & Medicine Section of the Max Planck Society. jbauer@biochem.mpg.de

Lutz Bornmann works as a sociologist of science at the Division for Science and Innovation Studies in the Administrative Headquarters of the Max Planck Society (Munich, Germany). He has been working on issues in the promotion of young academics and scientists in the sciences and on quality assurance in higher education since the late 1990s. His current research interests include research evaluation, peer review, and bibliometric indicators. He is a member of the editorial boards of the *Journal of Informetrics* and *Scientometrics* and is an advisory editorial board member of *EMBO Reports*. Since 2004, he has published more than 100 papers in journals covered by Thomson Reuters, with a total of more than 1,500 citations. Essential Science Indicators lists him in the top 1% of all scientists based on total citations over the last 10 years. Lutz.Bornmann@gv.mpg.de

Benjamin F. Bowman studied microbiology at the Technical University of Munich and completed his doctoral dissertation at the Max Planck Institute of Biochemistry. Since 1979 he has been responsible for establishing scientific information retrieval services for the institutes of the Biology & Medicine Section of the Max Planck Society.

These internal services are primarily directed to the research needs of scientists, but also include bibliometric analyses for science administrators and policymakers. He is a member of various advisory boards relating to information services and bibliometric activities in Germany. bowman@biochem.mpg.de

Kevin W. Boyack is president of SciTech Strategies, Inc., and has been with the company since the summer of 2007. After receiving a PhD in Chemical Engineering from Brigham Young University, he spent 17 years at Sandia National Laboratories, where he worked in various areas including combustion (experimental and modeling), transport processes, socioeconomic war gaming, and science mapping. Since joining SciTech his work has centered on developing more accurate global maps of science. He has published nearly 30 articles dealing with various aspects of science mapping and related metrics. Current interests include detailed mapping of the structure and dynamics of science and technology, application of full text to science mapping and bibliometrics, and the identification of emerging topics. kboyack@mapofscience.com

Ronald E. Day is an associate professor in the Department of Information and Library Science of the School of Informatics and Computing at Indiana University Bloomington. He is the author of *The Modern Invention of Information: Discourse, History, and Power*, coeditor and cotranslator of Suzanne Briet's *What Is Documentation?*, and coeditor of *Rethinking Knowledge Management: From Knowledge Objects to Knowledge Processes*, as well as having published numerous articles. He writes at the intersection of critical theory and information and knowledge studies, on politics, art, philosophy, and social theory. roday@indiana.edu

Nicola De Bellis graduated in philosophy at the University of Bari, Italy, and obtained a PhD in the history of science in 1998 with a doctoral dissertation on the communication structure of Renaissance natural history. Having joined a medical library in 2002, he has been working since then on the boundary line between information science on the one hand and the history and philosophy of science on the other. He currently acts as a consultant on bibliometric databases and techniques to the academic personnel of the University of Modena and Reggio Emilia. nicola.debellis@unimore.it

Jonathan Furner is an associate professor in the Department of Information Studies, as well as a faculty affiliate of the Center for Digital Humanities and of the Center for Information as Evidence, at the University of California, Los Angeles. He works on projects in cultural informatics and in the history and philosophy of information science. He teaches in UCLA's PhD program in Information Studies, the Master of Library and Information Science program, and the interdepartmental MA program in Moving Image Archive Studies, which he chairs. He has a PhD in information studies from the

University of Sheffield, and an MA in philosophy and social theory from the University of Cambridge. furner@gseis.ucla.edu

Yves Gingras is a professor in the Department of History of the Université du Québec à Montréal (UQAM), scientific director of the Observatoire des sciences et des technologies (OST), and Canada Research Chair in the History and Sociology of Science. He is also a member of the Centre Interuniversitaire de recherche sur la Science et la technologie (CIRST). His research interests include the transformation of universities, the role of mathematical analogies in physics, historical bibliometrics, and the evaluation of research. He has published many books and more than a hundred papers in various journals like *Social Studies of Science, History of Science, Social Epistemology, Journal of the American Society for Information Science and Technology* and *Scientometrics*. gingras.yves@uqam.ca

Stefanie Haustein studied history, American linguistics and literature, and information science at Heinrich Heine University in Düsseldorf, Germany. In 2012 she received a PhD for a dissertation titled *Multidimensional Journal Evaluation: Analyzing Scientific Periodicals beyond the Impact Factor*, for which she was awarded the Eugene Garfield Doctoral Dissertation Scholarship 2011. Haustein has worked on the bibliometrics team at Forschungszentrum Jülich and as a lecturer in the Department of Information Science at Heinrich Heine University. Currently she is a bibliometric analyst at Science-Metrix in Montréal, Canada, and a postdoctoral fellow at the Université de Montréal. stefanie.haustein@umontreal.ca

Edwin A. Henneken has been working as an IT specialist for the SAO/NASA Astrophysics Data System (ADS) since October 2002. In this role, he evaluates, integrates, and enhances existing software systems used by the ADS project. He participates in curating ADS' holdings and manages ADS' Education and Public Outreach as well as its digital preservation efforts. Henneken uses the ADS holdings for informetrics research. He holds a master's degree in astrophysics from Leiden Observatory, The Netherlands. He also did research in the Geophysics Department at the Vrije Universiteit in Amsterdam, and worked as an IT and systems specialist at a large telecommunications company. ehenneken@cfa.harvard.edu

Peter A. Hook is a doctoral candidate at the Indiana University School of Informatics and Computing. He has a juris doctor (JD) degree from the University of Kansas (1997) and an MS in library and information science from the University of Illinois (2000). Additionally, Hook has spent 10 years working as an academic law librarian. His primary research focus is information visualization and domain mapping. He is currently completing his dissertation (*The Structure and Evolution of the Academic Disci-*

pline of Law in the United States: Generation and Validation of Course-Subject Co-occurrence Maps), in which he evaluates whether course-coupling analysis (the aggregate of the same professor teaching multiple, different courses) produces accurate topic maps of an academic discipline. Pahook@indiana.edu

Judith Kamalski is Manager, Strategic Research Insights & Analytics in Elsevier's Academic and Government Institutional Markets. She focuses on demonstrating Elsevier's bibliometric expertise and capabilities by connecting with the research community. She is heavily involved in analyzing, reporting, and presenting commercial research performance evaluation projects for academic institutes as well as governments. She has worked within several key areas of Elsevier, including journal publishing, strategy, sales, and most recently Research and Academic Relations. Kamalski has a PhD from the Utrecht Institute of Linguistics and also holds master's degrees in Corporate Communications and French Linguistics & Literature. J.Kamalski@elsevier.com

Richard Klavans is the founder of SciTech Strategies, Inc., and has published extensively on the art and science of science mapping. He has created large-scale maps of science for research planning in industry (Abbott Labs, Astra Zeneca, Dupont, Glaxo, Kellogg, Kraft, SmithKline Beecham, and Unilever), government agencies (DOE, NSF, and NIH), and over 20 universities. His most recent research initiative is the prediction of scientific breakthroughs using a dynamic microstructural map of science. His educational background includes an engineering degree from Tufts, a master's degree from Sloan at MIT, and a PhD from Wharton at the University of Pennsylvania. rklavans@ mapofscience.com

Kayvan Kousha is a research associate at the Statistical Cybermetrics Research Group, University of Wolverhampton (UK). His broad research areas include web citation analysis and online scholarly impact assessment. His over 30 refereed publications include a range of investigations on the potential role of digitized web contents for impact assessment such as online books, syllabi, presentations, blogs, videos, and images. The findings suggest that web sources can potentially be helpful as supplementary impact indictors in some subject areas, where traditional citation metrics might be insufficient for research evaluation. k.kousha@wlv.ac.uk

Michael J. Kurtz is an astronomer and computer scientist at the Harvard-Smithsonian Center for Astrophysics in Cambridge, Massachusetts, which he joined after receiving a PhD in physics from Dartmouth College in 1982. Kurtz is the author or coauthor of over 300 technical articles and abstracts on subjects ranging from cosmology and extragalactic astronomy, to data reduction and archiving techniques, to information systems and text retrieval algorithms. He is a fellow of the American Physical Society in

Astrophysics and a fellow of the American Association for the Advancement of Science in Computer and Information Science. In 1988 Kurtz conceived what has now become the Smithsonian/NASA Astrophysics Data System, the core of the digital library in astronomy. He has been associated with the project since that time, and was awarded the 2001 Van Biesbroeck Prize of the American Astronomical Society for his efforts. kurtz@cfa.harvard.edu

Julia Lane is a senior managing economist at the American Institutes for Research in Washington, D.C., and professor of economics, BETA—University of Strasbourg, CNRS. She was previously the program director of the Science of Science and Innovation Policy program at the National Science Foundation, and the cochair of the federal interagency group on the Science of Science Policy. She is the developer of the STAR METRICS program, which is an interagency activity to document the results of science investments, and coeditor of the *Handbook of Science of Science Policy*. jlane@air.org

Mark Largent is an associate professor of Science Policy at James Madison College at Michigan State University and director of the Science, Technology, Environment, and Public Policy Specialization. He is a historian of science and technology and researches the role of scientists in American public policy. Largent@msu.edu

Vincent Larivière is an assistant professor at the École de bibliothéconomie et des sciences de l'information (EBSI) of the Université de Montréal, where he holds the Canada Research Chair on the transformations of scholarly communication. He is also an associate researcher at the Observatoire des sciences et des technologies (OST) and a member of the Centre interuniversitaire de recherche sur la science et la technologie (CIRST). His research has been published in venues such as the *Journal of the American Society for Information Science and Technology*, *Scientometrics*, and *PLoS ONE*. He currently serves as an academic editor for *PLoS ONE*. vincent.lariviere@umontreal.ca

Loet Leydesdorff (PhD, sociology; MA, philosophy; and MSc, biochemistry) is a professor in "Communication and Innovation in the Dynamics of Science and Technology" at the Amsterdam School of Communications Research of the University of Amsterdam. He is also a visiting professor at the Institute of Scientific and Technical Information of China in Beijing and an honorary professor at the Science and Technology Policy Research Unit of the University of Sussex. He has published extensively in systems theory, social network analysis, scientometrics, and the sociology of innovation and is a member of a number of editorial boards. He received the Derek de Solla Price Award for scientometrics and informetrics in 2003 and held "The City of Lausanne" Honor Chair at the School of Economics, Université de Lausanne, in 2005. loet@leydesdorff.net

Werner Marx studied chemistry at the University of Bonn, where he completed his diploma and doctoral dissertation in physical chemistry. In 1982 he joined the Information Service of the institutes of the Chemical Physical Technical Section of the Max Planck Society, located at the Max Planck Institute for Solid State Research in Stuttgart. This service offers support with respect to all kinds of scientific information. The target groups are researchers, administrators, and decision makers at the Max Planck Society. w.marx@fkf.mpg.de

Katherine W. McCain holds degrees in the life sciences and information studies and is a professor in the iSchool at Drexel University. She has been involved in bibliometric/scientometric research since the early 1980s and has a particular interest in issues relating to the importance of context in the analysis of the structure of fields and literatures—including contextual cocitation and codescriptor mapping and citation context analysis. She serves on the editorial boards of the *Journal of the American Society for Information Science and Technology*, *Scientometrics*, and the *Journal of Informetrics*. In 2007, *Scientometrics* awarded her the Derek de Solla Price Medal for her contributions to quantitative studies of science. mccainkw@drexel.edu

Margit Palzenberger is a member of the management team of the Max Planck Digital Library, the Max Planck Society's unit for the deployment of advanced scientific information services. She is an expert in the integration, quantitative analysis, and visualization of data describing and steering information processes in the academic environment. She has a background in biology (University of Salzburg), with 15 years' experience in adopting and teaching quantitative methods in ecology and environmental sciences. palzenberger@mpdl.mpg.de

Andrew Plume is Director of Scientometrics & Market Analysis in Research & Academic Relations at Elsevier. Plume studies information flows in the scholarly literature by analyzing patterns of publications and citations. His particular interest lies in the use (and abuse) of the Impact Factor and the emergence of alternative metrics for journal evaluation. Plume frequently presents on these topics, among others, to journal editors, learned and scholarly societies, and the publishing community. After receiving his PhD in plant molecular biology from the University of Queensland (Australia), and conducting postdoctoral research at Imperial College London, Plume joined Elsevier in 2004. He has coauthored research and review articles in the peer-reviewed literature and is a member of the editorial board of *Research Trends*. a.plume@elsevier.com

Jason Priem is a PhD student and Royster Fellow, studying information science at the University of North Carolina at Chapel Hill. Since coining the term *altmetrics*, he has remained active in the field, organizing the annual altmetrics workshop, giving invited talks, and publishing research. Priem is also a cofounder of ImpactStory (http://impactstory.it), an open-source webapp that helps scholars track and report the broader impacts of their research. He blogs at http://jasonpriem.org and tweets as @jasonpriem.

Rebecca Rosen is a senior researcher specializing in the development of an evidence-based science policy portfolio at the American Institutes for Research. She has worked as an AAAS Science Policy Fellow at both the National Science Foundation and the National Institutes of Health. Rosen was a member of the Federal Interagency group that developed the groundbreaking STAR METRICS project that aims to document the outcomes of federal science investments. Her work at NIH and NSF helped to build a policy-level community of practice around the emerging science of science policy field. Rosen has a doctoral degree in neuroscience from Emory University. rrosen@air.org

Hermann Schier works in the Information Service of the institutes of the Chemical Physical Technical Section of the Max Planck Society. He supports Max Planck scientists in the fields of materials science, chemistry, and physics with a special focus on patents and the development of new methods in science evaluation. He studied chemistry in Stuttgart, completed his diploma thesis in solid-state chemistry, moved to molecular electronics for his PhD, and worked as a materials scientist in industry. h.schier@fkf.mpg.de

Hadas Shema is a graduate student in the Department of Information Science of Bar-Ilan University in Israel. She studies the characteristics of online scientific discourse and is a member of the European Union's Academic Careers Understood through Measurement and Norms (ACUMEN) project. Shema coauthors a *Scientific American* blog called Information Culture (http://blogs.scientificamerican.com/information-culture) and tweets at @Hadas_Shema. dassysh@gmail.com

Mike Thelwall is a professor of information science and leader of the Statistical Cybermetrics Research Group at the University of Wolverhampton, UK, and a research associate at the Oxford Internet Institute. Thelwall has developed tools for gathering and analyzing web data, including hyperlink analysis, sentiment analysis, and content analysis for Twitter, YouTube, blogs, and the general Web. His publications include 152 refereed journal articles, 7 book chapters, and 2 books, including *Introduction to Webometrics*. He is an associate editor of the *Journal of the American Society for Information Science and Technology* and serves on three other editorial boards. m.thelwall@wlv.ac.uk

Daril A. Vilhena received a PhD from the Department of Biology at the University of Washington. He works at the intersection of semantic information and network structure to study the coevolution between structure and function. He is a member of Carl Bergstrom's lab and the Eigenfactor project. daril@uw.edu

Jevin D. West is an assistant professor in the iSchool at the University of Washington. He uses network models to measure the flow of information through social and biological systems. As a doctoral student, he founded Eigenfactor.org—a free website that ranks and maps science. Researchers, librarians, publishers, and administrators from around the world use Eigenfactor tools for better navigating the scholarly terrain. The metrics are included in Thomson Reuters' JCR, and the maps have been featured in the *Chronicle of Higher Education*, *Nature*, and *Science*. West has been invited to give more than 40 talks about his research at universities and conferences across the globe. jevinw@uw.edu

Paul Wouters is director of the Centre for Science and Technology Studies and professor of scientometrics at Leiden University. He is also a visiting professor of cybermetrics at the University of Wolverhampton. He has published on the history of the Science Citation Index, on scientometrics, and on the way the criteria of scientific quality have been changed by citation analysis. He is particularly interested in the role of information and information technologies in the creation of new scientific and scholarly knowledge. He coordinates the European project ACUMEN (http://research-acumen.eu) on research careers and evaluation of individual researchers. p.f.wouters@cwts.leidenuniv.nl

Index